Uday K. Chakraborty (Ed.)

Computational Intelligence in Flow Shop and Job Shop Scheduling

Studies in Computational Intelligence, Volume 230

Editor-in-Chief
Prof. Janusz Kacprzyk
Systems Research Institute
Polish Academy of Sciences
ul. Newelska 6
01-447 Warsaw
Poland
E-mail: kacprzyk@ibspan.waw.pl

Further volumes of this series can be found on our homepage: springer.com

Vol. 208. Roger Lee, Gongzu Hu, and Huaikou Miao (Eds.)
Computer and Information Science 2009, 2009
ISBN 978-3-642-01208-2

Vol. 209. Roger Lee and Naohiro Ishii (Eds.)
Software Engineering, Artificial Intelligence, Networking and Parallel/Distributed Computing, 2009
ISBN 978-3-642-01202-0

Vol. 210. Andrew Lewis, Sanaz Mostaghim, and Marcus Randall (Eds.)
Biologically-Inspired Optimisation Methods, 2009
ISBN 978-3-642-01261-7

Vol. 211. Godfrey C. Onwubolu (Ed.)
Hybrid Self-Organizing Modeling Systems, 2009
ISBN 978-3-642-01529-8

Vol. 212. Viktor M. Kureychik, Sergey P. Malyukov, Vladimir V. Kureychik, and Alexander S. Malyoukov
Genetic Algorithms for Applied CAD Problems, 2009
ISBN 978-3-540-85280-3

Vol. 213. Stefano Cagnoni (Ed.)
Evolutionary Image Analysis and Signal Processing, 2009
ISBN 978-3-642-01635-6

Vol. 214. Been-Chian Chien and Tzung-Pei Hong (Eds.)
Opportunities and Challenges for Next-Generation Applied Intelligence, 2009
ISBN 978-3-540-92813-3

Vol. 215. Habib M. Ammari
Opportunities and Challenges of Connected k-Covered Wireless Sensor Networks, 2009
ISBN 978-3-642-01876-3

Vol. 216. Matthew Taylor
Transfer in Reinforcement Learning Domains, 2009
ISBN 978-3-642-01881-7

Vol. 217. Horia-Nicolai Teodorescu, Junzo Watada, and Lakhmi C. Jain (Eds.)
Intelligent Systems and Technologies, 2009
ISBN 978-3-642-01884-8

Vol. 218. Maria do Carmo Nicoletti and Lakhmi C. Jain (Eds.)
Computational Intelligence Techniques for Bioprocess Modelling, Supervision and Control, 2009
ISBN 978-3-642-01887-9

Vol. 219. Maja Hadzic, Elizabeth Chang, Pornpit Wongthongtham, and Tharam Dillon
Ontology-Based Multi-Agent Systems, 2009
ISBN 978-3-642-01903-6

Vol. 220. Bettina Berendt, Dunja Mladenic, Marco de de Gemmis, Giovanni Semeraro, Myra Spiliopoulou, Gerd Stumme, Vojtech Svatek, and Filip Zelezny (Eds.)
Knowledge Discovery Enhanced with Semantic and Social Information, 2009
ISBN 978-3-642-01890-9

Vol. 221. Tassilo Pellegrini, Sören Auer, Klaus Tochtermann, and Sebastian Schaffert (Eds.)
Networked Knowledge - Networked Media, 2009
ISBN 978-3-642-02183-1

Vol. 222. Elisabeth Rakus-Andersson, Ronald R. Yager, Nikhil Ichalkaranje, and Lakhmi C. Jain (Eds.)
Recent Advances in Decision Making, 2009
ISBN 978-3-642-02186-2

Vol. 223. Zbigniew W. Ras and Agnieszka Dardzinska (Eds.)
Advances in Data Management, 2009
ISBN 978-3-642-02189-3

Vol. 224. Amandeep S. Sidhu and Tharam S. Dillon (Eds.)
Biomedical Data and Applications, 2009
ISBN 978-3-642-02192-3

Vol. 225. Danuta Zakrzewska, Ernestina Menasalvas, and Liliana Byczkowska-Lipinska (Eds.)
Methods and Supporting Technologies for Data Analysis, 2009
ISBN 978-3-642-02195-4

Vol. 226. Ernesto Damiani, Jechang Jeong, Robert J. Howlett, and Lakhmi C. Jain (Eds.)
New Directions in Intelligent Interactive Multimedia Systems and Services - 2, 2009
ISBN 978-3-642-02936-3

Vol. 227. Jeng-Shyang Pan, Hsiang-Cheh Huang, and Lakhmi C. Jain (Eds.)
Information Hiding and Applications, 2009
ISBN 978-3-642-02334-7

Vol. 228. Lidia Ogiela and Marek R. Ogiela
Cognitive Techniques in Visual Data Interpretation, 2009
ISBN 978-3-642-02692-8

Vol. 229. Giovanna Castellano, Lakhmi C. Jain, and Anna Maria Fanelli (Eds.)
Web Personalization in Intelligent Environments, 2009
ISBN 978-3-642-02793-2

Vol. 230. Uday K. Chakraborty (Ed.)
Computational Intelligence in Flow Shop and Job Shop Scheduling, 2009
ISBN 978-3-642-02835-9

Uday K. Chakraborty (Ed.)

Computational Intelligence in Flow Shop and Job Shop Scheduling

Springer

Uday K. Chakraborty
Mathematics & Computer Science Department
University of Missouri
Saint Louis, MO 63121
USA
E-mail: uday@cs.umsl.edu

ISBN 978-3-642-26036-0 e-ISBN 978-3-642-02836-6

DOI 10.1007/978-3-642-02836-6

Studies in Computational Intelligence ISSN 1860-949X

Typeset & Cover Design: Scientific Publishing Services Pvt. Ltd., Chennai, India.

Printed in acid-free paper

9 8 7 6 5 4 3 2 1

springer.com

Preface

For over fifty years now, the famous problem of flow shop and job shop scheduling has been receiving the attention of researchers in operations research, engineering, and computer science. Over the past several years, there has been a spurt of interest in computational intelligence heuristics and metaheuristics for solving this problem. This book seeks to present a study of the state of the art in this field and also directions for future research.

The ten chapters in this volume have been written by leading experts in the area. Chapter 1 provides a survey of the effect of the flow shop problem's structural properties on algorithm performance and analyzes the advantages of a structural property-based tabu search. In Chapter 2 a comprehensive review and evaluation of no-idle permutation flow shop scheduling is presented, with iterated greedy methods shown to outperform the other algorithms for this problem. Chapter 3 presents a new multi-objective ant-colony algorithm for minimizing makespan and total flowtime. A new, multi-objective simulated annealing approach for solving the permutation flow shop is introduced in Chapter 4. The blocking flow shop scheduling problem is considered in Chapter 5 where a new strategy is developed by combining an estimation of distribution algorithm with local search. Chapter 6 develops a scatter search-based strategy for multiobjective (average tardiness and the number of tardy jobs) fuzzy permutation flow shop and applies that to a real-world problem of engine piston manufacturing, producing results better than those obtained with a hybrid genetic algorithm. Chapter 7 presents new, genetic algorithm-based methods for job shop scheduling under uncertainty (fuzzy processing times, fuzzy due dates, stochastic processing times, and flexible job shop with fuzzy processing times). Classical and flexible job shop scheduling is also considered in Chapter 8 where Giffler-Thompson procedure-based genetic algorithms minimize makespan and also a weighted sum of makespan, total tardiness and total idle time. Chapter 9 presents a broad survey of recent research in flow shop and job shop scheduling. Chapter 10 proposes new ways of applying two continuous optimization heuristics, namely particle swarm optimization and differential evolution, to single-machine scheduling which is a discrete optimization problem. While single-machine scheduling does not belong to flow shop or job shop scheduling, this work has been included because of its novelty value and its potential for extension to flow shop scheduling.

I gratefully acknowledge the inspiration, advice and support that I received from Springer's Janusz Kacprzyk, Thomas Ditzinger and Heather King. I am grateful to

Charles Chui, Prabhakar Rao, Richard Friedlander and Nasser Arshadi, all of UMSL, for their encouragement and advice. Thanks to the contributing authors for being so patient during the long review process. I owe much to the following researchers for their help with reviewing the manuscripts: A. Agarwal, M. Chakraborty, C.Z. Janikow, B. Jarboui, C. Kahraman, A. Konar, D. Lei, C. Rajendran, R. Ruiz, P. Siarry, M.F. Tasgetiren, L. Wang, Q. Zhang, H. Ziegler.

St. Louis,
April 2009

Uday Kumar Chakraborty

Contents

Structural Property and Meta-heuristic for the Flow Shop Scheduling Problem

Feng Jin[1,2], Shiji Song[2], and Cheng Wu[2]

[1] Shanghai Baosight Software Co., Ltd., Shanghai 201203, China
jinfeng99@tsinghua.org.cn
[2] Department of Automation, Tsinghua University, Beijing 100084, China
{shijis,wuc}@tsinghua.edu.cn

Summary. According to the *No Free Lunch* Theorem, all algorithms equal to the randomly blind search if no problem information is known. Therefore, it is very important to study the problem properties (especially structural properties) and introduce them into algorithms so as to improve the algorithm performance (both solution quality and computational effort). For the flow shop scheduling problem (FSP) with makespan criterion, structural properties are wildly used in the existing literature, but there is no systematic review on it. This chapter surveys the existing structural properties, which are divided into two types: neighborhood properties (such as the famous block property) and solution space properties (such as the big-valley phenomenon).

This chapter also shows how to introduce the structural properties into meta-heuristic algorithms like tabu search (TS). By comparing the performance of structural property based TS with the simple version of TS, it is shown how much the meta-heuristic algorithm can benefit from the structural properties.

1 Introduction

Makespan minimization in *permutation flow shop scheduling problem* (PFSP) is an OR topic that has been intensively addressed in the last 50 years. Since the problem is known to be NP-complete for more than two machines, most of the research effort has been devoted to the development of heuristic procedures in order to provide good approximate solutions to the problem.

The currently reported approximation algorithms can be categorized into one of two types: constructive methods or improvement methods. Constructive methods include slope-index-based heuristics [1, 2], the CDS heuristic [3], the RA heuristic [4] and the NEH algorithm [5] (more refer to [6]). Most of the improvement approaches are based on modern meta-heuristics, such as simulated annealing [7, 8], tabu search [9, 10, 11] and genetic algorithms [12, 13, 14].

Among these algorithms, meta-heuristic algorithms perform very well. Since the framework of meta-heuristic algorithms are quite open and are problem-independent, they can be easily applied to various PFSPs. In the past 20 years, meta-heuristic algorithms were very popular in solving PFSPs and they did provide many good results. Encouraged by the meta-heuristic algorithms, more

U.K. Chakraborty (Ed.): Comput. Intel. in Flow Shop and Job Shop Sched., SCI 230, pp. 1–20.
springerlink.com

intelligent algorithms, such as *Ant Colony Optimization* (ACO) algorithm and *Particle Swarm Optimization* (PSO) algorithm are developed very quickly.

Because of the generality and portability of the meta-heuristic algorithm, many researchers tend to focus on the method innovation while ignoring the properties of the scheduling problem itself. As a result, many algorithms would catch one and lose another on optimization effect and efficiency. Many papers show that the meta-heuristic algorithm can obtain better solutions than the constructive method, and require less computational time than the exact method such as branch and bound algorithm. However, it also implies that the meta-heuristic algorithm requires more computational time than the constructive method and obtains worst results than the exact method.

Therefore, in some sense, comparing to the constructive method and the exact method, a meta-heuristic algorithm with on problem information is only a kind of compromise between optimization effect and efficiency. It conforms to the results in Kalczynski and Kamburowski [15], who find that many meta-heuristic algorithms are not better than the simple NEH algorithm after a fairer comparison. In fact, this validates the *No Free Lunch* (NFL) Theorem, which points out that all algorithms equal to the randomly blind search if no problem information is known.

On the other hand, NFL Theorem also suggests that algorithm performance can be improved by introducing problem information (problem property). For the continuous optimization problem, local search can be nicely guided by the gradient information. However, no such structural information is available for the combinatorial optimization problem to which PFSP belongs. This motivates us to study the structural properties (similar as the gradient information) and let the structural property guide the search process in solving PFSPs.

In fact, there are already a lot of structural properties studied for the PFSP in the previous research. Nowicki and Smutnicki [9] propose the block properties and successfully introduce them into the tabu search algorithm. Reeves and Yamada [13] study the distribution of local optimums in the solution space and introduce the *Big Valley* phenomenon into the genetic algorithm. These two algorithms are considered as the best two in the existing algorithms for PFSP [16, 17]. However, little attention has been devoted to establish a common framework for these properties so they can be effectively combined or extended. In this chapter, we review and classify the main contributions regarding this topic and discuss future research issues.

In this chapter, we divide the exiting structural properties into two types: neighborhood property and solution space property. The former considers the relation between two solutions, namely basic solution and its neighbor. The latter considers the statistic property of all solutions in the solution space.

The remainder of the chapter is organized as follows: Section 2 gives the definition of the problem. Structural properties of neighborhood and solution space are reviewed in Section 3 and Section 4, respectively. A structural property based tabu search is proposed and compared with the simple version of tabu search in Section 5. Section 6 concludes this chapter and gives some directions for future research.

2 Problem Definition

The *permutation flow shop scheduling problem* (PFSP) considered in this chapter is commonly defined as follows: a set $N = \{1, 2, \ldots, n\}$ of n jobs is to be processed through a set $M = \{1, 2, \ldots, m\}$ of m machines. Each job $i \in N$ is processed on machine 1 first, machine 2 second, ... , and machine m last. Thus, the work-flow in this shop is unidirectional. Associated with each job $i \in N$ and machine $j \in M$ is the known and deterministic processing time p_{ij}. All jobs are available at time zero. Each job can only be processed on at most one machine and each machine can process only one job at any time. Preemption is not allowed, i.e., once the processing of a job has started on a machine, it must be completed without interruption at that machine. Only permutation schedules are considered, i.e., different jobs have the same processing order on all machines. Let Π denote the set of all $n!$ possible permutation schedules in the solution space. Because of various simplifying assumptions about PFSPs stated above and found in the literature [18, 19], the completion time of job $\pi(i)$ at sequence position i in schedule $\pi = (\pi(1), \pi(2), \ldots, \pi(n))$ on machine j, $C_{\pi(i),j}$ can be expressed as:

$$C_{\pi(i),j} = \max\{C_{\pi(i-1),j}; C_{\pi(i),j-1}\} + p_{\pi(i),j} \tag{1}$$

with the boundary conditions $C_{\pi(i),0} = C_{\pi(0),j} = 0$ for all $i \in N$ and all $j \in M$.

Then the PFSP considered here is to find a permutation schedule $\pi \in \Pi$ such that its makespan $C_{\max}(\pi) = C_{\pi(n),m}$ is minimum. Note that expanding the recursive relation (1) above, the makespan of permutation schedule $\pi = (\pi(1), \pi(2), \ldots, \pi(n))$ can be written as either:

$$C_{\max}(\pi) = \max_{u_0 \le u_1 \le u_2 \le \ldots \le u_{m-1} \le u_m} \sum_{j=1}^{m} \sum_{i=u_{j-1}}^{u_j} p_{\pi(i),j} \tag{2}$$

where $u_0 = 1$ and $u_m = n$.

3 Neighborhood Property

For the continuous optimization problem, the relation between the basic solution and its neighbor is described as gradient information. In this section, we consider the relation between basic solution and its neighbor for the PFSP. We address the following questions: if we have evaluated the basic solution, can we know anything about its neighbor? By the relation, can we easily identify the non-improving neighbors so as to accelerate the local search process? Before we review the neighborhood property, the definition of neighborhood is given as follows.

3.1 Neighborhood

Usually, neighbors of π are generated by changing the positions of one or more jobs in π. In this chapter, we generate neighbors based on *shift* operation, which is commonly used in the existing literature. To describe the shift operation, let x, y ($x, y = 1, 2, \ldots, n$ and $x \neq y$) be two positions in π. With respect to π, a pair $v = (x, y)$ defines a shift operation, i.e., removing job $\pi(x)$ from position x and inserting it in position y. Then the shift operation v generates a neighbor of π as follows:

$$\pi_v = \begin{cases} (\pi(1), \ldots, \pi(x-1), \pi(x+1), \ldots, \pi(y), \pi(x), \pi(y+1), \ldots, \pi(n)) & \text{if } x < y \\ (\pi(1), \ldots, \pi(y-1), \pi(x), \pi(y), \ldots, \pi(x-1), \pi(x+1), \ldots, \pi(n)) & \text{if } x > y \end{cases}$$

For a given schedule π with n jobs, the original neighborhood can be expressed as

$$N_O(\pi) = \{\pi_v \mid v \in V_O\} \tag{3}$$

where $V_O = \{(x, y) \mid y \neq x, x-1;\ x, y = 1, 2, \ldots, n\}$. Generally, there are $(n-1)^2$ such neighbors in the neighborhood for a given schedule π with n jobs. It requires $O(n^3)$ time to evaluate all these neighbors as each schedule needs $O(n)$ time. This is quite time consuming especially when we repeatedly evaluate a neighborhood in a local search based algorithm.

In fact, such neighborhood is *knowledge-poor*. It is not necessary to evaluate all the $(n-1)^2$ schedules in $N_O(\pi)$. Structural properties shown in the next subsection will reveal that some neighbors are definitely not better than the basic solution. Obviously, if such non-improving neighbors are excluded from the neighborhood, the search process can be greatly accelerated.

3.2 Critical Path and Block

To describe the block property, we should introduce the definition of critical path and block first. Consider the following network $N(\pi)$ with vertex valuations for each permutation $\pi \in \Pi$. The vertex (i, j) represents the operation of job $\pi(i)$ on machine j and the valuation is the processing time $p_{\pi(i)j}$.

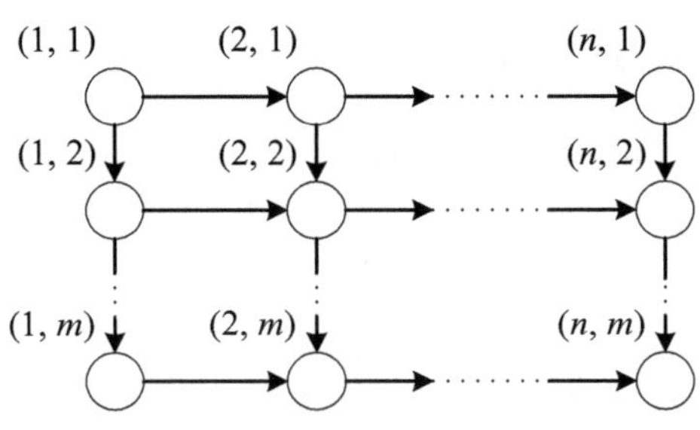

Fig. 1. Network $N(\pi)$

For any path in $N(\pi)$, its length is given by the sum of the valuations of all vertices of the path. To be convenient, let a sequence of integers $u = (u_1, u_2, \ldots, u_{m-1})$ satisfying $1 \leq u_1 \leq u_2 \ldots \leq u_{m-1} \leq n$ denote a *path* from $(1,1)$ to (m,n) in π, which contains vertices $(1,1), (2,1), \ldots, (u_1,1)$, $(u_1+1,2), \ldots, (u_2,2)$, …, $(u_{m-1}+1,m), \ldots, (n,m)$. Then the length of path u can be expressed as

$$l(u) = \sum_{j=1}^{m} \sum_{i=u_{j-1}}^{u_j} p_{\pi(i),j} \tag{4}$$

where $u_0 \equiv 1$ and $u_m \equiv n$.

Definition 1 (Critical Path). *A path $u^* = (u_1^*, u_2^*, \ldots, u_{m-1}^*)$ is called a* critical path *of π if it is the longest path in $N(\pi)$, i.e. $l(u^*) = \max_u l(u)$.*

Comparing to formulation (2), we know that the length of a critical path equals the makespan of π namely $C_{\max}(\pi)$. Then for a critical path u^* and any general path u of π, it has

$$C_{\max}(\pi) = l(u^*) \geq l(u) \tag{5}$$

Definition 2 (Block). *Based on the critical path u^*, a sequence of jobs $B_k = (\pi(u_{k-1}^*), \pi(u_{k-1}^*+1), \ldots, \pi(u_k^*))$ is called the kth* block *in π, $k = 1, 2, \ldots, m$. And the kth* internal block *is defined as a subsequence of B_k:*

$$B_k^* = \begin{cases} B_k - \{\pi(u_1^*)\} & \text{if } k = 1, \\ B_k - \{\pi(u_{k-1}^*), \pi(u_k^*)\} & \text{if } 1 < k < m, \\ B_k - \{\pi(u_{m-1}^*)\} & \text{if } k = m. \end{cases}$$

The use of these definitions is illustrated in the following example.

Example 1. Figure 2 shows a schedule of seven jobs on three machines. The permutation is $\pi = (1,7,3,2,4,5,6)$ and its critical path is $u^* = (2,6)$ which generates three blocks: $B_1 = (1,7)$, $B_2 = (7,3,2,4,5)$ and $B_3 = (5,6)$ and three relevant internal blocks: $B_1^* = (1)$, $B_2^* = (3,2,4)$ and $B_3^* = (6)$.

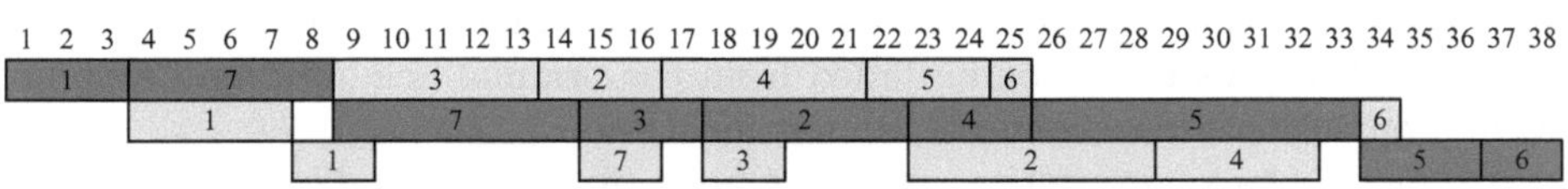

Fig. 2. Gantt chart of permutation $\pi = (1,7,3,2,4,5,6)$

3.3 Block Property

With the definition of block, Nowicki and Smutnicki [9] propose the neighborhood property.

Theorem 1 (Block Property 1 [9]). *Shifting a job within the internal block does not generate a better neighbor.*

Let permutation π_v be the neighbor generated by shifting a job within an internal block of permutation π. Theorem 1 holds because π_v has a general path containing exactly the same vertices as the critical path of π [9]. By formulation (5), we can know $C_{\max}(\pi_v) \geq C_{\max}(\pi)$ without evaluating π_v.

While Theorem 1 deals with shifting a job within the same block, the following theorem considers shifting a job from one block to another block.

Theorem 2 (Block Property 2 [10, 11]). *Suppose π_v is generated by move $v = (x, y)$, where jobs $\pi(x)$ and $\pi(y)$ are in the p-th and l-th internal blocks of π, respectively. Then it has*

$$C_{\max}(\pi_v) \geq C_{\max}(\pi) + p_{\pi(x)l} - p_{\pi(x)p} \tag{6}$$

Theorem 2 holds because π_v has a general path where there is only one vertex different from the critical path of π [10]. In fact, moving a job within the same internal block implies $l = p$ in Theorem 2. Therefore, Theorem 1 can be considered as a special situation of Theorem 2.

Obviously, Theorem 2 gives a lower bound of π_v. With the lower bound, a lot of non-improving neighbors can be identified and excluded from the neighborhood in $O(1)$ time (note that evaluating a neighbor requires $O(mn)$ time). It will greatly save the computational effort in a local search algorithm.

Example 2. For the data and permutation $\pi = (7, 1, 3, 2, 4, 5, 6)$ in Example 1, suppose job 3 is to be shifted, i.e. $\pi(x) = 3$. By Theorem 1, we can easily identify that neighbors (1,7,2,**3**,4,5,6) and (1,7,2,4,**3**,5,6) are not better than π, as they are generated by shifting job 3 within the same block. By Theorem 2, neighbors (**3**,1,7,2,4,5,6) and (1,**3**,7,2,4,5,6), which are generated by shifting job 3 to the first block, have lower bounds of $40 (= 38 + 5 - 3)$. Therefore, without evaluating the exact makespans of these two neighbors, we know they are not better than π either. Similarly, we can know lower bounds for (1,7,2,4,5,**3**,6) and (1,7,2,4,5,6,**3**) are $37 (= 38 + 2 - 3)$ without extra evaluation.

Lower bounds of neighbors of π are summarized in Table 1. There are totally 36 neighbors of π but we can obtain tight lower bounds for 26 of them. From Table 1, 14 neighbors can be excluded from the original neighborhood without evaluation as their lower bounds are not less than $C_{\max}(\pi)$.

Although block properties are developed for the standard PFSP with makespan criterion, they have been extended to PFSPs in more complex environment, such as PFSP with mixed no-wait/no-store [20], or with buffers /blocking/finite intermediate storage [21, 22, 23, 24]. The ideas of critical path structure and blocks of jobs have also been extended to other scheduling problem with other criterions [25, 26, 27, 28, 29, 30]. Block properties have successfully been introduced into various meta-heuristics for solving the PFSP, such as TS [9, 10, 11, 20, 21, 22, 23, 24, 25, 26, 27, 31, 32], GA [13, 33] and SA [34]. References [9] and [13], which successfully employ the block properties, are considered the

Table 1. Lower bound of π_v

$\pi(x)$	π_v	$LB(\pi_v)$
1	(7,**1**,3,2,4,5,6),(7,3,**1**,2,4,5,6),(7,3,2,**1**,4,5,6),(7,3,2,4,**1**,5,6)	39
	(7,3,2,4,5,**1**,6),(7,3,2,4,5,6,**1**)	37
3	(**3**,1,7,2,4,5,6),(1,**3**,7,2,4,5,6)	40
	(1,7,2,**3**,4,5,6),(1,7,2,4,**3**,5,6)	38
	(1,7,2,4,5,**3**,6),(1,7,2,4,5,6,**3**)	37
2	(**2**,1,7,3,4,5,6),(1,**2**,7,3,4,5,6)	36
	(1,7,3,4,**2**,5,6)	38
	(1,7,3,4,5,**2**,6),(1,7,3,4,5,6,**2**)	39
4	(**4**,1,7,3,2,5,6),(1,**4**,7,3,2,5,6)	40
	(1,7,**4**,3,2,5,6)	38
6	(**6**,1,7,3,2,4,5),(1,**6**,7,3,2,4,5)	37
	(1,7,**6**,3,2,4,5),(1,7,3,**6**,2,4,5),(1,7,3,2,**6**,4,5),(1,7,3,2,4,**6**,5)	37

best two papers in solving PFSP with makespan criterion [16, 17]. Block properties have also been used for improving the classic NEH algorithm [35] or for the worst-case analysis [36].

3.4 Statistic Analysis on the Block Property

Computational results from the above references show that block properties can greatly reduce the computational effort of meta-heuristics. However, it brings up a new question: how much effort be saved by the block properties from the view of statistics. Then the following two questions are of interest:

1) How large is the average number of neighbors in which there exists a general path with the same vertices as the critical path of π? These neighbors cannot lead to a cost improvement as $C_{\max}(\pi_v) \geq C_{\max}(\pi)$. Denote $E|U_0(m,n)|$ such average number for the PFSP with m machines and n jobs.

2) How large is the average number of neighbors in which there exists a general path that differs only by one vertex from the critical path of π? For such neighbors we can easily state a lower bound for the objective value. Denote $E|U_1(m,n)|$ such average number for the PFSP with m machines and n jobs.

In fact, as early as at the beginning of 1990s, the answers were given by Werner [37], who studies the path structure of PFSP. However, we should note that the following theorems are based on the random PFSP, in which all processing times are randomly generated from the same distribution[1]. Let $z(a,b) = \binom{a+b-2}{b-1}$.

Theorem 3 ([37]). *Let $n \geq 3$ and $m \geq 2$. Then*

$$E|U_0(m,n)| = \frac{2[(n-1)\cdot z(m+1,n-2) - z(m+2,n-3)] - z(n-2,m)}{z(n,m)} \tag{7}$$

[1] For more information about random PFSP and structured PFSP, please refer to [38, 39].

Theorem 4 ([37]). *Let $n \geq 2$ and $m \geq 3$. Then*

$$E|U_1(m,n)| = \frac{3(n-1)\cdot z(n-1,m-1) + 2\binom{n-1}{2}\cdot z(n-1,m)}{z(n,m)} - \frac{2(n-2)\cdot z(m+1,n-4) - 2z(m+2,n-5)}{z(n,m)} \tag{8}$$

Figure 3(a) and Figure 3(b) show the percentage of $E|U_0(m,n)|$ and $E|U_1(m,n)|$ with respect to neighborhood size $(n-1)^2$, respectively. For the given number of machines m, the expectation values of both $E|U_0(m,n)|$ and $E|U_1(m,n)|$ increase when the number of jobs n increases. Figure 3(a) indicates that the percentage of non-improving neighbors is rather large for large ratio of n/m. Figure 3(b) shows that it is possible to obtain lower bounds for more than half of the neighborhood in theory.

From Theorem 1 to Theorem 3, it is clear that meta-heuristics, such as TS or SA, can be greatly benefited from the adaptive neighborhoods which take the block properties into consideration.

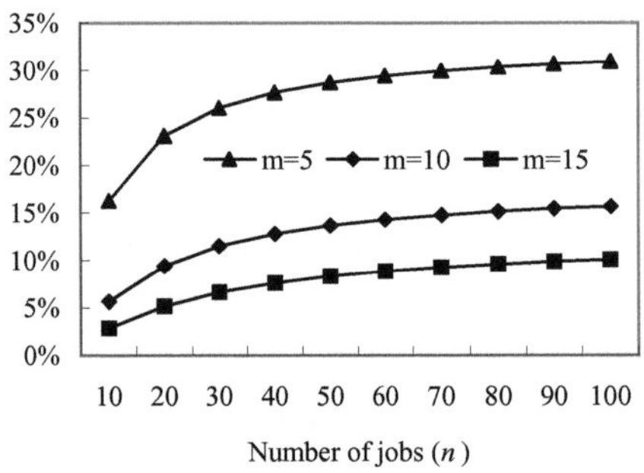

(a) Percentage of $E|U_0(m,n)|$

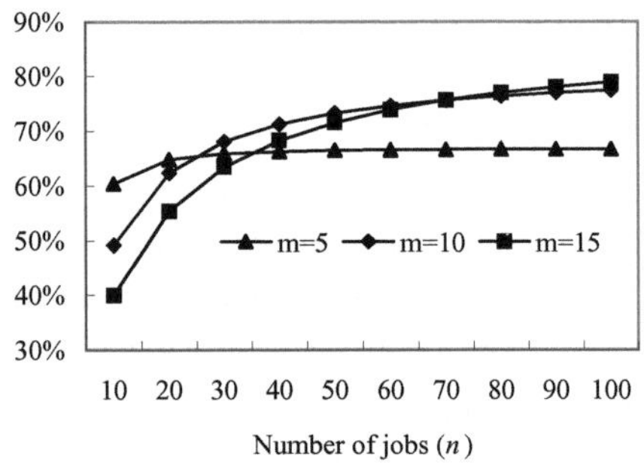

(b) Percentage of $E|U_1(m,n)|$

Fig. 3. Percentage of (a) $E|U_0(m,n)|$ and (b) $E|U_1(m,n)|$ with respect to neighborhood size $(n-1)^2$

4 Solution Space Property

It is clear that the search process should be adjusted to peculiar properties of the solution space. However, research on the solution space property is not so extensive as the neighborhood property. Till now, only some space phenomena have already been detected and reported, including the big valley phenomenon and the normality of makespan distribution.

4.1 Big Valley Phenomenon

In fact, the notion of *big valley* is not precisely defined. However, it visualizes the structure of solution space and implies: 1) local optima are radially distributed in the problem space relative to a global optimum at the center; 2) the more distant the local optima are from the center, the worse are their objective function values.

The concept of a big-valley structure was fist introduced by instances of the Traveling Salesman Problem (TSP) [40] using the 2-opt local search operator. For the PFSP, it has been empirically demonstrated that when the shift operator defined in Section 3.1 is applied to random FSPs (such as Taillard's benchmark suite), it yields a big valley structure [13, 41]. To show the big-valley structure of PFSP, an operator-independent precedence-based measure is defined as follows:

$$D(\pi, \pi') = \frac{n(n-1)}{2} - \sum_{i,j,i \neq j} pre(i, j, \pi, \pi') \tag{9}$$

where the function $pre(i, j, \pi, \pi')$ equals 1 if job i is scheduled before job j both in permutation π and in π'; otherwise it equals 0.

Figure 4, which is taken from [13], shows the correlation between distances and relative makespans. The x-axis in Figure 4(a) represents the average distance from other local optima (MEAND), and in Figure 4(b) represents the distance from the global optima (BESTD). The y-axis represents their makespans relative to the global optimum (OBJFN). Figure 4(a) indicates that local optima tend to be relatively close to other local optima and local optima near one another have similar evaluations. Figure 4(b) shows that better local optima tend to be closer to global optima. These two plots empirically verifies the big-valley hypothesis.

Many papers have demonstrated the existence of big-valley structure for PFSP's solution space [13, 41, 42, 43, 44].

Recently, a few more results on big-valley structure are presented. Notable results are: 1) for the random PFSP, big-valley structure holds for all solutions

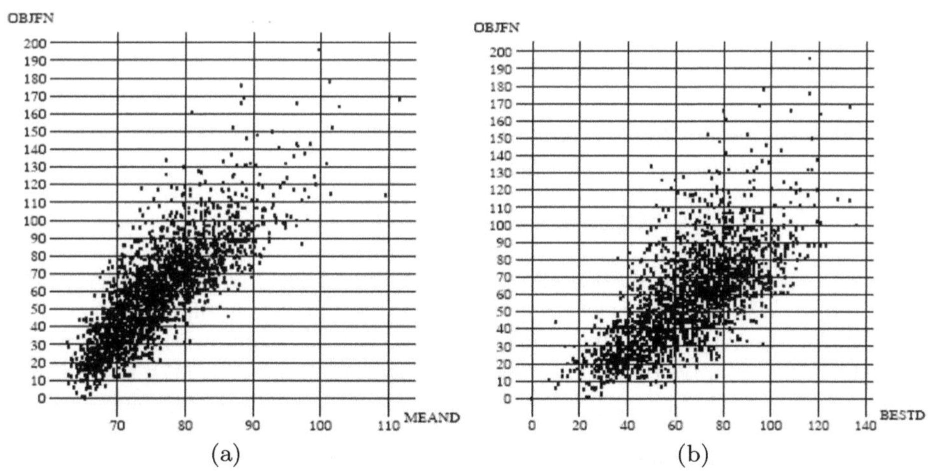

Fig. 4. 2313 distinct local optima for the ta021 (20×20) problem are plotted in terms of (a) average distance from other local optima and (b) distance from global optima (x-axis), against their relative makespans(y-axis)[13]

from the space, not only for local minima [42]; 2) for the structured PFSP, the big-valley structure degrades into a stepped valley structure composed of plateaus of equivalent fitness solutions [43, 44].

Such big-valley structure suggests that when going along trajectory linking two local optima, it is possible to find a new local optimum or even a global one. This is the foundation of scatter search and path relinking. Therefore, big-valley structure is widely applied in scatter search and path relinking, which is commonly employed as a part of GA or TS [13, 42].

4.2 Normality of the Makespan Distribution

Since FSP is NP-hard in the strong sense [45], numerous heuristic algorithms have proposed for finding optimal or near optimal schedules [16, 46]. An inherent shortcoming, which is common for most heuristic algorithms for combinatorial problems, is that it is difficult to evaluate the *goodness* of the heuristic solution, i.e., find the gap between the value of a heuristic solution and its corresponding optimal value.

A possible way to overcome this shortcoming is to study the makespan distribution in the solution space. If the distribution curve can be determined, it will be possible to determine, in the probabilistic sense, the number of better solutions that may still exist in the solution space. Because of this reason and the desire to use the developments in simulation techniques to solve scheduling problems, the makespan distribution of PFSP was first studied by Heller [47, 48] and was claimed to be asymptotically normal if the number of jobs is sufficiently large.

Heller's normality claim is very attractive. We know that a normal distribution is determined only by two parameters namely mean and variance, which can be obtained by sampling. If Heller's claim is right, the makespan distribution can be easily determined by sampling in the solution space, and then we can evaluate the goodness of a given heuristic solution of PFSP.

The normality phenomenon can be observed from the empirical results for the random PFSP. For example, we randomly sampled 200,000 schedules of Ta061 and calculated the corresponding makespan. In Figure 5, x-axis represents *relative makespan* $C^r_{\max}$, which is defined as:

$$C^r_{\max}(\pi) = C_{\max}(\pi) - C^*_{\max}(\pi)$$

where $C^*_{\max}(\pi)$ is the smallest makespan value among the 200,000 sampled schedules[2] and y-axis represents the frequency of makespan in the solution space. Figure 5 shows the makespan distribution fits the normal distribution quite well.

Besides Heller's observation [48] and the theoretical analysis [47], the normality phenomenon is also observed by several other researchers [49, 50, 51, 52][3].

[2] Such transformation shifts the makespan distribution curve to make it start from 0 but doesn't change its shape.

[3] However, Moras et al. [52] also imply that the minimum and the maximum makespan values for the PFSs are not symmetrical from the mean value of the normal distribution.

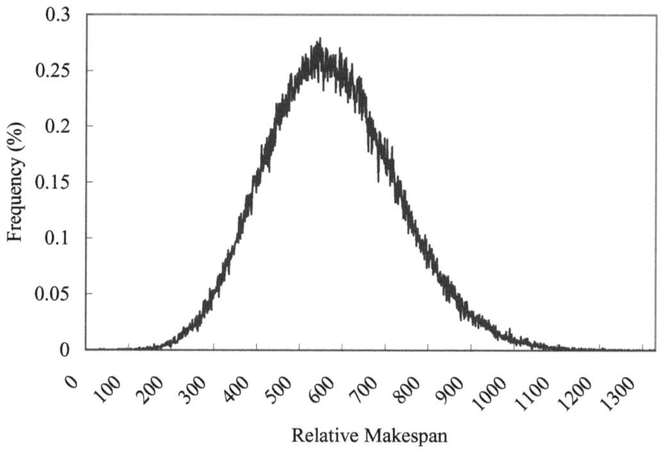

Fig. 5. Makespan distribution of random FSP (Ta061)

They support the normality claim and use it to develop solution procedures to find approximate solutions to FSPs. Elmaghraby [53] and Nowicki and Smutnicki [42] took the normality of makespan distribution as a doubtless result in their research.

However, since its first appearance, Heller's normality claim has been a topic of debate among researchers. For example, Giffler et al. [54], Nugent [55], Conway et al. [56], Gupta et al. [57] and Ashour [58] raise doubts about the validity of the claim that the makespan distribution of permutation flowshop schedules (PFSs) is normal even if the number of jobs is large. Analysis of the extreme value distribution of PFSs [59, 60] found that the left tail of the makespan distribution is different than the typical normal distribution, thus raising doubt about the validity of the normality claim. Taillard [61] mentioned the normality of makespan distribution in his experimental results. He neither confirmed nor refuted the normality claim.

While there has been considerable debate about the validity of the normality claim, there is no systematic and theoretical investigation of the makespan distribution of permutation flow shop schedules until Jin et al. [39]. They point out errors in Heller [47], which is supposed to give the proof of normality of the makespan distribution. Because of the errors, theoretical analysis in [47] can neither prove the normality nor prove the non-normality of makespan distribution. Then they theoretically and empirically investigate the makespan distribution of the structured PFSP. They show that the normality claim is *not* valid for the structured PFSPs such as job-dominated and machine-dominated PFSPs (Figure 6 shows makespan distribution for typical structured PFSPs). Therefore, Heller's claim is not right for all PFSPs, at least not right for the structural PFSPs.

However, considering the observation on random PFSP, it is still possible that makespans of random PFSP are normally distributed when the number of jobs is

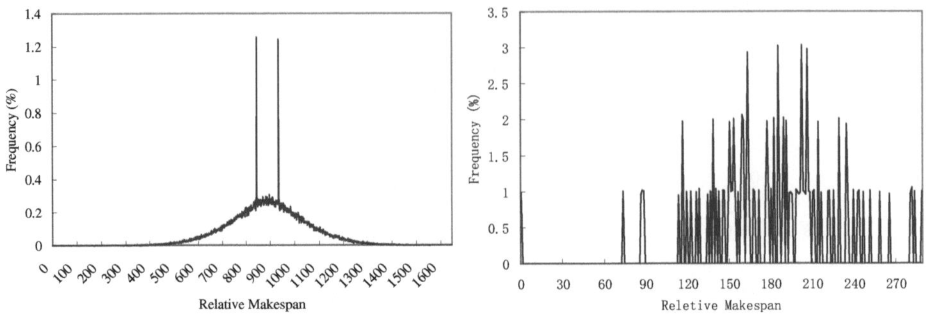

(a) Job-dominated PFSP (b) Machine-dominated PFSP

Fig. 6. Makespan distribution of structured PFSPs

very large. But the problem how to prove the normality of makespan distribution for random PFSPs remains open.

5 Case Study: Structural Property Based Tabu Search

As reviewed above, structural properties (especially the block property and big valley phenomenon) have been widely introduced into meta-heuristics. However, to reach the best performance, many other elements, that are often not explained thoroughly to the reader, are introduced the into the algorithm [62]. Therefore, it is difficult to know that how much efficiency is purely brought by the structural property[4].

Therefore, in this section, we do not attempt to propose an algorithm as good as the state-of-the-art best algorithm. Instead, we will first present a simple version of a tabu search algorithm and then extend it a little to include structural properties. By comparing the improvement in performance, it can be shown that how much meta-heuristics can benefit from the structural property.

5.1 Reduce Neighborhood by Structural Property

Consider the neighborhood generated by shift operator in Section 3.1. There are $(n-1)^2$ neighbors in the original neighborhood for a given schedule π with n jobs. Such large-sized neighborhood can assist local search methods to avoid being trapped in a bad local optimum. However, the number of neighbors in the neighborhood drastically increases with the number of jobs. For such a neighborhood, it requires $O(n^3)$ time to evaluate all the neighbors as each schedule needs $O(n)$ time. Although the computational complexity can be reduced to $O(n^2)$ by the fast computation technique [61], it is still quite time consuming especially when we repeatedly evaluate a neighborhood in the local search method.

[4] For the similar reason, Watson et al. [63] de-construct the algorithm proposed by Nowicki and Smutnicki [27] to determine the components that are integral to its performance, and the degree to which they share the responsibility.

However, we can obtain a lower bound of $C_{\max}(\pi_v)$ by Theorem 2. Denote $LB(\pi_v) = C_{\max}(\pi) + p_{\pi(x)l} - p_{\pi(x)p}$ the lower bound of π_v. Obviously, if $p_{\pi(x)l} \geq p_{\pi(x)p}$, $LB(\pi_v) \geq C_{\max}(\pi)$, which implies $C_{\max}(\pi_v) \geq C_{\max}(\pi)$. Then we can know that π_v is not better than π without evaluating π_v.

Therefore, we can reduce the original neighborhood by the structural properties as follows:

$$N(\pi, UB) = N_0(\pi) - \{\pi_v \mid LB(\pi_v) \geq UB\} \tag{10}$$

where UB is a given upper bound. Such definition excludes neighbors whose performance is worse than UB.

Let $UB = C_{\max}(\pi)$. Then the following tow points are worth noting.

(1) Note that y (the position to insert) does not appear in the right side of Formulation (6). It implies neighbors generated by shifting job $\pi(x)$ to any position in the lth block can be excluded. Since the average block size is n/m, a lot of non-improving neighbors can be excluded if $n \gg m$.

(2) According to Formulation (6), the larger $C_{\max}(\pi)$ is, the larger $LB(\pi_v)$ will be. It implies the worse the basic solution π is, the more non-promising neighbors can be excluded and then the smaller $N(\pi, UB)$ will be.

5.2 Algorithm Description and Computational Complexity

The tabu search (TS) algorithm is commonly used in solving combinatorial optimization problems. It starts from an initial basic solution and searches its neighborhood for a solution with the best performance. Then the search moves to this best one as a new basic solution, and then repeats the process until some stopping condition is satisfied. Obviously, TS algorithm is a local search based approach. It avoids being trapped at a local optimum by introducing a mechanism called tabu list, which defines some moves that are forbidden to be applied currently.

There are three basic elements in TS algorithm: initial solution, tabu list and neighborhood. The choice of neighborhood is very important as it affects the search effectiveness and efficiency. The details of elements are given as follows.

Initial Solution

The initial solution is generated by the famous NEH algorithm [5].

Tabu List

Let $T = (T_1, T_2, \ldots, T_{maxt})$ denote a tabu list where $T_i = (g, h)$ is a job pair and $maxt$ is the length of the tabu list. If search moves from a basic schedule π to its neighbor π_v through a move $v = (x, y)$, we add the job pair $(\pi(x), \pi(x+1))$ to the tabu list if $x < y$ and add $(\pi(x), \pi(x-1))$ otherwise. If the length of the tabu list exceeds $maxt$, remove the oldest element from the tabu list. During the search, move $v = (x, y)$ is forbidden if one of the following conditions is satisfied:

1) there are one or more job pairs $(\pi(j), \pi(x))$,$j = x+1, \ldots, y$ in the tabu list if $x < y$,

2)there are one or more job pairs $(\pi(x), \pi(j))$,$j = y, \ldots, x-1$ in the tabu list if $x > y$.

Neighborhood

In the simple version of TS, we employ the original neighborhood $N_0(\pi)$ and in the block property based TS, we employ the reduced neighborhood $N(\pi, C_{\max}(\pi))$.

Denote STS and BTS the simple version of TS and block property based TS, respectively.

Tabu Search Procedure

The main procedures of the two TS algorithms (STS and BTS) are almost the same. The only difference exists in the neighborhood selection. The procedure of STS algorithm is given as follows.

Algorithm STS: Simple Tabu Search

Input: Basic solution $\pi_{basic} = null$, best solution ever known $\pi^*_{cur} = null$, number of maximum iterations $maxIter$, length of tabu list $maxt$

Output: Best solution ever known π^*_{cur}

Step 1: Generate the initial solution π_0 and set $\pi_{basic} = \pi_0$ and $\pi^*_{cur} = \pi_0$;

Step 2: For 1 to $maxIter$

  Step 2.1: Generate the original neighborhood $N_0(\pi_{basic})$;

  Step 2.2: Find the best unforbidden neighbor and let it be π_{basic};

  Step 2.3: Update the tabu list;

  Step 2.4: If π_{basic} is better than π^*_{cur}, set $\pi^*_{cur} = \pi_{basic}$;

Step 3: Return π^*_{cur}.

Algorithm BST, as follows, is identical to algorithm STS except Step 2.1 (neighborhood generation). Therefore, only the modified Step 2.1 is given below.

Algorithm BTS: Block Property based Tabu Search

Step 2.1: Generate the original neighborhood $N_0(\pi_{basic})$ and reduce it to $N(\pi_{basic}, C_{\max}(\pi_{basic}))$;

Computational Complexity Analysis

In the procedure presented above, the computational complexity of Step 1 is $O(n^2)$. For Step 2 in algorithm STS, since there are $O(n^2)$ neighbors in the original neighborhood and each requires $O(n)$ times to be evaluated, the computational complexity to evaluate the original neighborhood is $O(n^3)$. However, it can be reduced to $O(n^2)$ by the fast computing technique [61]. Therefore, the computational complexity of STS is $O(n^2 * maxIter)$. Algorithm BTS only differs from STS in Step 2.1. As the reduction depends on the neighborhood structure, it is hard to exactly figure out how much computational effort can be saved by applying the structure property. However, we do know that the computational complexity BTS is not more than $O(n^2 * maxIter)$ either.

5.3 Computational Results

Algorithms STS and BTS were coded in C++ and run on a Pentium 4 computer (2.6 GHz) with 512M Bytes of memory. The two algorithms were tested on the largest 50 benchmark problems taken from Taillard [64](Ta071-Ta120). The benchmark set contains problems of various sizes, including 100, 200 and 500 jobs and ten and 20 machines respectively. There are ten problems in each problem size. Set the length of tabu list $maxt = 8$ and the maximum iteration number $maxIter = 1000$.

A measure called *PRD* (percentage relative difference) for each algorithm A is defined as follows:

$$PRD(A) = \frac{C^A_{\max} - C^*_{\max}}{C^*_{\max}} \times 100\% \tag{11}$$

where $C^A_{\max}$ is the makespan obtained by algorithm A and $C^*_{\max}$ is the optimal makespan or the best known lower bound, obtained from Taillard's homepage (http://mistic.heig-vd.ch/taillard/).

Figure 7(a) shows the average CPU time required for solving PFSPs in various sizes. For both TS algorithms, CPU time increases with the problem size. However, Figure 7(a) indicates that CPU time of STS algorithm increase much quicklier than BST algorithm. Figure 7(b) shows the CPU time required for solving the largest ten PFSPs in Taillard's benchmark suit. From Figure 7(b), it is clear that BTS algorithm requires only about half of the CPU time consumed by STS algorithm. It is because all neighbors in the neighborhood are evaluated in STS algorithm and only the promising neighbors, which are possible to yield improvement, are evaluated in BTS algorithm. Computational effort is greatly reduced by the structural property. While less time is required, PRD values in Figure 8 show that BTS algorithm can provide much better solutions than STS algorithm.

Algorithms STS and BTS are tested in the same environment and the procedure is almost the same. The only difference is the introduction of structural properties in BTS. Computational results shown in Figures 7 and 8 suggest

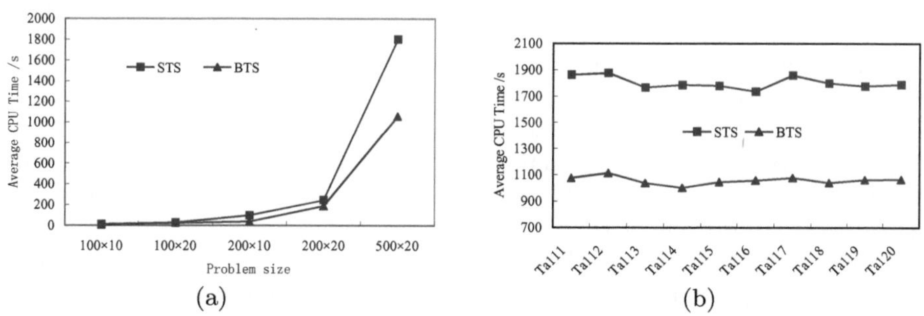

Fig. 7. Computational time for the two TS algorithms

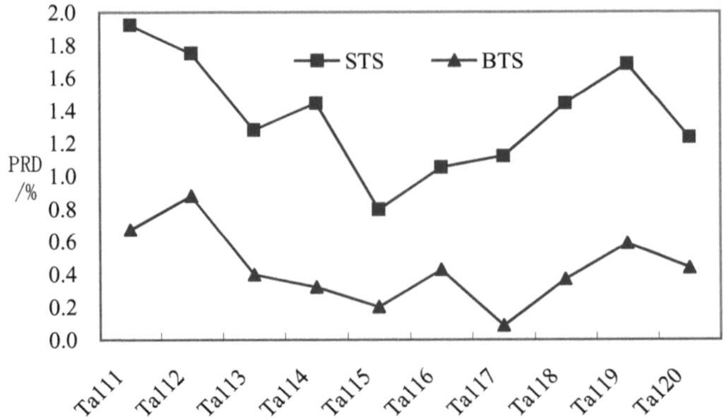

Fig. 8. PRD values for the largest ten instances

that structural properties can not only reduce the computational effort, but also enhance the solution quality.

6 Conclusions and Directions of Future Research

In this chapter, we reviewed the structural property of permutation flow shop scheduling problem with makespan criterion. We mainly considered two types of structural properties, including neighborhood properties (block properties) and solution space properties (big-valley phenomenon and normality of makespan distribution). For each part, we tried to give a brief literature review by noting contributions and gave a glimpse of meta-heuristics which employed the structural properties. We also gave an example to show how to introduce the structural properties into meta-heuristic algorithms like Tabu Search. By comparing the performance of the simple version of tabu search (STS) and block property based tabu search (BTS), we have shown how much the meta-heuristic can benefit from the structural property.

From the above discussion, it is clear that structural properties are important to meta-heuristics and require continued research. Based on our review of existing research work, we suggest the following fruitful directions for future research:

1) Since block property has successfully applied in meta-heuristic for solving PFSPs with makespan criterion, it is possible to extend the idea of critical path to PFSPs with other criterions and more realistic constraints.

2) The problem, whether the makespan distribution of random FSP is normal, remains open. As most test problems in the existing belongs to the random FSP, it is valuable to prove it or refute it.

3) The description of big-valley structure is intuitive but not precise. Therefore, we may mathematically formulate the big-valley structure and make the property more clear and powerful.

4) We may introduce structural properties into more meta-heuristics to develop more efficient algorithms.

Acknowledgments

This work is partially supported by 973 Program of China under 2002CB312205, National Science Foundation of China under 60574077 and 60874071, 863 Program of China under 2007AA04Z102.

References

[1] Palmer, D.S.: Sequencing jobs through a multistage process in the minimum total time: a quick method of obtaining a near optimum. Operational Research Quarterly 16(1), 101–107 (1965)
[2] Gupta, J.N.D.: A functional heuristic algorithm for the flowshop scheduling problem. Operational Research Quarterly 22(1), 39–47 (1971)
[3] Campbell, H.G., Dudek, R.A., Smith, M.L.: A heuristic algorithm for the n job, m machine sequencing problem. Management Science 16(10), 630–637 (1970)
[4] Dannenbring, D.G.: An evaluation of flow shop sequencing heuristics. Management Science 23(11), 1174–1182 (1977)
[5] Nawaz, M., Enscore Jr., E.E., Ham, I.: A heuristic algorithm for the m-machine, n-job flow-shop sequencing problem. Omega-International Journal of Management Science 11(1), 91–95 (1983)
[6] Laha, D., Chakraborty, U.K.: A constructive heuristic for minimizing makespan in no-wait flowshop scheduling. International Journal of Advanced Manufacturing Technology (2008), doi: 10.1007/s00170-008-1454-0
[7] Osman, I.H., Potts, C.N.: Simulated annealing for permutation flow-shop scheduling. Omega-International Journal of Management Science 17(6), 551–557 (1989)
[8] Ogbu, F.A., Smith, D.K.: Application of the simulated annealing algorithm to the solution of the cmax flowshop problem. Computers & Operations Research 17(3), 243–253 (1990)
[9] Nowicki, E., Smutnicki, C.: A fast tabu search algorithm for the permutation flowshop problem. European Journal of Operational Research 91(1), 160–175 (1996)
[10] Grabowski, J., Pempera, J.: New block properties for the permutation flow shop problem with application in tabu search. Journal of the Operational Research Society 52(2), 210–220 (2001)
[11] Grabowski, J., Wodecki, M.: A very fast tabu search algorithm for the permutation flow shop problem with makespan criterion. Computers & Operations Research 31(11), 1891–1909 (2004)
[12] Reeves, C.R.: A genetic algorithm for flowshop sequencing. Computers & Operations Research 22(1), 5–13 (1995)
[13] Reeves, C.R., Yamada, T.: Genetic algorithms, path relinking, and the flowshop sequencing problem. Evolutionary Computation 6(1), 45–60 (1998)
[14] Wang, L., Zheng, D.Z.: An effective hybrid heuristic for flow shop scheduling. International Journal of Advanced Manufacturing Technology 21(1), 38–44 (2003)
[15] Kalczynski, P.J., Kamburowski, J.: On the NEH heuristic for minimizing the makespan in permutation flow shops. Omega-International Journal of Management Science 35(1), 53–60 (2007)
[16] Framinan, J.M., Gupta, J.N.D., Leisten, R.: A review and classification of heuristics for permutation flow-shop scheduling with makespan objective. Journal of the Operational Research Society 55(12), 1243–1255 (2004)

[17] Ruiz, R., Maroto, C.: A comprehensive review and evaluation of permutation flowshop heuristics. European Journal of Operational Research 165(2), 479–494 (2005)
[18] Gupta, J.N.D.: A review of flowshop scheduling research. In: Ritzman, L.P., Krajewski, L.J., Berry, W.L., Goodman, S.M., Hardy, S.T., Vitt, L.D. (eds.) Disaggregation Problems in Manufacturing and Service Organizations, pp. 363–388. Martin Nijhoff Publishers, The Hague (1979)
[19] Gupta, J.N.D., Stafford Jr., E.F.: Flowshop scheduling research after five decades. European Journal of Operational Research 169(3), 699–711 (2006)
[20] Grabowski, J., Pempera, J.: Sequencing of jobs in some production system. European Journal of Operational Research 125(3), 535–550 (2000)
[21] Smutnicki, C.: A two-machine permutation flow shop scheduling problem with buffers. Or. Spektrum 20(4), 229–235 (1998)
[22] Grabowski, J., Pempera, J.: The permutation flow shop problem with blocking. A tabu search approach. Omega-International Journal of Management Science 35(3), 302–311 (2007)
[23] Nowicki, E.: The permutation flow shop with buffers: A tabu search approach. European Journal of Operational Research 116(1), 205–219 (1999)
[24] Li, S.H., Tang, L.X.: A tabu search algorithm based on new block properties and speed-up method for permutation flow-shop with finite intermediate storage. Journal of Intelligent Manufacturing 16(4-5), 463–477 (2005)
[25] Nowicki, E., Zdrzalka, S.: Single machine scheduling with major and minor setup times: a tabu search approach. Journal of the Operational Research Society 47(8), 1054–1064 (1996)
[26] Nowicki, E., Smutnicki, C.: A fast taboo search algorithm for the job shop problem. Management Science 42(6), 797–813 (1996)
[27] Nowicki, E., Smutnicki, C.: New algorithm for the job shop problem. Technical report, Institute of Engineering Cybernetics, Wroclaw University of Technology (2003)
[28] Bozejko, W., Grabowski, J., Wodecki, M.: Block approach - tabu search algorithm for single machine total weighted tardiness problem. Computers & Industrial Engineering 50(1-2), 1–14 (2006)
[29] Bozejko, W., Wodecki, M.: A new inter-island genetic operator for optimization problems with block properties. In: Rutkowski, L., Tadeusiewicz, R., Zadeh, L.A., Żurada, J.M. (eds.) ICAISC 2006. LNCS, vol. 4029, pp. 334–343. Springer, Heidelberg (2006)
[30] Jin, F., Song, S.J., Wu, C.: A simulated annealing algorithm for single machine scheduling problems with family setups. Computers & Operations Research (2008), `http://dx.doi.org/10.1016/j.cor.2008.08.001`
[31] Nowicki, E., Smutnicki, C.: The flow shop with parallel machines: A tabu search approach. European Journal of Operational Research 106(2-3), 226–253 (1998)
[32] Negenman, E.G.: Local search algorithms for the multiprocessor flow shop scheduling problem. European Journal of Operational Research 128(1), 147–158 (2001)
[33] Tseng, L.-Y., Lin, Y.-T.: A hybrid genetic algorithm for the flow-shop scheduling problem. In: Ali, M., Dapoigny, R. (eds.) IEA/AIE 2006. LNCS (LNAI), vol. 4031, pp. 218–227. Springer, Heidelberg (2006)
[34] Wodecki, M., Bozejko, W.: Solving the flow shop problem by parallel simulated annealing. In: Wyrzykowski, R., Dongarra, J., Paprzycki, M., Waśniewski, J. (eds.) PPAM 2001. LNCS, vol. 2328, pp. 236–244. Springer, Heidelberg (2002)

[35] Jin, F., Song, S.J., Wu, C.: An improved version of the NEH algorithm and its application to large-scale flow-shop scheduling problems. IIE Transactions 39(2), 229–234 (2007)
[36] Smutnicki, C.: Some results of the worst-case analysis for flow shop scheduling. European Journal of Operational Research 109(1), 66–87 (1998)
[37] Werner, F.: On the combinatorial structure of the permutation flow shop problem. ZOR, Methods and Models of Operations Research 35(4), 273–289 (1991)
[38] Watson, J.P., Barbulescu, L., Whitley, L.D., Howe, A.E.: Contrasting structured and random permutation flow-shop scheduling problems: Search-space topology and algorithm performance. Informs Journal on Computing 14(2), 98–123 (2002)
[39] Jin, F., Gupta, J.N.D., Song, S.J., Wu, C.: Makespan distribution of permutation flowshop schedules. Journal of Scheduling 11(6), 421–432 (2008)
[40] Boese, K.D., Kahng, A.B., Muddu, S.: A new adaptive multi-start technique for combinatorial global optimizations. Operations Research Letters 16(2), 101–113 (1994)
[41] Reeves, C.R.: Landscapes, operators and heuristic search. Annals of Operations Research 86, 473–490 (1999)
[42] Nowicki, E., Smutnicki, C.: Some aspects of scatter search in the flow-shop problem. European Journal of Operational Research 169(2), 654–666 (2006)
[43] Watson, J.P., Barbulescu, L., Howe, A.E., Whitley, L.D.: Algorithm performance and problem structure for flow-shop scheduling. In: Proceedings of the National Conference on Artificial Intelligence, pp. 688–696 (1999)
[44] Barbulescu, L., Watson, J.P., Whitley, L.D., Howe, A.E.: Problem structure and flow-shop scheduling. In: Proceedings of the Sixteenth Congreso de Ecuaciones Diferencialesy Aplicaciones, pp. 27–38 (1999)
[45] Garey, M.R., Johnson, D.S.: Computers and intractability. A guide to the theory of NP-completeness. Freeman, New York (1979)
[46] Reza Hejazi, S., Saghafian, S.: Flowshop-scheduling problems with makespan criterion: A review. International Journal of Production Research 43(14), 2895–2929 (2005)
[47] Heller, J.: Combinatorial, probabilistic, and statistical aspects of an mxj scheduling problem. Technical Report NYO-2540, Institute of Mathematical Sciences. New York University, New York (1959)
[48] Heller, J.: Some numerical experiments for mxj flow shop and its decision-theoretical aspects. Operations Research 8(2), 178–184 (1960)
[49] Pulle, C.V.: An analysis of Inter-relationship of multiple criteria in a flowshop with set-up sequence dependence. PhD thesis, Texas Tech University (1976)
[50] Azim, M.A., Moras, R.G., Smith, M.L.: Antithetic sequences in flow shop scheduling. Computers & Industrial Engineering 17(1-4), 353–358 (1989)
[51] Caffrey, J., Hitchings, G.: Makespan distributions in flow shop scheduling. International Journal of Operations & Production Management 15(3), 50–58 (1995)
[52] Moras, R., Smith, M.L., Kumar, K.S., Azim, M.A.: Analysis of antithetic sequences in flowshop scheduling to minimize makespan. Production Planning and Control 8(8), 780–787 (1997)
[53] Elmaghraby, S.E.: The machine sequencing problem review and extensions. Technical report (1968)
[54] Giffler, B., Thompson, G.L., Van Ness, V.: Numerical experience with linear and monte carlo algorithms for solving scheduling problems. In: Muth, J.F., Thompson, G.L. (eds.) Industrial Scheduling. Prentice Hall, Englewood Cliffs (1963)

[55] Nugent, C.E.: On sampling approaches to the solution of n-by-m static sequencing problem. PhD thesis, Cornell University (1964)
[56] Conway, R.W., Maxwell, W.L., Miller, L.W.: Theory of Scheduling. John Wiley and Sons Inc., New York (1967)
[57] Gupta, J.N.D., Smith, M.L., Martz, H.F., Dudek, R.A.: Monte carlo experimentation with flowshop scheduling problem. Technical Report QT-103-68, Department of Industrial Engineering, Texas Technological College (1968)
[58] Ashour, S.: Sequencing Theory. Springer, New York (1972)
[59] Dannenbring, D.G.: Procedures for estimating optimal solution values for large combinatorial problems. Management Science 23(12), 1273–1283 (1977)
[60] Panwalker, S.S., Charles, O.E.: Analysis of the left tail for the makespan distribution in flowshop problems. Journal of Operational Research Society of India 18(4), 215–220 (1981)
[61] Taillard, E.: Some efficient heuristic methods for the flow shop sequencing problem. European Journal of Operational Research 47(1), 65–74 (1990)
[62] Ruiz, R., Stutzle, T.: A simple and effective iterated greedy algorithm for the permutation flowshop scheduling problem. European Journal of Operational Research 177(3), 2033–2049 (2007)
[63] Watson, J.P., Howe, A.E., Whitley, L.D.: Deconstructing nowicki and smutnicki's i-tsab tabu search algorithm for the job-shop scheduling problem. Computers & Operations Research 33(9), 2623–2644 (2006)
[64] Taillard, E.: Benchmarks for basic scheduling problems. European Journal of Operational Research 64(2), 278–285 (1993)

Scheduling in Flowshops with No-Idle Machines

Rubén Ruiz, Eva Vallada, and Carlos Fernández-Martínez

Grupo de Sistemas de Optimización Aplicada. Instituto Tecnológico de Informática (ITI). Ciudad Politécnica de la Innovación, Edificio 8G. Acceso B. Universidad Politécnica de Valencia. Camino de Vera s/n, 46022 Valencia, Spain
rruiz@eio.upv.es, evallada@eio.upv.es, cfernandez@iti.upv.es

Summary. This chapter deals with an interesting and not so well studied variant of the classical permutation flowshop problem with makespan criterion. In the studied variant, no idle time is allowed on machines. In order to ensure this no-idle constraint, the start times of jobs on machines must be delayed until all assigned jobs can be processed without incurring in idle times. This is a real situation arising in practice when expensive machinery is operated or when specific machines cannot be easily started and stopped due to technological constraints.

We provide a comprehensive characterization and modelization of the no-idle permutation flowshop, along with a detailed literature review. Existing methods are critically evaluated. We propose several improvements over existing approaches as well as adaptations of state-of-the-art algorithms that were proposed for related problems. An extensive computational campaign is conducted. Results are carefully analyzed by means of sound statistical techniques. The results indicate that the recent Iterated Greedy methods outperform existing algorithms by a significant margin.

1 Introduction

Flowshop scheduling is a very active field of research with close to 55 years of history. Flowshop problems are easy to formulate yet remarkably complex, both from a mathematical as well as from a computational point of view. In a flowshop, there is a set $N = 1, 2, \ldots, n$ of n unrelated product orders to produce. These are usually referred to as "jobs". The production shop is composed of a set $M = 1, \ldots, m$ of m machines that are disposed in series. Each job visits each machine in order. This order might be, without loss of generality, machine 1, machine 2 and so on until machine m. As a result of this, each job $j, j \in N$ is composed of m serial tasks, each one to be performed on a machine $i, i \in M$. The processing time of the tasks is referred to as $p_{j,i}$ which basically denotes the non-negative, known and deterministic processing time of job j at machine i.

The flowshop problem then consists of finding a production sequence of the n jobs in the m machines so that a given performance criterion is optimized. The total number of feasible solutions to this problem is derived from the possible job's arrangements on machines. For each machine, we have $n!$ possible job permutations. Thus, the total number of feasible solutions or schedules is $(n!)^m$.

U.K. Chakraborty (Ed.): Comput. Intel. in Flow Shop and Job Shop Sched., SCI 230, pp. 21–51.
springerlink.com

However, this general case is seldom considered in the flowshop research field. Instead, a simplification is to consider a single permutation of jobs for all the machines. This brings the overall number of solutions down to $n!$ Under this simplification, the problem is referred to as permutation flowshop scheduling problem or PFSP in short.

In this chapter we study an interesting variant of this problem where no idle time is allowed on machines. As we will see, this results in a different problem. The chapter continues with detailed characterizations of both the regular as well as the no-idle flowshops. Later, the chapter also provides a detailed literature review in Section 2, along with a discussion of existing approaches, improvements over published methods and adaptations of high performing state-of-the-art algorithms in Section 3. A complete computational and statistical campaign is performed in Section 4. Finally, conclusions and suggestions for future research are given in Section 5.

1.1 Regular Flowshop Problem

Before going into details, a more formal definition of the PFSP is given. First of all, a number of assumptions are usually considered: (Baker, 1974):

- All jobs are independent and available for processing at time 0.
- All machines are continuously available.
- Each machine can process at most one job at a time and each job can be processed only on one machine at a time.
- The processing of a given job at a machine cannot be interrupted once started, i.e, no preemption is allowed.
- Setup times are sequence independent and are included in the processing times or are otherwise ignored.
- An infinite in-process storage buffer is assumed. If a given job needs an unavailable machine then it joins a queue of unlimited size waiting for that machine.

Most optimization criteria are based on the completion times of the jobs at the different machines which are denoted by $C_{j,i}$. Similarly, C_j denotes the time at which job j is completed at the last machine. The completion times $C_{j,i}$ can be easily calculated as follows:

Given a permutation π of n jobs, where $\pi_{(j)}$ denotes the job in the j-th position, the completion times are calculated in $\mathcal{O}(nm)$ with the following recursive expression:

$$C_{\pi_{(j)},i} = \max\left\{C_{\pi_{(j)},i-1}, C_{\pi_{(j-1)},i}\right\} + p_{\pi_{(j)},i} \tag{1}$$

where $C_{\pi_{(j)},0} = 0$ and $C_{\pi_{(0)},i} = 0$, $\forall i \in M, \forall j \in N$. The most common optimization criterion is the minimization of the maximum completion time or makespan (C_{max}) where $C_{max} = C_{\pi_{(n)},m}$. Under this objective, the PFSP is denoted as $F/prmu/C_{max}$ following the well known $\alpha|\beta|\gamma$ notation for scheduling problems given in Graham et al. (1979).

The earliest research papers on the PFSP focused on makespan minimization. The seminal paper of Johnson (1954) is widely recognized as the first study. However, a closer look brings even earlier papers, like the one of Salveson (1952).

Johnson mainly studied the PFSP with only two machines ($m = 2$) and provided a polynomial algorithm of $\mathcal{O}(n \log n)$ steps to solve this special case to optimality. The three or more machines problem is known to be $\mathcal{NP}$-Complete in the strong sense (Garey et al., 1976).

Exact approaches for the PFSP under makespan criterion (PFSP-C_{max}) are fairly effective, but only for a small number of jobs, and specially, machines. For the sake of completeness, and in order to completely characterize the PFSP, we introduce the following Mixed Integer Programming (MIP) model. Note that this is a well known model and certainly not the only possible one.

Decision variables:

$$X_{j,k} = \begin{cases} 1, & \text{if job } j \text{ occupies position } k \text{ of the sequence} \\ 0, & \text{otherwise} \end{cases}$$
$$j, k = \{1, \dots, n\}$$
$$C_{k,i} = \text{Completion time of job at position } k \text{ on machine } i$$
$$k = \{1, \dots, n\}, i = \{1, \dots, m\}$$

Objective function:

$$\min C_{max} = C_{n,m} \tag{2}$$

Constraints:

$$\sum_{k=1}^{n} X_{j,k} = 1, \quad j = \{1, \dots, n\} \tag{3}$$

$$\sum_{j=1}^{n} X_{j,k} = 1, \quad k = \{1, \dots, n\} \tag{4}$$

$$C_{k,1} \geq \sum_{j=1}^{n} X_{j,k} \cdot p_{j,1}, \quad k = \{1, \dots, n\} \tag{5}$$

$$C_{k,i} \geq C_{k,i-1} + \sum_{j=1}^{n} X_{j,k} \cdot p_{j,i}, \quad k = \{1, \dots, n\}, i = \{2, \dots, m\} \tag{6}$$

$$C_{k,i} \geq C_{l,i} + \sum_{j=1}^{n} X_{j,k} \cdot p_{j,i}, \quad k = \{2, \dots, n\}, l = \{1, \dots, k-1\}, i = \{1, \dots, m\} \tag{7}$$

$$C_{k,i} \geq 0, \quad k = \{1, \dots, n\}, i = \{1, \dots, m\} \tag{8}$$

$$X_{j,k} \in \{0, 1\}, \quad j, k = \{1, \dots, n\} \tag{9}$$

We can see that minimizing C_{max} is equivalent to minimizing the completion time of the job in the last position of the sequence and on the last machine.

Constraint sets (3) and (4) ensure that each position is occupied by exactly one job. Sets (5) and (6) control the completion times of all jobs in the first and on subsequent machines, making sure that these completion times are larger than those of previous machines. With constraint set (7) we ensure that completion times also take into account jobs in preceding positions on all machines. Finally, sets (8) and (9) define the nature of the decision variables.

The PFSP-C_{max} has been thoroughly studied in the literature. Here we provide just some of the most cited papers, like the heuristics by Page (1961), Palmer (1965), Campbell et al. (1970) or Dannenbring (1977). By far, the most known heuristic for the $F/prmu/C_{max}$ problem is the NEH by Nawaz et al. (1983). NEH is considered as the champion among heuristics, according to many studies like Turner and Booth (1987), Taillard (1990) and more recently, Ruiz and Maroto (2005). As a matter of fact, in the study of Ruiz and Maroto, NEH is confronted against more modern –and complex– heuristics like the ones of Koulamas (1998), Suliman (2000) and Davoud Pour (2001) and NEH is proved to perform better.

Apart from the review and computational evaluation of Ruiz and Maroto (2005), the reader might find additional valuable information in the reviews of Framinan et al. (2004) and Hejazi and Saghafian (2005).

Of course, C_{max} is not the only criterion studied. Total completion time, defined as $TCT = \sum_{j=1}^{n} C_j$, results in a $\mathcal{NP}$-Hard problem already for $m \geq 2$ (Gonzalez and Sahni, 1978). Furthermore, if there are no release times for the jobs, i.e., if $r_j = 0, \forall j \in N$, then the total or average completion time equals the total or average flowtime, denoted as F in the literature. Other studied criteria are those based in due dates. Given a due date d_j for job j, T_j denotes the tardiness of job j, which is defined as $T_j = \max\{C_j - d_j, 0\}$. Total tardiness minimization results in a $\mathcal{NP}$-Hard problem in the strong sense for $m \geq 2$ as shown in Du and Leung (1990). A recent review for the total tardiness version of the PFSP (the $F/prmu/\sum T_j$ problem) is given by Vallada et al. (2008). Lastly, there is a recent trend in which several objectives are jointly considered. A comprehensive review and evaluation of multiobjective approaches for PFSP is provided by Minella et al. (2008).

1.2 No-Idle Flowshop Variant

In this chapter we are interested in a variant of the PFSP that arises when no idle time is allowed at machines. This constraint models an important practical situation that arises when expensive machinery is employed. Idling on such expensive equipment is often not desired. Clear examples are the steppers used in the production of integrated circuits by means of photolithography. Other examples come from sectors where less expensive machinery is used but where machines cannot be easily stopped and restarted. Ceramic roller kilns, for example, consume a large quantities of natural gas when in operation. Idling is not an option because it takes several days to stop and to restart the kiln due to a very large thermal inertia. In all such cases, idling must be avoided.

Table 1. Processing times for a PFSP example with four machines and five jobs

	jobs (j)				
machines (i)	1	2	3	4	5
1	31	39	23	23	33
2	22	25	22	22	41
3	25	41	47	14	27
4	30	34	22	13	19

In order to better understand the no-idle constraint, we make use of an example problem with five jobs and four machines. The processing times $p_{j,i}$ are given in Table 1. The optimum solution, easily obtainable by complete enumeration or by solving the corresponding instance of the model given in Section 1.1 is $\pi^*_{idle} = \{3, 1, 2, 5, 4\}$ and is depicted in Figure 1. It is straightforward to see that all machines, with exception of machine 1, have idle times. For example, there is an idle time on the second machine of 18 time units between the completion time of the job 1 in second position and the beginning of job 2 in the third position. These idle times are necessary because by the time job 1 is finished in machine 2, job 2 cannot start processing since it is still being processed in machine 1. Despite idle times, the makespan value is of 226 time units.

In the no-idle flowshop problem with makespan criterion, denoted as $F/prmu, no{-}idle/C_{max}$, these idle times are not allowed. In order to ensure this, and following the previous example, the start times of jobs 3, 1 and 2 on machine 2 need to be delayed so that no idle time is present in the schedule. The same previous sequence $\{3, 1, 2, 5, 4\}$ results in a makespan of value 258 if the no-idle constraint is enforced. As a result, the makespan value is more than 14% worse.

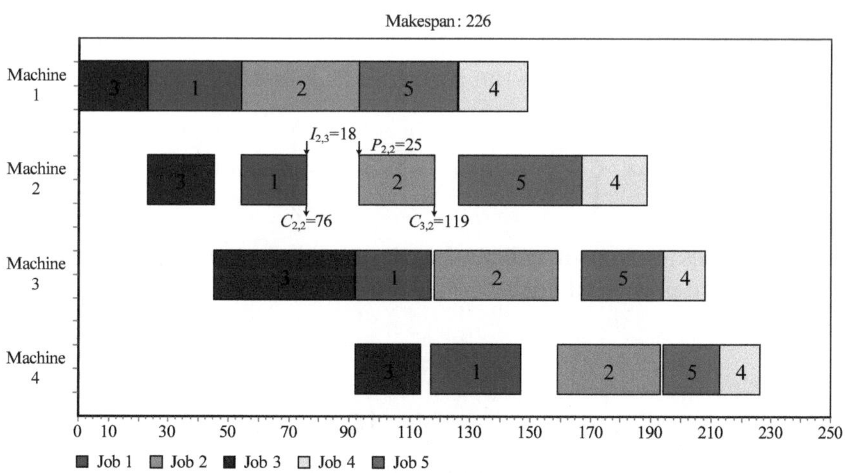

Fig. 1. Optimum solution for the PFSP example. Idle time allowed.

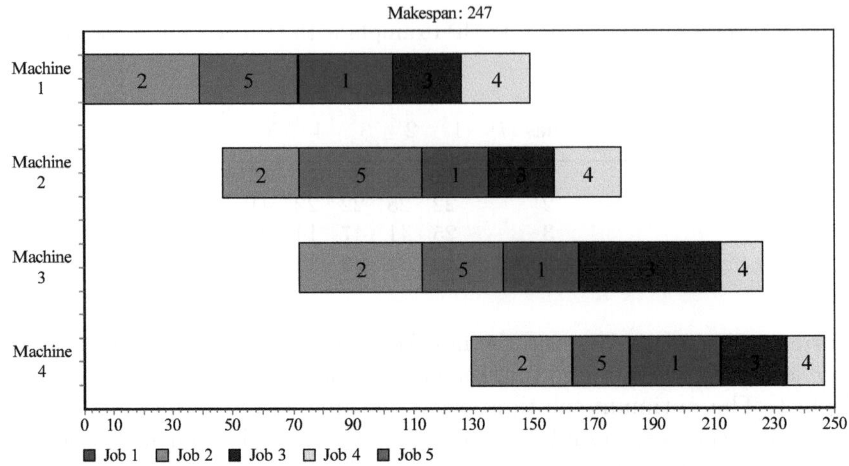

Fig. 2. Optimum solution for the PFSP example. No idle time allowed.

Although related, the no-idle PFSP and the regular PFSP are very different. As a matter of fact, the optimum solution for the example problem of Table 1 is $\pi^*_{no-idle} = \{2, 5, 1, 3, 4\}$, with a makespan value of 247 and is shown in Figure 2. We can see that π^*_{idle} and $\pi^*_{no-idle}$ are very different and only job 4 is located in the same position in both sequences. Similarly, the optimum makespan with the no-idle constraint is better than the makespan obtained by enforcing the no-idle constraint to π^*_{idle}.

Calculating the completion times $C_{j,i}$ in a no-idle flowshop is not trivial. Following the previous examples of Figures 1 and 2 we see that for each machine, jobs are delayed until we are sure that they can be processed without idle time. Therefore, we first need to calculate when a given machine can start processing with no needed idle time. We denote this as $S_i, i = \{1, \ldots, m\}$. Obviously, $S_1 = 0$. With this in mind, we calculate the S_i values as follows:

$$S_i = S_{i-1} + \max_{1 \leq h \leq n} \left\{ \sum_{j=1}^{h} p_{\pi(j),i-1} - \sum_{j=1}^{h-1} p_{\pi(j),i} \right\}, \quad i = \{2, \ldots, m\} \tag{10}$$

Once the S_i values are known, calculating the completion times is straightforward since the jobs are processed with no-idle time:

$$C_{\pi(1),i} = S_i + p_{\pi(1),i}, \quad i = \{1, \ldots, m\} \tag{11}$$

$$C_{\pi(j),i} = C_{\pi(j-1),i} + p_{\pi(j),i}, \quad j = \{2, \ldots, n\}, i = \{1, \ldots, m\} \tag{12}$$

As a result, $C_{max} = C_{\pi(n),m}$. However, it is interesting to mention that the completion times are not really needed for makespan criterion as $C_{max} = S_m + \sum_{j=1}^{n} p_{j,m}$. As we can see, calculating C_{max} for the no-idle PFSP has the same complexity as for the regular PFSP ($\mathcal{O}(nm)$) although more calculations are needed at each step. Note that in order to come up with a $\mathcal{O}(nm)$ complexity, the summations inside the max term in expression (10) have to be stored at each step. For example:

$$\sum_{j=1}^{h} p_{\pi(j),i-1} = \sum_{k=1}^{h-1} p_{\pi(k),i-1} + p_{\pi(h),i-1}$$

Similarly to the PFSP, it is easy to come up with a MIP model to obtain the optimum solution for the no-idle PFSP. We propose an adaptation of the model presented in Saadani et al. (2005) here. The variable definition, objective function and constraint sets (3), (4) and (8), (9) are unchanged.

Constraint sets (5)-(7) are changed by:

$$C_{1,1} = \sum_{j=1}^{n} X_{j,1} \cdot p_{j,1} \tag{13}$$

$$C_{k+1,i} = C_{k,i} + \sum_{j=1}^{n} X_{j,k+1} \cdot p_{j,i}, \quad k = \{1, \ldots, n-1\}, i = \{1, \ldots, m\} \tag{14}$$

$$C_{k,i+1} \geq C_{k,i} + \sum_{j=1}^{n} X_{j,k} \cdot p_{j,i+1}, \quad k = \{1, \ldots, n\}, i = \{1, \ldots, m-1\} \tag{15}$$

We see that the structure of these constraints has changed. The most important aspect is constraint set (14) where we enforce that the completion time of a job in position $k+1$ is exactly equal to the completion time of job in position k plus the processing time of the job in position $k+1$. This ensures the no-idle constraint.

The computational complexity of the $F/prmu, no-idle/C_{max}$ problem is briefly commented in Tanaev et al. (1994) which in turn refers to an older communication in Russian. In any case, the $\mathcal{NP}$-Hardness of the $F3/prmu, no-idle/C_{max}$ was proved in Baptiste and Hguny (1997). Similarly, and according to Adiri and Pohoryles (1982), when Garey et al. (1976) proved the $\mathcal{NP}$-Completeness in the strong sense of the problem $F2/prmu/\sum C_j$ they did so with a no-idle instance, and therefore, the problem $F2/prmu, no-idle/\sum C_j$ is also $\mathcal{NP}$-Complete.

2 Literature Review

To the best of our knowledge, Adiri and Pohoryles (1982) where the first to address the no-idle PFSP. They studied also the no-wait flowshop. The main contribution is a polynomial algorithm for solving the $F2/prmu, no-idle/\sum C_j$ problem to optimality. They also provided results for $m > 2$ but for special cases with dominating machines only.

Vachajitpan (1982) was the first to study the makespan objective. He proposed a MIP model with the additional characteristic that non-permutation sequences are allowed. Of course, the proposed model is shown to be impracticable even for small problem sizes. A Branch and Bound (B&B) method is also presented, but in this case for the permutation case. No computational results are provided beyond a small example.

Heuristics for the general m-machine no-idle PFSP were first examined by Woollam (1986) for the makespan objective. Basically, several heuristics were taken from the literature, including some of the aforementioned ones like the NEH. From the solution given in those heuristics (idle time allowed), a no-idle sequence was calculated, followed by a series on $n-1$, adjacent pairwise exchange moves. Computational results were carried out with five heuristics and instances of up to 25 jobs and 25 machines in size (25×25). Nowadays, such sizes are deemed as small. However, for such cases, NEH produced the best results.

Baptiste and Hguny (1997) proposed a B&B method for the general m-machine no-idle PFSP with makespan criterion. They also proved the $\mathcal{NP}$-Hardness of the problem.

Čepek et al. (2000) pointed out some errors found in the paper by Adiri and Pohoryles (1982). Furthermore, they demonstrated that in the case of total completion time criterion and two machines, it suffices to search permutation schedules only.

In a fairly unknown paper, Narain and Bagga (2003) study the $F3/prmu, no-idle/C_{max}$ problem. They provide a MIP model and a B&B algorithm along with some rather limited computational results. The same problem with three machines is studied by Saadani et al. (2003). They proposed a lower bound and an effective heuristic. This heuristic compared favorably against an earlier method by the authors (Saadani et al., 2001). Notice that this work was later published in Saadani et al. (2005).

Kamburowski (2004) elaborates over Saadani et al. (2003) paper. The author proposes a network representation and identifies some paradoxes by which reducing some processing times might result in a prolongation of the makespan and viceversa.

Saadani et al. (2005) propose a Traveling Salesman Problem (TSP)-based heuristic for the $F/prmu, no-idle/C_{max}$. Basically, the authors modelize the distance between any two possible jobs as the resulting no-idle makespan value when sequencing these two jobs in all m machines. Starting from the minimum distance, the Nearest Insertion Rule (NRI) heuristic is applied by inserting, one by one, and in all positions, all pending jobs. The heuristic has a complexity of $\mathcal{O}(n^3)$ and is easily implementable. The authors tested the proposed heuristic against a MIP model and its optimum solution provided by LINGO in problems of sizes up to 17×30.

In two similar papers, Narain and Bagga (2005a,b), study the $F2/prmu, no-idle/\sum C_j$ and $F/prmu, no-idle/C_{max}$ problems, respectively. However, in the second case, only special variants with dominating machines are studied and heuristics are presented.

Kalczynski and Kamburowski (2005) proposed a heuristic for the $F/prmu, no-idle/C_{max}$ problem with a reported computational complexity of $\mathcal{O}(n^2m)$. The heuristic is compared against that of Saadani et al. (2005) with instances of size up to 100×40. Better results are reported on most instance sizes. The authors also present an adaptation of the NEH for the no-idle problem. Their proposed method is also shown to outperform this NEH heuristic.

As of late, no-idle flowshop has received renewed interest. Kalczynski and Kamburowski (2007) study special situations and problem combinations between the no-idle and no-wait flowshops.

Recently, Baraz and Mosheiov (2008) have proposed a simple two stage heuristic for the $F/prmu, no-idle/C_{max}$. In the first stage, pending jobs are added, one at a time, at the end of an incomplete sequence, and the job resulting in the least no-idle added makespan, is appended to the sequence. This phase carries out $\mathcal{O}(n^2)$ steps. In the second phase, all possible job interchanges are tested and the best moves are performed. There are $n(n-1)$ possible job pairs. Therefore, the authors conclude that the running time of their proposed heuristic is $\mathcal{O}(n^2)$. However, we want to point out a very important mistake here. The authors are not considering the added complexity of calculating the no-idle makespan at each step. Since this calculation has a computational complexity of $\mathcal{O}(nm)$, we conclude that the correct total computational complexity of their proposed heuristic is actually $\mathcal{O}(n^3m)$. In any case, the authors demonstrate the superiority of their proposed heuristic against that of Saadani et al. (2005) but bypass other important papers like the one of Kalczynski and Kamburowski (2005). The size of the instances tested go all the way up to 400×8.

Also recently, in two similar papers, Pan and Wang (2008a,b) propose discrete differential evolution and a discrete particle swarm algorithms for the same problem. In both papers, an acceleration for the insertion neighborhood is proposed. This reduces the computational complexity of a single insertion neighborhood scan from $\mathcal{O}(n^3m)$ to $\mathcal{O}(n^2m)$ if the insertion is done in order. This acceleration is based on the very well known accelerations presented in Taillard (1990) for the same neighborhood but for the PFSP. Both algorithms use a form of advanced local search called Iterated Greedy (Ruiz and Stützle, 2007) that will be discussed later. The authors use the also well known benchmark of Taillard (1993) –extended to the no-idle flowshop– to test the results. In both papers, the authors test the proposed methods against the heuristics of Baraz and Mosheiov (2008) and Kalczynski and Kamburowski (2005). The results indicate that both the differential evolution and the particle swarm methods provide state-of-the-art results. However, these two methods are not compared between them.

As we can see, not many approaches have been suggested for the general m-machine $F/prmu, no-idle/C_{max}$ problem. However, it seems that this trend is reversing as several papers have appeared recently. It is one of the objectives of this paper to quantitatively compare these last approaches in order to identify the state-of-the-art.

There are other related papers that consider no-idle times, although not as a hard constraint or on related settings. For example, Liao (1993) relaxes the no-idle constraint and tries to minimize the number of idle intervals instead. This number is treated as a goal, that is subject to minimum makespan. The author presents a MIP model and a heuristic. A similar problem is studied in Saadani and Baptiste (2002) where the no-idle constraint is relaxed. In this case, a B&B algorithm is proposed for the three machine case where an optimal placement for one or more idle intervals is sought. A different paper is that by Giaro (2001) where "compact" open and flowshop problems are studied. Compact means that both no-idle and no-wait constraints exist. The author shows the incredible complexity of these problems where even proving the existence of a feasible compact schedule is already $\mathcal{NP}$-Hard.

Other shop settings are also studied in the literature. Narasimhan and Panwalkar (1984) and Narasimhan and Mangiameli (1987) study a two stage hybrid flowshop with no-idle parallel machines in the first stage. The symmetric problem is studied by Wang et al. (2005) where some heuristics are proposed for the case where the no-idle machines are on the second stage.

Niu and Gu (2006) study a no-idle PFSP with the additional consideration of fuzzy processing times. In this case, the mean makespan along with the makespan spread are studied with a mixture between particle swarm optimization and genetic algorithms. Deteriorating jobs on no-idle dominant machines –a very special case– is studied in two related papers, Cheng et al. (2007a,b).

3 New Approaches, Discussion and Adaptation of Existing State-of-the-Art Methods

It is frequent in the scheduling literature to propose new algorithms for a given specific problem and to compare against existing approaches for that problem only. While this is reasonable, sometimes it is worthwhile to look for state-of-the-art methods in related problems. As we have seen in Sections 1.1 and 1.2, the regular and no-idle flowshop problems are different from a mathematical point of view. However, the search space is based on permutations and much of the existing knowledge –specially in the field of metaheuristics– might be applicable.

Therefore, in this Section we discuss improvements to some of the earlier reviewed methods that were specifically proposed for the no-idle flowshop. We also propose adaptations of high performing existing methods that where proposed for related problems like the regular PFSP.

Let us first analyze the recent proposal of Baraz and Mosheiov (2008). We refer to this heuristic as GH_BM. For the sake of complexity, we detail the heuristic here.

1. STEP 1 (greedy). Perform n iterations. At each iteration, append to the current sequence the unscheduled job yielding the least additional no-idle makespan.
2. STEP 2 (pairwise job interchange). From the sequence obtained at STEP 1, perform a single pass in the interchange neighborhood, testing all possible pairs of job exchanges. Accept those exchanges improving makespan.

At STEP 1, n jobs are tested in the first iteration, $n-1$ in the second and so on until the last job. Therefore, we have $n(n+1)/2$ steps. At each step, the makespan has to be calculated, with a cost of nm. Therefore the computational complexity, as discussed before, is $\mathcal{O}(n^3m)$. STEP 2 is essentially similar, as it is well known that the cardinality of the interchange neighborhood is also $n(n+1)/2$.

We want to focus our attention to this heuristic. The choice of the constructive heuristic in STEP 1 is probably not the best one. It has been long known that the NEH (Nawaz et al., 1983) heuristic is the best performer for flowshop problems in many different scenarios. See for example Framinan et al. (2003) or Ruiz and Maroto (2005) for recent results on this and for different objectives. Furthermore, NEH was shown very recently to be an excellent performer for the regular PFSP (see Rad et al., 2009). NEH considers lengthy jobs early in the sequence and then carries out insertions as shown in the following steps:

1. Sum the processing times of all jobs on all machines: $P_j = \sum_{i=1}^{m} p_{j,i}$.
2. Sort jobs in descending order of P_j.
3. Take job j, $j = 1, \ldots, n$ from the sorted list, insert it in all possible j positions of the partial incumbent sequence and place it in the position that results in the lowest C_{max}.

Even without the accelerations of Taillard (1990), the complexity of the NEH heuristic is $\mathcal{O}(n^3m)$, which is the same as STEP 1 in the GH_BM heuristic. Considering the good performance of the NEH, it seems reasonable to substitute STEP 1 by the NEH heuristic. Furthermore, using Taillard (1990) accelerations reduces the complexity of the NEH to $\mathcal{O}(n^2m)$. The extension of these accelerations to the no-idle flowshop have been proposed, as already mentioned, by Pan and Wang (2008a,b). Accelerations for the PFSP are extremely effective. As shown in Rad et al. (2009), a very efficient NEH implementation results in CPU times of only 77 milliseconds for instances as large as 500×20 in a modern desktop computer.

As regards STEP 2, it has been long known that for the PFSP, insertion neighborhoods give better results than adjacent interchange and general interchange neighborhoods. This was tested in many domains as early as in the work of Osman and Potts (1989). This is also true even for genetic mutation operators as an insert mutation performs much better than a swap or interchange mutation as shown in Reeves (1995) or more recently, in Ruiz et al. (2006). Furthermore, when scanning all the insertion neighbors of a single job, the same accelerations discussed before can be applied. This effectively brings down the application of a single pass of the insertion neighborhood to $\mathcal{O}(n^2m)$. As a result, a better alternative is to apply the insertion local search in STEP 2.

Considering the previous discussion, we propose an improvement of the GH_BM heuristic –which we call GH_BM2– that uses NEH with accelerations in STEP 1 and that uses a single pass insertion local search, also with accelerations, in STEP 2.

The second heuristic we want to draw our attention upon is the TSP-based SGM method by Saadani et al. (2005). This heuristic is composed of several steps:

1. Calculate the distance D_{jk} between any possible pair of jobs $j, k=\{1,\ldots,n\}$, $j \neq k$. D_{jk} is actually equivalent to S_m from expression (10) if just jobs j and k are scheduled in all m machines considering no-idle constraints.
2. Take the minimum D_{jk} and schedule jobs j and k in this order. Store the scheduled jobs in a partial sequence π_p
3. For every unscheduled job l, insert the job in every possible position of the incomplete sequence, this is, insert job l in the first position, in the second and so on until position $|\pi_p|+1$. Among all pending jobs and all possible positions, insert the job in the position that resulted in the minimum tour length increase.
4. Go back to step 3 until all jobs are scheduled.

At first, SGM heuristic might look expensive. However, it is actually very fast if implemented with care. As the heuristic is no more than an adaptation of the Nearest Insertion Rule (NRI) from the TSP to the no-idle PFSP, we do not have to calculate any makespan value. For example, the different "tour lengths" when inserting a given job 5 into all positions of $\pi_p = \{1,2,3\}$ are obtained by two simple summations and a single substraction. For this example, the current tour length is $L = D_{12} + D_{23} + D_{31}$. As a result, when inserting job 5 in the second position the new length is calculated as follows: $L' = L - D_{12} + D_{15} + D_{52}$. This does not reduce the theoretical complexity of the heuristic, as step 1 has a complexity of $\mathcal{O}(n^2)$ and step 3 has a complexity of $\mathcal{O}(n^3)$[1]. However, the amount of work per step is very low and the result is a very fast heuristic. For our tests, we have implemented such a fast version of the SGM algorithm.

Another heuristic considered in this chapter is the one proposed by Kalczynski and Kamburowski (2005). This heuristic is referred to as KK and consists of the following steps:

1. Calculate the sequence π_i by applying Johnson's (1954) algorithm on machines i and $i+1$, for $i = \{1,\ldots,m-1\}$. Set $\omega = \varnothing$, $K = |\omega| = 0$, and $\sigma_i = \pi_i$ for $i = \{1,\ldots,m-1\}$.
2. Assume that the current sequences are $\pi_i = (\sigma_i, \omega)$, where ω is the subsequence of scheduled jobs. For every unscheduled job s and position $k = \{1,\ldots,\lceil n/(n-K)\rceil\}$ in ω, compute $R(s,k) = \sum_{i=1}^{m-1} C_{max}((\sigma_i - \{s\}, \omega(s,k))$;

[1] In computational complexity, constants multiplying large numbers are normally overlooked. However, in the case of scheduling, n and m are usually not measured in the thousands. Therefore, constants might be relevant. More specifically, in the first iteration of the third step of the SGM heuristic, $n-2$ pending jobs are inserted in 3 possible positions. In the second iteration, $n-3$ jobs are inserted in 4 positions and so on. This gives a total number of insertions of $\sum_{j=2}^{n-1}\{(n-j)\cdot(j+1)\} = \frac{1}{6}n^3 + \frac{1}{2}n^2 - \frac{8}{3}n + 2$. As we have seen, at each insertion, only three basic operations are carried out. The result is that although the heuristic is $\mathcal{O}(n^3)$, its performance is practice is very good.

M_i, M_{i+1}). The sequence $\sigma_i - \{s\}, \omega(s,k)$ is that obtained by deleting job s from σ_i and inserting it into the k-th position of ω. The objective is to find s^* and k^* that minimize $R(s,k)$. Set $\omega = \omega(s^*,k^*), \sigma_i = \sigma_i - \{s^*\}$ for $i = \{1,\ldots,m-1\}$ and $K = K+1$.

3. If $K < n$, return Step 2. Otherwise, $\pi_{kk} = \omega$ is the final sequence.

In this case, the method starts from m-1 sequences built with Johnson's algorithm. For each sequence, Step 2 chooses the best job to be deleted from the current Johnson's sequence and to be inserted in the subsequence of scheduled jobs. The complexity of this method is $\mathcal{O}(n^3m)$. However, the authors state that the complexity of the heuristic is $\mathcal{O}(n^2m)$ since a speed up procedure based on the Critical Path Method (CPM) can be applied to compute the makespan value in $\mathcal{O}(1)$. Nevertheless, the cost to obtain the CPM is $\mathcal{O}(n)$ and it is necessary to compute it again when the sequence changes. On the other hand, the value of m is a very strong factor to compute the $R(s,k)$ values. Moreover, we have to consider that several searches and additional operations have to be carried out before the makespan value can be computed, since it is necessary to know the position of the deleted job in the Johnson's sequence. So, despite great efforts, and after coding all the speed-ups and formulae from Kalczynski and Kamburowski (2005), the complexity of the implemented KK heuristic remains at $\mathcal{O}(n^3m)$. We want to remark that in the original paper, the authors do not report CPU times. Additionally, we contacted Dr. Quan-Ke Pan for help as he did code the KK heuristic in the papers Pan and Wang (2008a,b). The code we received did not include the alleged accelerations either.

Recently, Rad et al. (2009) have proposed some algorithms based on the NEH procedure. From the proposed heuristics, here we adapt the three best performing ones to the no-idle PFSP. Namely, we consider the following heuristics:

- FRB3. It is an extension of the NEH method. After inserting a job in a given position, all jobs inserted in previous iterations are again reinserted in all possible positions. This reinsertion is motivated by the fact that after inserting a new job, existing jobs could be moved in order to better accommodate the new one. FRB3 was shown in Rad et al. (2009) to improve the performance of the NEH method almost a 300% on average. However, this comes at a cost, since the worst case computational complexity rises to $\mathcal{O}(n^3m)$.
- FRB4$_k$. It is a simplification of the FRB3 method. After inserting a job j, only the jobs at positions $\pm k$ from the position where job j has been finally inserted are reinserted. Since typically $k \ll n$, the computational complexity of FRB4$_k$ is $\mathcal{O}(n^2m)$. In any case, the empirical observed running time will be much higher that the accelerated NEH.
- FRB5. It is an extension of FRB3. Basically, after placing a given job j, a full local search in the insertion neighborhood until local optima is carried out. Furthermore, jobs are extracted at each step at random and without repetition to enforce an unbiased and powerful search. Given that the number of local search steps cannot be derived, the worst case complexity cannot be calculated. FRB5 was shown in Rad et al. (2009) to be much slower than NEH on average. However, it also produced the best results.

FRB3, FRB4$_k$ and FRB5 are adapted to the no-idle PFSP by simply calculating no-idle C_{max} values at each step. We employ the mentioned accelerations in the insertions. Furthermore, for FRB4$_k$, we test two k values, namely 4 and 12.

Lastly, we are interested in a recent state-of-the-art algorithm proposed for the regular PFSP-C_{max}. The Iterated Greedy method with local search (IG$_{LS}$) was shown in Ruiz and Stützle (2007) to produce better results than other much more complex state-of-the-art methods. Iterated Greedy (IG) has been successfully applied to other shop environments, like the SDST flowshop in Ruiz and Stützle (2008), no-wait flowshops (Pan et al., 2008), hybrid flowshops (Ying, 2008) and many others. Recently, Vallada and Ruiz (2009) have devised cooperative IG methods that have improved the results even further. There have been even approaches for multiobjective PFSP like the one shown in Framinan and Leisten (2008). It is clear that there is a strong recent trend in the application of IG methods to flowshop problems.

As the name implies, IG iterates over greedy constructive heuristics. IG$_{LS}$ starts from the solution given by the NEH. Then a local search step is carried out (this local search will be explained later). Then, three phases are iteratively applied until a termination criterion is met. First, we have the destruction. During this phase, some jobs are extracted from the incumbent sequence, at random. The second phase is construction, were the removed jobs are inserted, one by one, in all positions of the partially destructed sequence. Each job is placed in the position resulting in the lowest C_{max} increase. The reconstruction phase is applied until a new complete sequence is obtained. Lastly, this new sequence undergoes a local search step. After the application of these three steps, the new solution is considered for replacing the incumbent one. More details can be seen in Ruiz and Stützle (2007). IG$_{LS}$ uses the principle and accelerations of the NEH. Of special mention is the local search step, which is detailed in Figure 3. As we can see, it is not a straightforward local search. First of all, in the inner

```
procedure LocalSearch_Insertion(π)
  improve := true;
  while (improve = true) do
    improve := false;
    for i := 1 to n do
      remove a job k at random from π without repetition
      π' := best permutation after inserting k in all positions of π;
      if C_max(π') < C_max(π) then
        π := π';
        improve := true;
      endif
    endfor
  endwhile
  return π
end
```

Fig. 3. Local search employed in IG$_{LS}$ (Ruiz and Stützle, 2007)

loop, all jobs are extracted one by one, but instead of doing this in order, it is done at random, to avoid a biased result. Each job is inserted in all possible positions (using accelerations) and if a better C_{max} value is found, the solution is replaced. We continue until all jobs have been reinserted and if an improvement is found, the search starts again for all jobs. The local search is finished when no improvements are found after reinserting all jobs. Note that this is also the local search employed in the FRB5 heuristic. A very similar local search is applied in the second step of the GH_BM2 method. However, we apply a single pass, i.e., the while loop is eliminated.

As mentioned in Section 2, Pan and Wang (2008a,b) employ IG_{LS} as a local search method inside their proposed approaches. More specifically, they carry out a few iterations of IG_{LS} to good solutions found during the search. In this chapter we propose the application of the pure IG_{LS} method and not as a surrogate local search step.

4 Computational Evaluation

In this Section we aim at comparing all existing heuristics that we have reviewed with detail in the previous Section. We will comment first on the benchmark employed.

The vast majority of the flowshop literature concentrates on the well known benchmark of Taillard (1993). This benchmark is composed of 12 groups of 10 instances each, totalling 120 instances. Each group is characterized by a combination of n and m values ($n \times m$). The groups are $\{20, 50, 100\} \times \{5, 10, 20\}$, $200 \times \{10, 20\}$ and 500×20. However, this benchmark was proposed as a difficult set of instances for the PFSP-C_{max} only. Despite of this, it is common in the literature to "adapt" this benchmark to other objectives and to other problem variants. We have, however, several concerns with this approach. First of all, Taillard's benchmark is not complete, in the sense that some combinations of n and m are missing. For example, there are no instances in the sets 200×5, and $500 \times \{5, 10\}$. Apart from not being complete, the different values of n and m are not equidistant. These two facts make statistical testing complicated as one cannot easily analyze the factor effect of n and m. Second, a benchmark of only 120 instances is not enough if small differences on performance are to be detected with some statistical significance. Third, there is no guarantee that a set of hard instances for the PFSP-C_{max} will be also hard for other problem variants. Last but not least, Taillard's benchmark is already aging and most instances have been already solved to optimality (at least for the problem for which they were originally devised). As a result of all of the above discussion, we propose an extended benchmark of instances specifically designed for the no-idle PFSP.

In the proposed benchmark we have 250 instances where we have all combinations of $n = \{50, 100, 150, 200, 250, 300, 350, 400, 450, 500\}$ and $m = \{10, 20, 30, 40, 50\}$. There are five replicates per combination. The processing times are uniformly distributed in the range $[1, 99]$ as usual in the literature. As we can see, there are more values of n and m and all of them are equidistant, with all combinations present. Such a larger benchmark is easier on statistical testing. The proposed benchmark, along with the best known solutions is available for download at `http://soa.iti.es`. Notice that we even have a second smaller benchmark for calibration and testing so that calibration and final results are not carried out over the same set of instances.

In the evaluation, we will test the following heuristic methods. All of them are deterministic:

1. NEH from Nawaz et al. (1983) with the accelerations published in Pan and Wang (2008a,b). Computational complexity $\mathcal{O}(n^2m)$.
2. Original NEH with no accelerations, referred to as NEH_{na}. Computational complexity $\mathcal{O}(n^3m)$
3. SGM from Saadani et al. (2005). Computational complexity $\mathcal{O}(n^3)$. Highly efficient version.
4. KK from Kalczynski and Kamburowski (2005) Computational complexity $\mathcal{O}(n^3m)$.
5. GH_BM from Baraz and Mosheiov (2008). Computational complexity $\mathcal{O}(n^3m)$.
6. New proposed algorithm GH_BM2 with accelerations. Based on the two phases of GH_BM. Computational complexity $\mathcal{O}(n^2m)$.
7. GH_BM2 without accelerations, referred to as GH_BM2_{na}. Computational complexity $\mathcal{O}(n^3m)$.
8. FRB3 from Rad et al. (2009). Computational complexity $\mathcal{O}(n^3m)$.
9. FRB4_k from Rad et al. (2009) with k values of 4 and 12. (FRB4_4 and FRB4_{12}). Computational complexity $\mathcal{O}(kn^2m)$ or $\mathcal{O}(n^2m)$.

As we can see, we also wanted to test NEH and GH_BM2 without accelerations (NEH_{na} and GH_BM2_{na}). This way we can assess the impact of accelerations in the CPU times.

We will also test the following metaheuristics. These are stochastic and do not provide the same result after each run. Most of them also have a stopping criterion that will be discussed later.

1. HDPSO from Pan and Wang (2008a).
2. DDELS from Pan and Wang (2008b). We will simply refer to this method as "DDE".
3. FRB5 from Rad et al. (2009).
4. IG_{LS} from Ruiz and Stützle (2007).

It has to be reminded that the local search step in HDPSO and DDE is actually a few applications of the IG_{LS} method. In turn, IG_{LS} and FRB5 share the same local search step which was detailed in Figure 3 before.

All methods have been coded in Delphi 2007. All algorithms share most code and especially the critical functions that evaluate the no-idle C_{max} as well as accelerations. Therefore, results are completely and fully comparable. For the tests we have used a cluster of 12 PC/AT computers with Intel Core 2 Duo E6600 processors running at 2.4 GHz and with 1 GB of RAM. There is no multi-core or multi-threading programming so a single core on each computer is actually used.

The performance measure that we will be using is the Relative Percentage Deviation (RPD) over the best known solution for each instance:

$$\text{Relative Percentage Deviation } (RPD) = \frac{Heu_{sol} - Best_{sol}}{Best_{sol}} \times 100 \qquad (16)$$

where Heu_{sol} is the solution given by any of the tested heuristics for a given instance and $Best_{sol}$ is the best known solution for each instance. These best known solutions are available from `http://soa.iti.es`.

It has to be noted that all metaheuristic methods (HDPSO, DDE, FRB5 and IG_{LS}) are stochastic and therefore five different runs are carried out. Furthermore, these methods –with the exception of FRB5– have a natural stopping criterion. Following previous works like Ruiz and Maroto (2005), Ruiz et al. (2006), Ruiz and Stützle (2007), Vallada et al. (2008) and others, we set a stopping time based on elapsed CPU time (not wall time). This elapsed CPU time is accurately measured inside each method in order to stop it whenever the maximum allowed CPU time has passed. Moreover, this maximum elapsed CPU time is set with the following formula: $n \cdot (m/2) \cdot t$ milliseconds. Setting the time limit in this way allows more computational effort as the number of jobs and/or the number of machines increases. This helps in lessening the effect of the instance size on the results and on the statistical analysis. Lastly, in order to test the effect of CPU time, these three methods (HDPSO, DDE and IG_{LS}) are tested with three different elapsed CPU time termination criteria, where t=10, 20 and 30. This means that for the largest instances of 500×50 and the highest value of $t = 30$, a maximum elapsed time of $500 \cdot (50/2) \cdot 30$ =375,000 milliseconds or 6.25 minutes are allowed. Results are separated in values of t. For example, HDPSO^{10} refers to the same method where t has been set to 10.

All in all, we have 10 heuristics that are run a single time and four metaheuristics that are run five times, three of them are run for three different stopping criteria. As a result we have a total of 15,000 data points. As we will see, with such a large dataset and comprehensive computational campaign, we are able to draw strong and statistically sound conclusions.

4.1 Heuristic Results

We first comment on the results of the 10 tested heuristics. The Average Relative Percentage Deviations ($\overline{RPD}$), grouped by n and m values, are given in Table 2. The elapsed CPU times (in seconds) needed by each method are given in Table 3. Note that first we comment on averages but afterwards we will provide statistical analyses.

As expected, NEH and NEH_{na} give the same exact results. The same applies to GH_BM2 and GH_BM2_{na}. The only difference is the CPU time employed. We can see that NEH_{na} is about 76 times slower, on average, than NEH. Similarly, GH_BM2_{na} is about 104 times slower than GH_BM2. Clearly, the accelerations

Table 2. Average Relative Percentage Deviation ($\overline{RPD}$) over the best solution known obtained by the tested heuristics

n	m	NEH	NEH_{na}	SGM	KK	GH_BM	GH_BM2	GH_BM2_{na}	FRB3	$FRB4_4$	$FRB4_{12}$
50	10	8.24	8.24	19.62	3.80	5.15	3.04	3.04	2.51	3.43	3.29
	20	10.78	10.78	24.03	6.90	8.94	5.18	5.18	4.64	5.74	4.74
	30	12.15	12.15	27.20	7.72	8.63	7.04	7.04	4.65	6.49	5.58
	40	12.00	12.00	27.09	7.51	9.40	6.57	6.57	4.21	6.54	5.15
	50	11.75	11.75	29.85	9.39	11.25	7.73	7.73	5.21	7.18	6.89
100	10	5.54	5.54	14.03	1.61	4.21	2.04	2.04	1.55	2.32	1.53
	20	7.95	7.95	23.85	2.29	5.55	3.34	3.34	2.06	4.09	3.76
	30	10.32	10.32	29.83	5.78	6.54	6.11	6.11	3.63	6.25	4.52
	40	11.40	11.40	34.66	6.14	11.30	6.93	6.93	5.28	8.01	6.74
	50	11.60	11.60	30.68	6.64	9.57	7.43	7.43	4.59	7.87	6.16
150	10	2.54	2.54	11.19	0.69	1.26	0.60	0.60	0.06	0.59	0.42
	20	6.54	6.54	21.07	2.52	4.73	2.71	2.71	2.18	3.12	2.96
	30	7.60	7.60	26.08	2.66	5.13	3.33	3.33	1.95	4.54	3.28
	40	11.13	11.13	32.25	4.87	9.21	6.15	6.15	3.93	6.03	5.43
	50	10.35	10.35	31.87	6.15	8.92	5.89	5.89	4.01	7.07	4.52
200	10	2.38	2.38	9.87	0.55	1.09	0.42	0.42	0.34	0.40	0.33
	20	4.08	4.08	18.53	1.43	3.26	1.70	1.70	1.02	2.24	2.03
	30	6.62	6.62	25.60	1.73	4.77	2.80	2.80	1.90	3.66	2.65
	40	9.24	9.24	30.57	3.31	7.53	4.10	4.10	2.38	5.18	3.96
	50	8.70	8.70	33.72	4.24	7.10	4.97	4.97	2.94	5.82	4.57
250	10	1.42	1.42	8.00	0.52	0.35	0.18	0.18	0.22	0.21	0.21
	20	4.53	4.53	19.83	1.35	3.09	1.59	1.59	0.74	2.20	1.34
	30	6.02	6.02	26.72	1.49	4.51	2.20	2.20	1.27	2.83	2.18
	40	8.13	8.13	30.34	1.89	5.95	3.80	3.80	1.51	4.70	2.99
	50	9.46	9.46	34.40	2.81	7.17	4.81	4.81	3.13	5.75	4.79
300	10	1.57	1.57	8.00	0.23	0.64	0.18	0.18	0.08	0.19	0.15
	20	4.08	4.08	18.85	1.06	2.43	1.75	1.75	0.53	1.44	1.22
	30	5.77	5.77	24.73	1.16	3.40	1.59	1.59	1.64	2.61	2.36
	40	6.48	6.48	27.91	1.43	5.03	2.70	2.70	1.71	3.39	2.72
	50	8.49	8.49	32.28	2.46	6.91	3.93	3.93	2.15	4.90	4.23
350	10	1.18	1.18	7.69	0.24	0.58	0.16	0.16	0.17	0.29	0.16
	20	3.16	3.16	15.83	0.79	2.33	0.89	0.89	0.43	1.05	0.90
	30	4.79	4.79	23.43	1.16	3.57	2.07	2.07	0.95	2.30	2.01
	40	5.60	5.60	26.39	1.24	4.29	2.63	2.63	1.48	3.20	2.89
	50	7.31	7.31	29.88	1.27	5.35	3.43	3.43	1.38	4.09	3.16
400	10	1.05	1.05	7.84	0.33	0.49	0.20	0.20	0.04	0.11	0.10
	20	3.12	3.12	16.07	0.76	2.16	1.02	1.02	0.62	1.29	1.00
	30	4.26	4.26	21.36	0.87	3.08	1.63	1.63	1.04	1.95	1.56
	40	5.32	5.32	24.61	1.36	3.19	1.67	1.67	0.57	2.26	1.72
	50	6.66	6.66	28.99	1.44	5.89	2.90	2.90	1.55	3.72	3.19
450	10	1.04	1.04	8.72	0.28	0.49	0.18	0.18	0.12	0.26	0.19
	20	3.01	3.01	17.18	0.75	1.95	0.96	0.96	0.44	1.08	0.71
	30	4.29	4.29	21.41	0.61	2.90	1.36	1.36	0.89	2.18	1.81
	40	4.60	4.60	25.82	0.98	3.42	1.85	1.85	0.78	2.28	1.86
	50	6.67	6.67	28.37	1.24	5.15	2.62	2.62	1.74	3.05	2.39
500	10	1.04	1.04	7.13	0.34	0.50	0.16	0.16	0.03	0.12	0.18
	20	1.89	1.89	13.22	0.47	0.96	0.47	0.47	0.23	0.51	0.44
	30	3.25	3.25	20.70	0.67	1.95	1.16	1.16	0.60	1.30	1.00
	40	4.90	4.90	23.96	1.04	3.54	2.22	2.22	0.99	2.83	2.05
	50	6.11	6.11	29.19	1.25	4.51	2.61	2.61	1.20	3.21	2.30
Average		6.12	6.12	22.61	2.35	4.59	2.82	2.82	1.75	3.24	2.61

proposed by Pan and Wang (2008a,b) should be applied at all costs. Additionally, our initial hypothesis from Section 1.2 that calculating the C_{max} value for the no-idle PFSP is costlier than for the regular PFSP is confirmed. Observing the results from Rad et al. (2009), we see that for the largest instances tested

Table 3. Elapsed CPU times needed by the tested heuristics (in seconds)

n	m	NEH	NEH$_{na}$	SGM	KK	GH_BM	GH_BM2	GH_BM2$_{na}$	FRB3	FRB4$_4$	FRB4$_{12}$
50	10	0.001	0.009	0.001	0.047	0.041	0.003	0.028	0.022	0.001	0.013
	20	0.003	0.016	0.001	0.088	0.059	0.006	0.056	0.044	0.013	0.022
	30	0.001	0.019	0.003	0.131	0.088	0.009	0.084	0.066	0.016	0.034
	40	0.003	0.031	0.003	0.184	0.106	0.009	0.109	0.088	0.016	0.047
	50	0.006	0.034	0.001	0.228	0.128	0.009	0.134	0.109	0.028	0.053
100	10	0.001	0.050	0.006	0.325	0.228	0.016	0.197	0.172	0.028	0.069
	20	0.003	0.103	0.003	0.681	0.406	0.016	0.403	0.344	0.044	0.100
	30	0.016	0.153	0.006	1.034	0.569	0.022	0.597	0.509	0.069	0.138
	40	0.016	0.203	0.009	1.391	0.756	0.034	0.791	0.684	0.088	0.191
	50	0.016	0.253	0.006	1.753	0.938	0.044	1.000	0.863	0.109	0.250
150	10	0.003	0.163	0.009	1.075	0.684	0.022	0.647	0.563	0.050	0.113
	20	0.016	0.331	0.016	2.241	1.275	0.038	1.303	1.147	0.106	0.231
	30	0.019	0.494	0.016	3.425	1.856	0.056	1.981	1.713	0.156	0.356
	40	0.028	0.666	0.016	4.609	2.475	0.072	2.638	2.284	0.197	0.444
	50	0.031	0.838	0.022	5.788	3.075	0.094	3.313	2.869	0.259	0.578
200	10	0.009	0.375	0.025	2.516	1.572	0.031	1.500	1.334	0.094	0.203
	20	0.031	0.781	0.025	5.297	2.950	0.066	3.100	2.691	0.181	0.416
	30	0.031	1.172	0.031	8.116	4.353	0.103	4.656	4.047	0.281	0.641
	40	0.044	1.563	0.038	10.906	5.769	0.125	6.284	5.403	0.366	0.819
	50	0.047	1.991	0.038	13.691	7.147	0.163	7.881	6.756	0.469	1.075
250	10	0.022	0.747	0.041	4.953	2.994	0.063	3.009	2.597	0.153	0.334
	20	0.031	1.519	0.047	10.441	5.738	0.109	6.159	5.256	0.291	0.650
	30	0.050	2.294	0.053	15.950	8.428	0.150	9.228	7.928	0.428	0.969
	40	0.075	3.100	0.063	21.419	11.238	0.200	12.475	10.559	0.575	1.338
	50	0.081	3.853	0.063	26.847	13.922	0.244	15.525	13.272	0.722	1.622
300	10	0.028	1.309	0.066	8.691	5.219	0.078	5.288	4.528	0.216	0.484
	20	0.050	2.638	0.078	18.269	9.959	0.141	10.747	9.172	0.413	0.959
	30	0.069	4.000	0.084	27.863	14.769	0.216	16.300	13.728	0.628	1.428
	40	0.094	5.431	0.097	37.572	19.650	0.281	22.019	18.256	0.850	1.950
	50	0.122	6.719	0.106	47.197	24.375	0.353	27.275	22.803	1.047	2.450
350	10	0.034	2.091	0.103	14.166	8.341	0.100	8.578	7.222	0.284	0.634
	20	0.063	4.288	0.116	29.753	16.109	0.188	17.431	14.506	0.584	1.394
	30	0.103	6.481	0.131	45.559	23.872	0.291	26.328	21.731	0.853	1.978
	40	0.128	8.675	0.150	61.197	31.822	0.375	35.369	28.931	1.153	2.719
	50	0.166	10.884	0.159	76.753	39.809	0.494	44.147	36.300	1.431	3.391
400	10	0.050	3.138	0.156	21.666	12.497	0.134	12.916	10.666	0.375	0.850
	20	0.084	6.447	0.175	45.522	24.291	0.266	26.316	21.594	0.763	1.784
	30	0.125	9.697	0.188	69.759	36.263	0.375	39.753	32.359	1.119	2.625
	40	0.166	13.122	0.209	93.481	48.697	0.500	53.541	43.313	1.491	3.513
	50	0.213	16.494	0.225	117.806	60.347	0.628	67.084	54.031	1.906	4.438
450	10	0.053	4.513	0.228	31.372	17.734	0.163	18.516	15.213	0.481	1.109
	20	0.109	9.272	0.244	66.203	34.856	0.319	37.803	30.672	0.959	2.256
	30	0.163	14.034	0.266	101.606	52.313	0.475	57.481	46.175	1.450	3.434
	40	0.216	18.959	0.284	136.275	69.600	0.628	77.481	61.731	1.931	4.494
	50	0.269	23.953	0.309	172.106	87.506	0.788	97.922	78.138	2.403	5.853
500	10	0.069	6.228	0.303	43.463	24.322	0.206	25.525	20.881	0.591	1.366
	20	0.134	12.859	0.334	91.925	47.778	0.394	52.341	42.216	1.213	2.906
	30	0.194	19.509	0.359	141.178	73.450	0.584	79.722	63.297	1.772	4.250
	40	0.263	26.309	0.391	189.394	96.788	0.788	107.088	84.556	2.384	5.759
	50	0.325	33.894	0.422	240.416	120.406	0.972	137.872	106.172	2.972	7.088
Average		0.077	5.834	0.114	41.447	21.551	0.229	23.759	19.190	0.680	1.596

there of size 500 × 20, the regular PFSP NEH needed 0.0773 seconds to give a solution. For the instances of the same size in this chapter, the NEH tested here for the no-idle PFSP needs 0.134 seconds, which is almost two times more costly. In any case, NEH is the fastest heuristic tested here. If coded with accelerations, it needs about 77 milliseconds, on average, to obtain a solution.

The next fastest method is SGM. As we hypothesized, its complexity of $\mathcal{O}(n^3)$ is compensated with the little work that is needed at each iteration. As a matter of fact, SGM is many times faster, on average, than GH_BM2 or both FRB4 methods, that have a complexity of $\mathcal{O}(n^2m)$. This also confirms the hypothesis that for scheduling problems, high constants in computational complexity calculations should not be overlooked.

As far as the $\overline{RPD}$ goes, we have that SGM gives rather poor results when compared to the other methods. Clearly, SGM is not recommended even if CPU time is considered as it is both slower and worse performing than NEH. The fact that SGM is not a good performer was already observed by Kalczynski and Kamburowski (2005), Baraz and Mosheiov (2008) and Pan and Wang (2008a,b). Although we have carried out a very effective coding, it does not suffice.

KK gives good results, better than NEH, which confirms the original findings of Kalczynski and Kamburowski (2005). However, our results in our implementation of KK are much better (when compared against NEH) than those reported by Pan and Wang (2008a,b). In these two papers, KK is tested against NEH and is shown only marginally better. Our results show that KK improves NEH on almost all instances. This is a very interesting result since as has been mentioned, NEH is unbeatable for the regular PFSP-C_{max}. It seems that this is not true for the no-idle PFSP. In any case, these results have to be considered only when CPU time is also accounted for. As mentioned, we have been unable to reproduce, despite our best efforts, the speed-ups reported in Kalczynski and Kamburowski (2005). As a matter of fact, our implementation of the KK heuristic results in a very slow method –the slowest among all tested heuristics–. Actually, and as we will see in next section, the average CPU time required by our KK implementation is very similar to that used by HDPSO^{10}, DDE^{10} or IG$^{10}_{LS}$ while the results are much worse. We do not claim here that KK cannot be implemented more efficiently, but after so much effort, it is clear that something is amiss in Kalczynski and Kamburowski (2005) paper and that additional information might be needed in order to code the speed-ups.

Another interesting result comes after comparing GH_BM and GH_BM2. As we stated initially, GH_BM2 is much faster, as using the insertion neighborhood allows important speed-ups. Actually, GH_BM is about 94 times slower than GH_BM2. Notice that Baraz and Mosheiov (2008) reported CPU times of 2.94 seconds for instances of size 200×8 on a Pentium IV 2.8 GHz. Our closer results are of 1.572 seconds on average for instances of size 200×10. Our Core 2 Duo processor running at 2.4 GHz is actually faster than a Pentium IV 2.8 GHz (even if running at a lower frequency clock). Therefore, we can safely state that we have a good implementation of GH_BM. If we compare the $\overline{RPD}$ values, GH_BM2 is about 63% better. Consequently, we can easily conclude that GH_BM2 is preferable to GH_BM. Of course, it could be argued that GH_BM could have been also accelerated. However, this is only true for step 1 as no accelerations are known for reducing the complexity of scanning the interexchange neighborhood in the PFSP. In any case, GH_BM2$_{na}$ is only a bit slower than GH_BM and also a 63% better.

The last three methods in the comparison offer very good results. FRB3, for example, gives the lowest $\overline{RPD}$ among all tested heuristics. The CPU times needed, however, are the third highest after KK and GH_BM. FRB4$_4$ is dominated, both from a CPU time and $\overline{RPD}$ by GH_BM2. FRB4$_{12}$ gives better $\overline{RPD}$ than GH_BM2 but at a significantly larger CPU time.

From the heuristics tested, we can conclude that FRB3 and GH_BM2 are the best performers, the first one as regards $\overline{RPD}$ and the second one as the best compromise between quality of results and CPU time. While KK gives results that are a bit better than GH_BM2, improving its speed to match that of GH_BM2 is certainly a challenge.

Of course, comparing average results could be misleading. We need to carefully test the statistical significance of these observed average differences. This will be done in later sections.

4.2 Metaheuristic Results

We now provide the results of the four tested metaheuristics. Recall that for three of them we have tested three different stopping criteria. The ($\overline{RPD}$) values and CPU times, also grouped by n and m values, are given in Tables 4 and 5.

As expected, the CPU times employed by all three methods that stop at $t = 10$ are almost identical. The same applies to $t = 20$ and $t = 30$. FRB5 has a rather erratic stopping time. This is because the method stops when the local search of Figure 3 reaches a local optimum and this depends on the stochastic order in which the jobs are inserted and on the instance data. Also, the local search is applied after each job is inserted in the NEH method which is extremely lengthy for larger instances. As a matter of fact, for instances with 500 jobs, FRB5 is actually slower than most methods. In any case, when comparing FRB5 with FRB3 we see that the added CPU time produces better results as the $\overline{RPD}$ of FRB5 is 1.36 versus that of FRB3 at 1.75.

A striking outcome are the results of DDE (DDELS as named in Pan and Wang, 2008b). The $\overline{RPD}$ does not improve from the original 2.65 given by DDE^{10} as DDE^{30} results in 2.65 as well. We checked our implementation carefully and found no errors. Dr. Quan-Ke Pan did send us his full source code. However, we decided to implement this DDE method following the details given in Pan and Wang, 2008b to the letter. In any case, DDE shares about 90% of the code with the HDPSO method by the same authors, since both use the NEH for initialization, PTL crossover, insertion mutation and IG for local search. Our hypothesis is that the problem is that, at each generation, two populations of size PS are generated. One by applying mutation to the original population, and another one by applying crossover. Then the original population and the two newly created ones undergo selection so to create a single population with PS individuals for the next generation. Only better individuals are passed over. After this, the best individual undergoes several iterations of the IG local search. Our observations indicate that when a certain level of evolution has been

Table 4. Average Relative Percentage Deviation ($\overline{RPD}$) over the best solution known obtained by the tested metaheuristics

n	m	HDPSO^{10}	HDPSO^{20}	HDPSO^{30}	DDE^{10}	DDE^{20}	DDE^{30}	FRB5	IG$^{10}_{LS}$	IG$^{20}_{LS}$	IG$^{30}_{LS}$
50	10	0.97	0.79	0.58	3.86	4.17	4.17	2.64	0.54	0.41	0.25
	20	0.99	0.61	0.52	4.77	4.88	4.89	3.11	0.59	0.39	0.33
	30	1.15	1.11	1.19	5.60	5.67	5.56	4.15	0.97	0.61	0.64
	40	1.20	1.12	1.16	5.81	5.70	5.11	3.51	1.09	0.96	0.78
	50	2.32	1.62	1.47	6.33	6.36	6.22	5.51	1.92	1.42	1.52
100	10	0.26	0.21	0.25	2.34	2.43	2.60	0.90	0.23	0.13	0.17
	20	0.74	0.62	0.58	2.95	2.85	3.04	1.72	0.57	0.44	0.33
	30	1.22	0.90	0.83	4.82	4.81	4.55	2.92	0.87	0.54	0.46
	40	1.65	1.15	1.23	6.49	6.40	6.30	4.56	1.49	0.87	0.87
	50	1.85	1.30	0.93	5.95	5.80	5.85	4.47	1.47	1.09	0.73
150	10	0.10	0.03	0.02	0.83	0.73	0.89	0.08	0.03	0.01	0.01
	20	0.84	0.61	0.54	2.94	2.73	2.61	1.39	0.59	0.45	0.34
	30	0.82	0.75	0.70	3.05	3.24	3.28	1.58	0.78	0.51	0.42
	40	1.95	1.16	1.27	6.25	6.38	6.19	3.02	1.52	0.91	0.73
	50	1.72	1.19	0.81	4.58	4.70	4.51	2.62	1.51	0.96	0.68
200	10	0.15	0.10	0.11	0.60	0.70	0.61	0.24	0.14	0.13	0.06
	20	0.42	0.25	0.25	1.50	1.52	1.62	0.60	0.36	0.22	0.12
	30	0.63	0.56	0.37	3.08	2.83	2.86	1.26	0.50	0.33	0.21
	40	1.24	0.91	0.49	4.07	4.24	4.03	2.05	0.83	0.49	0.44
	50	1.55	0.89	0.75	4.56	4.17	4.18	2.68	1.11	0.63	0.42
250	10	0.05	0.02	0.03	0.37	0.33	0.37	0.13	0.02	0.01	0.01
	20	0.37	0.28	0.23	1.55	1.80	1.69	0.52	0.22	0.21	0.17
	30	0.77	0.56	0.45	2.29	2.25	2.52	0.83	0.52	0.41	0.31
	40	1.02	1.01	0.66	3.60	3.62	3.65	1.45	1.04	0.64	0.54
	50	1.70	0.90	0.68	4.14	4.39	4.45	2.66	1.54	0.91	0.56
300	10	0.05	0.06	0.02	0.32	0.35	0.37	0.09	0.04	0.02	0.01
	20	0.36	0.33	0.30	1.46	1.68	1.75	0.54	0.31	0.28	0.23
	30	0.45	0.41	0.29	2.03	2.02	2.05	0.67	0.47	0.28	0.23
	40	0.75	0.57	0.43	2.93	2.81	2.73	0.87	0.76	0.45	0.26
	50	1.05	0.87	0.67	3.69	3.61	3.60	1.63	1.12	0.60	0.42
350	10	0.09	0.04	0.05	0.30	0.38	0.37	0.14	0.05	0.04	0.03
	20	0.40	0.28	0.28	1.28	1.23	1.33	0.33	0.32	0.21	0.23
	30	0.68	0.53	0.42	2.13	1.83	2.02	0.71	0.59	0.44	0.33
	40	1.01	0.71	0.49	2.33	2.44	2.42	0.97	0.85	0.48	0.39
	50	1.20	0.65	0.57	2.90	2.96	2.79	1.11	1.05	0.68	0.40
400	10	0.05	0.04	0.02	0.22	0.19	0.21	0.08	0.04	0.03	0.01
	20	0.26	0.21	0.23	1.06	1.09	1.06	0.53	0.25	0.17	0.14
	30	0.57	0.51	0.47	1.76	1.79	1.73	0.57	0.60	0.46	0.25
	40	0.57	0.39	0.35	1.87	1.84	2.08	0.51	0.41	0.29	0.33
	50	0.99	0.62	0.66	2.50	2.49	2.38	0.86	0.96	0.57	0.37
450	10	0.07	0.05	0.05	0.32	0.31	0.33	0.05	0.06	0.03	0.02
	20	0.25	0.18	0.19	1.00	1.00	1.13	0.28	0.22	0.18	0.12
	30	0.41	0.29	0.21	1.52	1.72	1.67	0.47	0.37	0.24	0.20
	40	0.58	0.48	0.44	1.66	1.77	1.59	0.48	0.58	0.44	0.36
	50	0.95	0.74	0.57	2.51	2.39	2.59	0.72	0.76	0.68	0.58
500	10	0.07	0.04	0.04	0.25	0.26	0.25	0.08	0.06	0.04	0.03
	20	0.20	0.15	0.13	0.60	0.65	0.72	0.15	0.20	0.15	0.10
	30	0.34	0.30	0.30	1.07	1.08	1.11	0.34	0.28	0.31	0.20
	40	0.82	0.62	0.46	2.08	2.00	2.11	0.65	0.69	0.45	0.35
	50	1.02	0.66	0.46	2.30	2.38	2.31	0.57	0.80	0.55	0.45
Average		0.78	0.57	0.48	2.65	2.66	2.65	1.36	0.65	0.43	0.34

achieved, mutation and crossover only deteriorate individuals and after selection, the new PS population is exactly equal to the original population and the algorithm stalls. Obviously, this is a design shortcoming. Our proposal for fixing this is that a selective local search should be applied to good individuals in the

Table 5. Elapsed CPU times needed by the tested metaheuristics (in seconds)

n	m	HDPSO^{10}	HDPSO^{20}	HDPSO^{30}	DDE^{10}	DDE^{20}	DDE^{30}	FRB5	IG$^{10}_{LS}$	IG$^{20}_{LS}$	IG$^{30}_{LS}$
50	10	2.52	5.02	7.51	2.50	5.00	7.50	0.10	2.50	5.00	7.50
	20	5.02	10.01	15.02	5.00	10.00	15.00	0.16	5.00	10.00	15.00
	30	7.51	15.02	22.51	7.50	15.00	22.50	0.22	7.50	15.00	22.50
	40	10.01	20.01	30.01	10.00	20.00	30.00	0.29	10.00	20.00	30.00
	50	12.50	25.01	37.51	12.50	25.00	37.50	0.34	12.50	25.00	37.50
100	10	5.03	10.03	15.02	5.00	10.00	15.00	0.54	5.00	10.00	15.00
	20	10.02	20.02	30.02	10.00	20.00	30.00	1.08	10.00	20.00	30.00
	30	15.02	30.02	45.01	15.00	30.00	45.00	1.74	15.00	30.00	45.00
	40	20.02	40.01	60.02	20.00	40.00	60.00	2.32	20.00	40.00	60.00
	50	25.01	50.01	75.01	25.00	50.00	75.00	2.93	25.00	50.00	75.00
150	10	7.56	15.03	22.54	7.50	15.00	22.50	1.36	7.50	15.00	22.50
	20	15.02	30.03	45.01	15.00	30.00	45.00	3.36	15.00	30.00	45.00
	30	22.52	45.02	67.54	22.50	45.00	67.50	5.49	22.50	45.00	67.50
	40	30.01	60.02	90.01	30.00	60.00	90.00	7.91	30.00	60.00	90.00
	50	37.51	75.03	112.50	37.50	75.00	112.50	10.34	37.50	75.00	112.50
200	10	10.06	20.06	30.07	10.00	20.00	30.00	3.09	10.00	20.00	30.00
	20	20.05	40.03	60.04	20.00	40.00	60.00	7.29	20.00	40.00	60.00
	30	30.02	60.04	90.03	30.00	60.00	90.00	12.71	30.00	60.00	90.00
	40	40.02	80.01	120.03	40.00	80.00	120.00	18.30	40.00	80.00	120.00
	50	50.01	100.02	150.03	50.00	100.00	150.00	24.74	50.00	100.00	150.00
250	10	12.55	25.11	37.62	12.50	25.00	37.50	5.25	12.50	25.00	37.50
	20	25.05	50.06	75.05	25.00	50.00	75.00	13.65	25.00	50.00	75.00
	30	37.52	75.02	112.54	37.50	75.00	112.50	23.72	37.50	75.00	112.50
	40	50.03	100.03	150.03	50.00	100.00	150.00	35.69	50.00	100.00	150.00
	50	62.51	125.02	187.52	62.50	125.00	187.50	48.19	62.50	125.00	187.50
300	10	15.04	30.08	45.05	15.00	30.00	45.00	8.56	15.00	30.00	45.00
	20	30.05	60.09	90.06	30.00	60.00	90.00	23.18	30.00	60.00	90.00
	30	45.01	90.03	135.07	45.00	90.00	135.00	39.63	45.00	90.00	135.00
	40	60.02	120.03	180.02	60.00	120.00	180.00	57.49	60.00	120.00	180.00
	50	75.02	150.01	225.03	75.00	150.00	225.00	84.26	75.00	150.00	225.00
350	10	17.57	35.12	52.57	17.50	35.00	52.50	13.42	17.50	35.00	52.50
	20	35.09	70.03	105.02	35.00	70.00	105.00	30.86	35.00	70.00	105.00
	30	52.56	105.05	157.52	52.50	105.00	157.50	60.53	52.50	105.00	157.50
	40	70.03	140.04	210.04	70.00	140.00	210.00	85.70	70.00	140.00	210.00
	50	87.56	175.00	262.52	87.50	175.00	262.50	126.35	87.50	175.00	262.50
400	10	20.11	40.08	60.10	20.00	40.00	60.00	18.75	20.00	40.00	60.00
	20	40.09	80.05	120.07	40.00	80.00	120.00	49.25	40.00	80.00	120.00
	30	60.03	120.05	180.06	60.00	120.00	180.00	87.66	60.00	120.00	180.00
	40	80.03	160.05	240.05	80.00	160.00	240.00	132.54	80.00	160.00	240.00
	50	100.00	200.05	300.02	100.00	200.00	300.00	182.26	100.00	200.00	300.00
450	10	22.60	45.07	67.58	22.50	45.00	67.50	26.36	22.50	45.00	67.50
	20	45.04	90.04	135.03	45.00	90.00	135.00	70.96	45.00	90.00	135.00
	30	67.53	135.03	202.57	67.50	135.00	202.50	121.36	67.50	135.00	202.50
	40	90.09	180.05	270.04	90.00	180.00	270.00	179.74	90.00	180.00	270.00
	50	112.55	225.03	337.51	112.50	225.00	337.50	249.74	112.50	225.00	337.50
500	10	25.13	50.11	75.12	25.00	50.00	75.00	36.54	25.00	50.00	75.00
	20	50.01	100.02	150.05	50.00	100.00	150.00	84.77	50.00	100.00	150.00
	30	75.03	150.00	225.10	75.00	150.00	225.00	156.72	75.00	150.00	225.00
	40	100.06	200.01	300.04	100.00	200.00	300.00	229.99	100.00	200.00	300.00
	50	125.05	250.04	375.04	125.00	250.00	375.00	344.56	125.00	250.00	375.00
Average		41.29	82.54	123.79	41.25	82.50	123.75	54.64	41.25	82.50	123.75

mutated and crossed populations, before actually applying selection or that a certain elitist strategy should be employed. In any case, such improvements are not necessary as the overall performance of DDE is not high. Much faster and simpler methods like GH_BM2 and $FRB4_{12}$ are comparable as far as $\overline{RPD}$ is concerned and are faster in return. Also, FRB3 is both faster and better performing.

As a result, DDE is hard to recommend over other algorithms for the no-idle PFSP.

Lastly, we comment on the performance of HDPSO and IG_{LS}. First of all, it must be reminded that IG_{LS} is used as a subroutine in HDPSO. Since we are stopping both algorithms at the same elapsed CPU time and since both share most code, the results are completely comparable. As we can see, the stand alone IG_{LS} gives significantly better results than HDPSO. Measuring the average percentage deviation between HDPSO and IG_{LS} we have that $HDPSO^{10}$ is a full 20% worse than IG_{LS}^{10} since the two $\overline{RPD}$ values are 0.78 and 0.65, respectively. What is more, the performance lead of IG_{LS} widens as more CPU time is allowed. For example, $HDPSO^{20}$ is 32.56% worse than IG_{LS}^{20} and $HDPSO^{30}$ is 41.18% worse than IG_{LS}^{30}. From the 1250 available results (250 instances and 5 replicates), IG_{LS}^{30} produces better results than $HDPSO^{30}$ in 736 cases, equal results in 150 cases and worse results in 364 cases. Most importantly, we want to strongly draw our attention to these last type of measurements. Counting "the number of times" a given method is better, equal or worse than another is not an indicator of performance. It is a strongly biased measure and can mislead conclusions. The average percentage deviation of IG_{LS}^{30} over $HDPSO^{30}$ in these 364 cases in which IG_{LS}^{30} gives a worse solution is a mere 0.24%. Therefore, it is easy to see that IG_{LS}^{30} is many times better (and by large) than $HDPSO^{30}$ and for the times where it is worse, it is by a small amount.

Summing up, the Particle Swarm part of the algorithm is actually hindering results. A simple, easy to code and straightforward IG method works much better by itself.

4.3 Statistical Analysis of Results

As mentioned in previous Sections, careful statistical testing is necessary to really ascertain the observed differences in average values. While it is expected, for example, that SGM will be statistically worse than all other methods, concluding the same when comparing two methods of similar performance like GH_BM2 and $FRB4_{12}$ is risky to say the least.

One of the most powerful and tested methodologies is the Design of Experiments (DOE), Montgomery (2005). DOE is a structured and organized method for determining the relationship between factors affecting the output of a process. In our case, we are interested in studying the effect on the response variable RPD. From the 15,000 data points available from the computational evaluation of the previous Section, we carry out a full factorial analysis where the effect of the following factors is studied:

- Number of jobs n
- Number of machines m
- Algorithm, (NEH, KK, GH_BM, GH_BM2, FRB3, $FRB4_4$, $FRB4_{12}$, HDPSO, DDE, FRB5 and IG_{LS})
- Stopping criterion t

Note that the last factor can only be studied in conjunction with the algorithms HDPSO, DDE and IG_{LS}. We have eliminated from the tests NEH_{na} and GH_BM2_{na} as they give the same results than the accelerated versions. Also, SGM is not tested as it is clear that its results are far worse than the others.

The initial means plot with Tukey 95% confidence intervals is shown in Figure 4. Recall that overlapping intervals for means indicates that the observed means are statistically equivalent. It has to be noted that the means plot of Figure 4 is not explicitly considering the interactions of the algorithms with the different n and m values. Therefore, it is an "overall" picture. For example, the means plots of IG^{30}_{LS} and IG^{20}_{LS} overlap. This means that for the overall observed $\overline{RPD}$, there is no statistically significant difference. However, zooming-in for

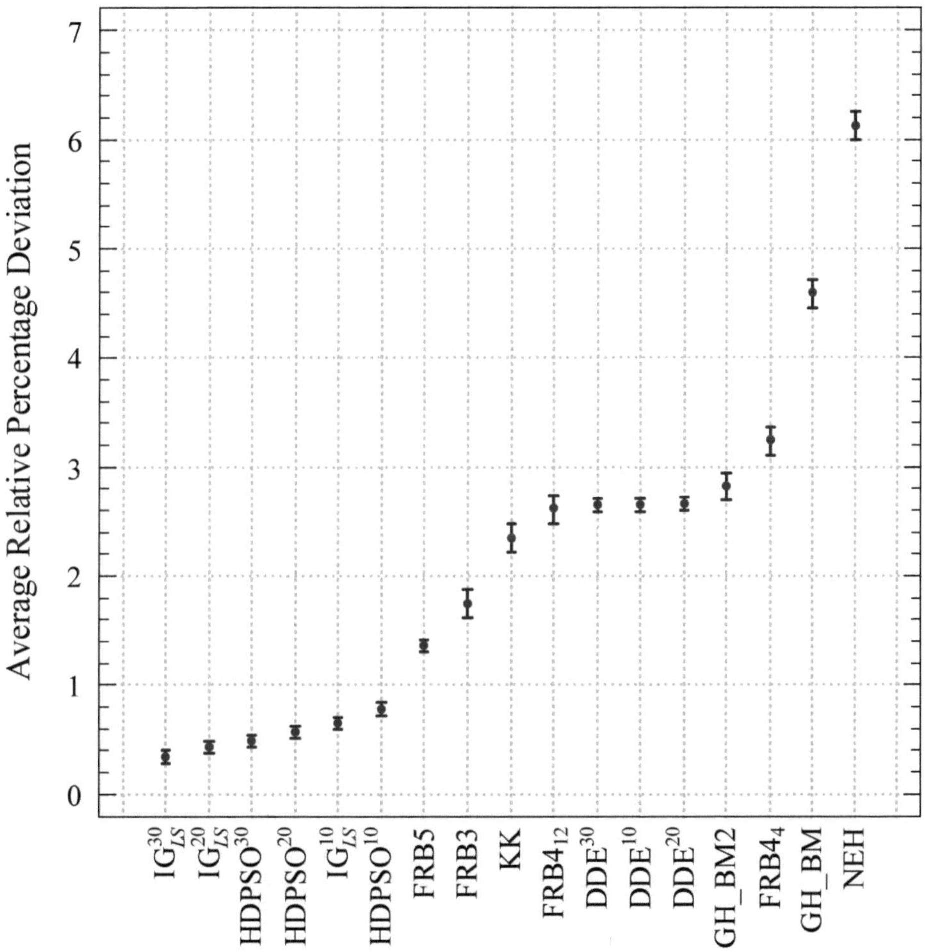

Fig. 4. Means plot for the algorithms $\overline{RPD}$ and 95% Tukey confidence intervals

different levels of n and m we observe statistically significant differences. We will provide some additional plots later.

What can be concluded is that most observed differences are statistically significant. For example NEH is statistically worse than all other methods. GH_BM is the second worst and there is a statistically significant difference between $FRB4_4$ and GH_BM2. However, all DDE methods are statistically equivalent to GH_BM2 and $FRB4_{12}$. We also see how IG^{30}_{LS} is indeed statistically better than $HDPSO^{30}$. All in all, most observed averages are statistically different.

Of high interest is to study which instance sizes affect algorithms the most. This information is not easy to see from large tables full of numbers. Instead, we give an interaction plot of factors n and m in Figure 5. We have to proceed with caution in the analysis of this plot. Since we do not know the optimum solution, we cannot state which instances are harder in an absolute way. Instead, we can

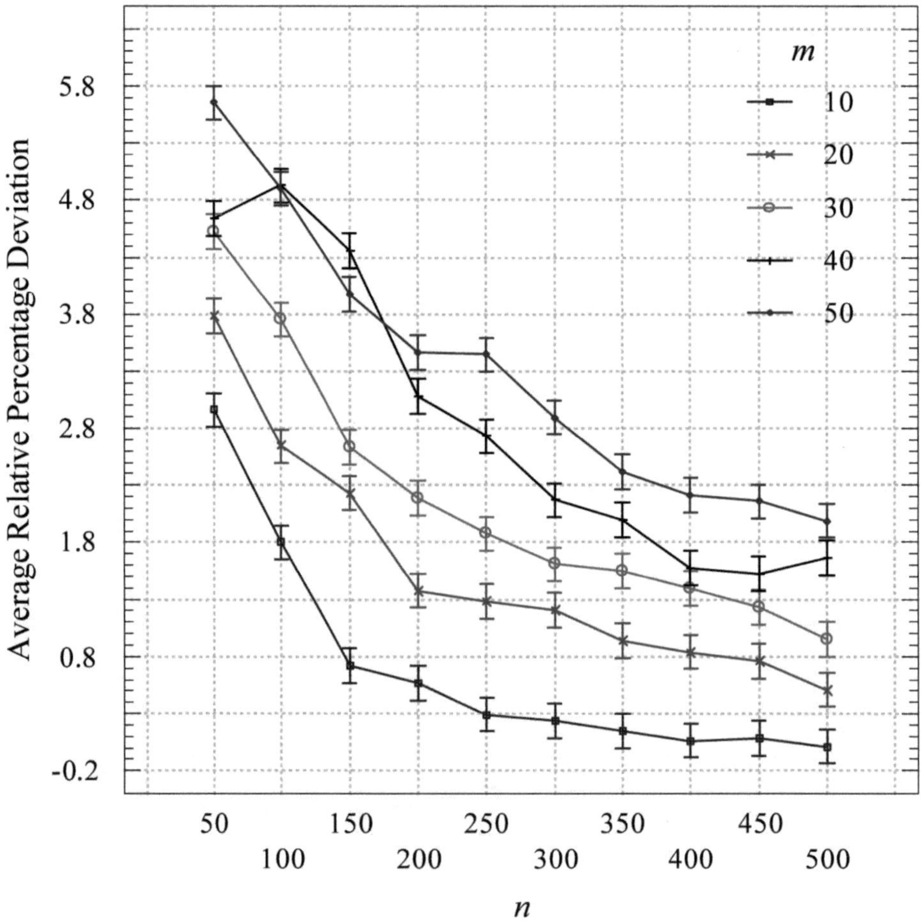

Fig. 5. Interaction plot between factors n and m with 95% Tukey confidence intervals

point which combinations of n and m result in higher and statistically significant $\overline{RPD}$ values for all algorithms. It is clear that increasing the number of machines results in higher percentage deviations. Interestingly, increasing the number of jobs results in lower percentage deviations. We hypothesize that with a larger number of jobs there are more options to come up with a better schedule even though the search space becomes larger. By far, the "hardest" instances are those with 50 jobs and 50 machines. There are not so many jobs and therefore fitting 50 no-idle machines becomes daunting for all methods.

Lastly, we zoom-in the performance of the two best methods, IG_{LS} and HDPSO. We plot the average performance against the different values of t in Figure 6. As can be seen, there is a clear statistically significant difference between IG_{LS} and HDPSO for all tested t values. As a matter of fact, for some instances

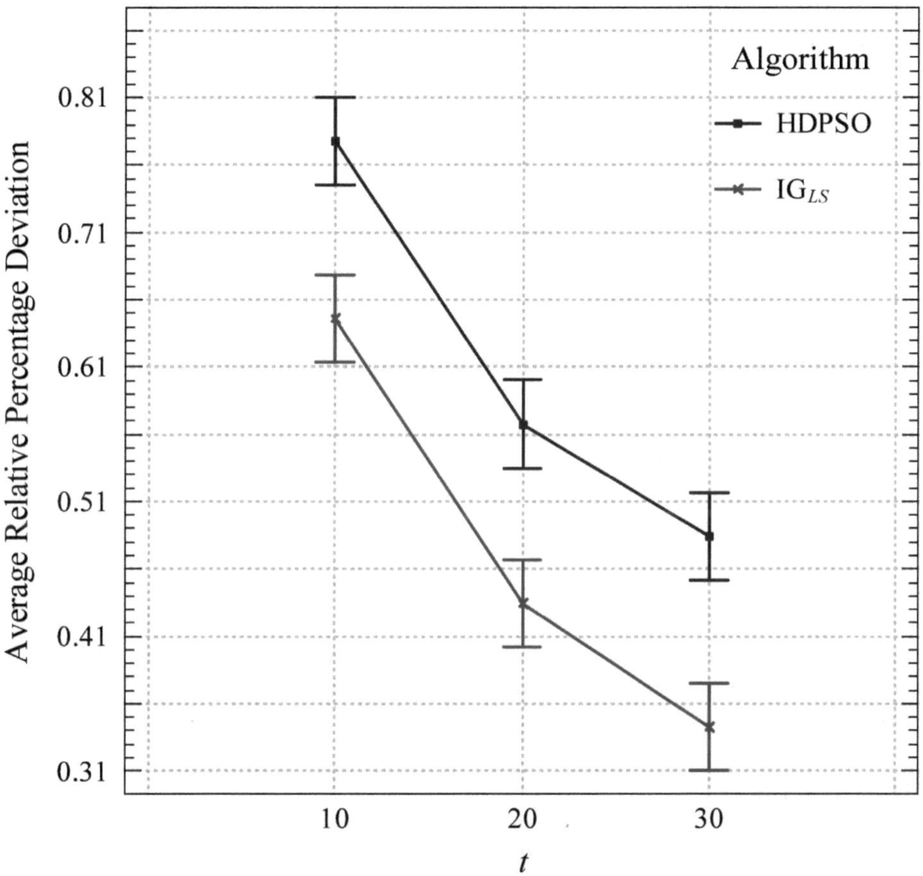

Fig. 6. Interaction plot between the algorithms IG_{LS} and HDPSO and t with 95% Tukey confidence intervals

sizes –not shown here– IG_{LS}^{20} is statistically better than $HDPSO^{30}$ which effectively means that IG_{LS} is able to reach better quality results when given one third less CPU time than HDPSO.

5 Conclusions and Future Research

This chapter has focused in a flowshop problem variant where idle times are not allowed on machines. This problem, known as the no-idle permutation flowshop, has been much less studied than the regular counterpart. We have provided a critical review of the existing literature, where each proposed algorithm has been carefully studied, and in some cases, improved. Namely, we have discussed a very effective implementation of the SGM method by Saadani et al. (2005) that despite not having improved its computational complexity of $\mathcal{O}(n^3)$, it has shown much lower empirical CPU running times when compared to other methods with better theoretical computational complexities. We have also provided an enhanced version of the GH_BM heuristic of Baraz and Mosheiov (2008). This improved version –referred to as GH_BM2– is much faster and effective than the original. Along with these improvements, we have made adaptations of methods that have been published very recently for the regular permutation flowshop problem. More specifically, we have adapted the Iterated Greedy (IG) metaheuristic from Ruiz and Stützle (2007) to the no-idle version. Some of the recent heuristics proposed in Rad et al. (2009) have also been adapted.

A total of 14 methods have been evaluated in a comprehensive computational campaign. State-of-the-art algorithms have been identified and validated through thorough statistical analyses. As the results indicate, adapted IG algorithms, as well as the proposed improved GH_BH2, together with the recent heuristics from Rad et al. (2009) constitute the best existing methods up to date for the no-idle permutation flowshop problem with makespan criterion.

There are many open research lines as this interesting problem variant has been seldom studied in the literature. No metaheuristic approaches have been proposed for other objectives apart from makespan. Furthermore, no-idle constraints in other environments like hybrid flowshops or job shops have not been studied. Additionally, more research is needed in exact methodologies, bounds and mathematical approaches for no-idle constraints. This way, researchers would benefit from a better understanding and characterization of this interesting problem.

References

Adiri, I., Pohoryles, D.: Flowshop no-idle or no-wait scheduling to minimize the sum of completion times. Naval Research Logistics 29(3), 495–504 (1982)

Baker, K.R.: Introduction to Sequencing and Scheduling. John Wiley & Sons, New York (1974)

Baptiste, P., Hguny, L.K.: A branch and bound algorithm for the $F/no-idle/C_{max}$. In: Proceedings of the International Conference on Industrial Engineering and Production Management, IEPM 1997, Lyon, France, vol. 1, pp. 429–438 (1997)

Baraz, D., Mosheiov, G.: A note on a greedy heuristic for flow-shop makespan minimization with no machine idle-time. European Journal of Operational Research 184(2), 810–813 (2008)

Campbell, H.G., Dudek, R.A., Smith, M.L.: A heuristic algorithm for the n- job, m-machine sequencing problem. Management Science 16(10), 630–637 (1970)

Cheng, M.B., Sun, S.J., He, L.M.: Flow shop scheduling problems with deteriorating jobs on no-idle dominant machines. European Journal of Operational Research 183(1), 115–124 (2007a)

Cheng, M.B., Sun, S.J., Yu, Y.: A note on flow shop scheduling problems with a learning effect on no-idle dominant machines. Applied Mathematics and Computation 184(2), 945–949 (2007b)

Dannenbring, D.G.: An evaluation of flow shop sequencing heuristics. Management Science 23(11), 1174–1182 (1977)

Davoud Pour, H.: A new heuristic for the n-job, m-machine flow-shop problem. Production Planning & Control 12(7), 648–653 (2001)

Du, J., Leung, J.Y.-T.: Minimizing total tardiness on one machine is NP-hard. Mathematics of Operations Research 15(3), 483–495 (1990)

Framinan, J.M., Gupta, J.N.D., Leisten, R.: A review and classification of heuristics for permutation flow-shop scheduling with makespan objective. Journal of the Operational Research Society 55(1), 1243–1255 (2004)

Framinan, J.M., Leisten, R.: A multi-objective iterated greedy search for flowshop scheduling with makespan and flowtime criteria. OR Spectrum 30(4), 787–804 (2008)

Framinan, J.M., Leisten, R., Rajendran, C.: Different initial sequences for the heuristic of Nawaz, Enscore and Ham to minimize makespan, idletime or flowtime in the static permutation flowshop sequencing problem. International Journal of Production Research 41(1), 121–148 (2003)

Garey, M.R., Johnson, D.S., Sethi, R.: The complexity of flowshop and jobshop scheduling. Mathematics of Operations Research 1(2), 117–129 (1976)

Giaro, K.: NP-hardness of compact scheduling in simplified open and flow shops. European Journal of Operational Research 130(1), 90–98 (2001)

Gonzalez, T., Sahni, S.: Flowshop and jobshop schedules: Complexity and approximation. Operations Research 26(1), 36–52 (1978)

Graham, R.L., Lawler, E.L., Lenstra, J.K., Rinnooy Kan, A.H.G.: Optimization and approximation in deterministic sequencing and scheduling: A survey. Annals of Discrete Mathematics 5, 287–326 (1979)

Hejazi, S.R., Saghafian, S.: Flowshop-scheduling problems with makespan criterion: A review. International Journal of Production Research 43(14), 2895–2929 (2005)

Johnson, S.M.: Optimal two- and three-stage production schedules with setup times included. Naval Research Logistics Quarterly 1(1), 61–68 (1954)

Kalczynski, P.J., Kamburowski, J.: A heuristic for minimizing the makespan in no-idle permutation flow shops. Computers & Industrial Engineering 49(1), 146–154 (2005)

Kalczynski, P.J., Kamburowski, J.: On no-wait and no-idle flow shops with makespan criterion. European Journal of Operational Research 178(3), 677–685 (2007)

Kamburowski, J.: More on three-machine no-idle flow shops. Computers & Industrial Engineering 46(3), 461–466 (2004)

Koulamas, C.: A new constructive heuristic for the flowshop scheduling problem. European Journal of Operational Research 105(1), 66–71 (1998)

Liao, C.J.: Minimizing the number of machine idle intervals with minimum makespan in a flowshop. Journal of the Operational Research Society 44(8), 817–824 (1993)

Minella, G., Ruiz, R., Ciavotta, M.: A review and evaluation of multi-objective algorithms for the flowshop scheduling problem. INFORMS Journal on Computing 20(3), 451–471 (2008)

Montgomery, D.: Design and Analysis of Experiments, 6th edn. John Wiley & Sons, New York (2005)

Narain, L., Bagga, P.C.: Minimizing total elapsed time subject to zero total idle time of machines in $n \times 3$ flowshop problem. Indian Journal of Pure & Applied Mathematics 34(2), 219–228 (2003)

Narain, L., Bagga, P.C.: Flowshop/no-idle scheduling to minimise the mean flowtime. Anziam Journal 47, 265–275 (2005a)

Narain, L., Bagga, P.C.: Flowshop/no-idle scheduling to minimize total elapsed time. Journal of Global Optimization 33(3), 349–367 (2005b)

Narasimhan, S.L., Mangiameli, P.M.: A comparison of sequencing rules for a two stage hybrid flowshop. Decision Sciences 18(2), 250–265 (1987)

Narasimhan, S.L., Panwalkar, S.S.: Scheduling in a two stage manufacturing process. International Journal of Production Research 22(4), 555–564 (1984)

Nawaz, M., Enscore Jr., E.E., Ham, I.: A heuristic algorithm for the m-machine, n-job flow-shop sequencing problem. OMEGA, The International Journal of Management Science 11(1), 91–95 (1983)

Niu, Q., Gu, X.S.: An improved genetic-based particle swarm optimization for no-idle permutation flow shops with fuzzy processing time. In: Yang, Q., Webb, G. (eds.) PRICAI 2006. LNCS, vol. 4099, pp. 757–766. Springer, Heidelberg (2006)

Osman, I., Potts, C.: Simulated annealing for permutation flow-shop scheduling. OMEGA, The International Journal of Management Science 17(6), 551–557 (1989)

Page, E.S.: An approach to the scheduling of jobs on machines. Journal of the Royal Statistical Society, B Series 23(2), 484–492 (1961)

Palmer, D.S.: Sequencing jobs through a multi-stage process in the minimum total time - a quick method of obtaining a near optimum. Operational Research Quarterly 16(1), 101–107 (1965)

Pan, Q.-K., Wang, L.: No-idle permutation flow shop scheduling based on a hybrid discrete particle swarm optimization algorithm. International Journal of Advanced Manufacturing Technology 39(7-8), 796–807 (2008a)

Pan, Q.-K., Wang, L.: A novel differential evolution algorithm for no-idle permutation flow-shop scheduling problems. European Journal of Industrial Engineering 2(3), 279–297 (2008b)

Pan, Q.-K., Wang, L., Zhao, B.-H.: An improved iterated greedy algorithm for the no-wait flow shop scheduling problem with makespan criterion. International Journal of Advanced Manufacturing Technology 38(7-8), 778–786 (2008)

Rad, S.F., Ruiz, R., Boroojerdian, N.: New high performing heuristics for minimizing makespan in permutation flowshops. OMEGA, the International Journal of Management Science 37(2), 331–345 (2009)

Reeves, C.R.: A genetic algorithm for flowshop sequencing. Computers & Operations Research 22(1), 5–13 (1995)

Ruiz, R., Maroto, C.: A comprehensive review and evaluation of permutation flowshop heuristics. European Journal of Operational Research 165(2), 479–494 (2005)

Ruiz, R., Maroto, C., Alcaraz, J.: Two new robust genetic algorithms for the flowshop scheduling problem. OMEGA, the International Journal of Management Science 34(5), 461–476 (2006)

Ruiz, R., Stützle, T.: A simple and effective iterated greedy algorithm for the permutation flowshop scheduling problem. European Journal of Operational Research 177(3), 2033–2049 (2007)

Ruiz, R., Stützle, T.: An iterated greedy heuristic for the sequence dependent setup times flowshop problem with makespan and weighted tardiness objectives. European Journal of Operational Research 187(3), 1143–1159 (2008)

Saadani, N.E.H., Baptiste, P.: Relaxation of the no-idle constraint in the flow-shop problem. In: Proceedings of the International Conference on Industrial Engineering and Production Management, IEPM 1997, Lyon, France, pp. 305–309 (2002)

Saadani, N.E.H., Guinet, A., Moalla, M.: A travelling salesman approach to solve the $F/no-idle/C_{max}$ problem. In: Proceedings of the International Conference on Industrial Engineering and Production Management, IEPM 2001, Quebec, Canada, vol. 2, pp. 880–888 (2001)

Saadani, N.E.H., Guinet, A., Moalla, M.: Three stage no-idle flow-shops. Computers & Industrial Engineering 44(3), 425–434 (2003)

Saadani, N.E.H., Guinet, A., Moalla, M.: A travelling salesman approach to solve the $F/no{-}idle/C_{max}$ problem. European Journal of Operational Research 161(1), 11–20 (2005)

Salveson, M.E.: On a quantitative method in production planning and scheduling. Econometrica 20(4), 554–590 (1952)

Suliman, S.M.A.: A two-phase heuristic approach to the permutation flow-shop scheduling problem. International Journal of Production Economics 64(1-3), 143–152 (2000)

Taillard, E.: Some efficient heuristic methods for the flow shop sequencing problem. European Journal of Operational Research 47(1), 67–74 (1990)

Taillard, E.: Benchmarks for basic scheduling problems. European Journal of Operational Research 64(2), 278–285 (1993)

Tanaev, V.S., Sotskov, Y.N., Strusevich, V.A.: Scheduling Theory. Multi-Stage Systems. Kluwer Academic Publishers, Dordrecht (1994)

Turner, S., Booth, D.: Comparison of heuristics for flow shop sequencing. OMEGA, The International Journal of Management Science 15(1), 75–78 (1987)

Vachajitpan, P.: Job sequencing with continuous machine operation. Computers & Industrial Engineering 6(3), 255–259 (1982)

Vallada, E., Ruiz, R.: Cooperative metaheuristics for the permutation flowshop scheduling problem. European Journal of Operational Research 193(2), 365–376 (2009)

Vallada, E., Ruiz, R., Minella, G.: Minimising total tardiness in the m-machine flowshop problem: A review and evaluation of heuristics and metaheuristics. Computers & Operations Research 35(4), 1350–1373 (2008)

Čepek, O., Okada, M., Vlach, M.: Note: On the two-machine no-idle flowshop problem. Naval Research Logistics 47(4), 353–358 (2000)

Wang, Z.B., Xing, W.X., Bai, F.S.: No-wait flexible flowshop scheduling with no-idle machines. Operations Research Letters 33(6), 609–614 (2005)

Woollam, C.R.: Flowshop with no- idle machine time allowed. Computers & Industrial Engineering 10(1), 69–76 (1986)

Ying, K.-C.: An iterated greedy heuristic for multistage hybrid flowshop scheduling problems with multiprocessor tasks. In press at Journal of the Operational Research Society (2008)

A Multi-Objective Ant-Colony Algorithm for Permutation Flowshop Scheduling to Minimize the Makespan and Total Flowtime of Jobs

Chandrasekharan Rajendran[1] and Hans Ziegler[2]

[1] Department of Management Studies
Indian Institute of Technology Madras, Chennai - 600 036, India
craj@iitm.ac.in

[2] Faculty of Business Administration and Economics
Department of Production and Logistics
University of Passau, D-94032 Passau, Germany
ziegler@uni-passau.de

Summary. The problem of scheduling in permutation flowshops is considered with the objectives of minimizing the makespan and total flowtime of jobs. A multi-objective ant-colony algorithm (MOACA) is proposed. The salient features of the proposed multi-objective ant-colony algorithm include the consideration of two ants (corresponding to the number of objectives considered) that make use of the same pheromone values in a given iteration; use of a compromise objective function that incorporates a heuristic solution's makespan and total flowtime of jobs as well as an upper bound on the makespan and an upper bound on total flowtime of jobs, coupled with weights that vary uniformly in the range [0, 1]; increase in pheromone intensity of trails by reckoning with the best solution with respect to the compromise objective function; and updating of pheromone trail intensities being done only when the ant-sequence's compromise objective function value is within a dynamically updated threshold level with respect to the best-known compromise objective function value obtained in the search process. In addition, every generated ant sequence is subjected to a concatenation of improvement schemes that act as local search schemes so that the resultant compromise objective function is improved upon. A sequence generated in the course of the ant-search process is considered for updating the set of heuristically non-dominated solutions. We consider the benchmark flowshop scheduling problems proposed by Taillard (1993), and solve them by using twenty variants of the MOACA. These variants of the MOACA are obtained by varying the values of parameters in the MOACA and also by changing the concatenation of improvement schemes. In order to benchmark the proposed MOACA, we rely on two recent research reports: one by Minella *et al.* (2008) that reported an extensive computational evaluation of more than twenty existing multi-objective algorithms available up to 2007; and a study by Framinan and Leisten (2007) involving a multi-objective iterated greedy search algorithm, called MOIGS, for flowshop scheduling. The work by Minella concluded that the multi-objective simulated annealing algorithm by Varadharajan and Rajendran (2005), called MOSA, is the best performing multi-objective algorithm for permutation flowshop scheduling. Framinan and Leisten found that their MOIGS performed better than the MOSA in terms of generating more heuristically non-dominated solutions. They also obtained a set of heuristically non-dominated solutions for every benchmark problem instance provided by Taillard (1993) by consolidating the solutions obtained by them and the solutions reported by Varadharajan and Rajendran. This set of heuristically non-dominated solutions (for every problem instance, up to

U.K. Chakraborty (Ed.): Comput. Intel. in Flow Shop and Job Shop Sched., SCI 230, pp. 53–100.
springerlink.com

100 jobs, of Taillard's benchmark flowshop scheduling problems) forms the reference or benchmark for the present study. By considering this set of heuristically non-dominated solutions with the solutions given by the twenty variants of the MOACA, we form the net heuristically non-dominated solutions. It is found that most of the non-dominated solutions on the net non-dominated front are yielded by the variants of the MOACA, and that in most problem instances (especially in problem instances exceeding 20 jobs), the variants of the MOACA contribute more solutions to the net non-dominated front than the corresponding solutions evolved as benchmark solutions by Framinan and Leisten, thereby proving the effectiveness of the MOACA. We also provide the complete set of heuristically non-dominated solutions for the ninety problem instances of Taillard (by consolidating the solutions obtained by us and the solutions obtained by Framinan and Leisten) so that researchers can use them as benchmarks for such research attempts.

1 Introduction

Flowshop scheduling problem involves the determination of an order of processing n jobs over m machines, arranged in series, to meet a desired objective or a measure of performance. The static permutation flowshop scheduling problem has been widely investigated over the years by considering separately the objectives of minimizing the makespan and total flowtime of jobs, and with the consideration of developing exact or heuristic methods (e.g. Johnson (1954), Ignall and Schrage (1965), Campbell *et al.* (1970), Gelders and Sambandam (1978), Miyazaki *et al.* (1978), Miyazaki and Nishiyama (1980), Nawaz *et al.* (1983), Rajendran (1993), Ho (1995), Wang *et al.* (1997), Woo and Yim (1998), Liu and Reeves (2001), Chung et al. (2002), Allahverdi and Aldowaisan (2002), Framinan and Leisten (2003), Framinan *et al.* (2005), Ruiz and Stuetzle (2007), Kalczynski and Kamburowski (2007) and (2008), Dong *et al.* (2008), Laha and Chakraborty (2008)). The use of metaheuristics such as simulated annealing, genetic algorithm and tabu search has been frequently resorted to solve flowshop scheduling problems (e.g. Widmer and Hertz (1989), Ben-Daya and Al-Fawzan (1998), and Ruiz *et al.* (2006)). In recent times, attempts are being made to solve combinatorial optimization problems by making use of swarm-intelligence algorithms. An important algorithm in this class is the ant-colony-optimization algorithm (or simply, ant-colony or ACO algorithm). The pioneering work has been done by Dorigo (1992), and an introduction to the ACO algorithms had been dealt with in Dorigo *et al.* (1996). Attempts have been made to solve the permutation flowshop scheduling problem with the objective of minimizing the makespan / total flowtime of jobs using ACO algorithms (e.g. Stuetzle (1998) dealing with the permutation flowshop scheduling problem with the objective of minimizing the makespan; Merkle and Middendorf (2000) dealing with the single-machine scheduling problem; T'kindt *et al.* (2002) considering the two-machine flowshop scheduling problem; and Rajendran and Ziegler (2004) and (2005) considering the m-machine permutation flowshop scheduling problem). Another swarm intelligence algorithm is the particle swarm algorithm which has shown promising results to solve flowshop scheduling problems (e.g., Tasgetiren *et al.* (2007) and Liao *et al.* (2007)).

While many attempts have been made to minimize separately makespan and total flowtime, only some attempts have been made to simultaneously minimize such measures of performance. In such a case, it is common to develop algorithms to obtain a set of Pareto-optimal solutions (or at least a set of heuristically non-dominated solutions). Two approaches to multi-objective scheduling are widely followed, namely, *a priori* approach in which the objectives are combined in the form of a weighted compromise function (mostly linear), and *a posteriori* approach in which a set of efficient or Pareto-optimal solutions (in the case of optimality being guaranteed) or a set of heuristically-efficient or heuristically non-dominated solutions (in the case of optimality being not guaranteed) is obtained. In the following, the term 'non-dominated solutions' or 'non-dominated sequences' refers to heuristically-efficient or heuristically non-dominated solutions or sequences, without the guarantee of efficiency or Pareto optimality. Some attempts in these directions are due to Rajendran (1994) and (1995), Sridhar and Rajendran (1996), Murata *et al.* (1996), Ishibuchi and Murata (1998), Bagchi (1999), Chang *et al.* (2002), Framinan *et al.* (2002), and Arroyo and Armentano (2005). In addition, attempts have also been done with the consideration of a lexicographical approach of optimizing a set of objectives (e.g., Daniels and Chambers (1990), Rajendran (1992), Chakravarthy and Rajendran (1999), T'kindt *et al.* (2002), Allahverdi (2004), and Framinan and Leisten (2006)).

Varadharajan and Rajendran (2005) developed a multi-objective simulated-annealing algorithm (with two variants, called MOSA-I and MOSA-II) for flowshop scheduling to minimize the makespan and total flowtime of jobs. The MOSA aims at discovering non-dominated solutions through the use of a simple probability function that is varied in such a way that the entire objective space is covered uniformly, thereby obtaining many non-dominated and well-dispersed solutions. The authors considered the benchmark flowshop problems of Taillard (1993), and obtained the non-dominated solution set for every problem, yielded by existing multi-objective flowshop scheduling algorithms, namely, the algorithms by Ishibuchi and Murata (1998), Bagchi (1999), Chang *et al.* (2002), and Framinan *et al.* (2002), as well as those by MOSA-I and MOSA-II. Subsequently they obtained the net non-dominated front by consolidating all the non-dominated fronts. They found that, in most cases, the MOSA contributes the most to the net non-dominated solution set, in comparison to the existing algorithms. Framinan and Leisten (2007) proposed a multi-objective iterated greedy search, called MOIGS, that is based on a partial enumeration heuristic. The MOIGS uses a parameter, called *d*, and the authors tried with the values of *d* ranging from 3 to 10. They found that the MOIGS discovers more non-dominated solutions than those discovered by Varadharajan and Rajendran (2005). In addition, they consolidated the solutions yielded by their MOIGS (with different values for *d*) and the solutions obtained by Varadharajan and Rajendran. It is be noted that Varadharajan and Rajendran consolidated the solutions yielded by MOSA-I, MOSA-II, and the solutions yielded by the algorithms of Ishibuchi and Murata (1998), Bagchi (1999), Chang *et al.* (2002), and Framinan *et al.* (2002). It is therefore evident that the final non-dominated solutions obtained by Framinan and Leisten are drawn from the implementations of MOIGS with eight different values of *d*, and from the implementations of MOSA-I, MOSA-II, and other algorithms considered by Varadharajan and Rajendran. These non-dominated solutions consolidated by Framinan and Leisten for a problem instance of Taillard (1993) could serve as the benchmark for researchers in the area of flowshop scheduling.

It is to be noted that most of the multi-objective flowshop scheduling algorithms were evaluated by the respective authors by comparing with the previously available literature, and that too, with the related objectives. It also appears that many researchers did not attempt to consider the possible adaptation of the generic multi-objective algorithms (such as NSGA by Srinivas and Deb (1994), SPEA by Zitzler and Thiele (1999), PESA by Corne *et al.* (2000), PESA-II by Corne *et al.* (2001), and NSGA-II by Deb *et al.* (2002)) to flowshop scheduling problems. A recent study by Minella *et al.* (2008) is possibly the first significant attempt to perform a comprehensive analysis by considering a number of flowshop scheduling algorithms (such as the multi-objective genetic algorithm by Murata *et al.* (1996), multi-objective tabu search (MOTS) by Armentano and Arroyo (2004), multi-objective genetic local search by Arroyo and Armentano (2005), MOSA by Varadharajan and Rajendran (2005), multi-objective genetic algorithm by Pasupathy *et al.* (2006), and PILS by Geiger (2007)), and also a number of generic multi-objective algorithms such as the NSGA, SPEA, PESA, PESA-II and NSGA-II. In all, a total of twenty three multi-objective algorithms were considered, and performance analyses were carried out. The authors consolidated the solutions for Taillard's benchmark problem instances. It was found that the MOSA by Varadharajan and Rajendran is the best performer among these twenty three algorithms with respect to multi-objective flowshop scheduling.

In the following, the problem of scheduling in permutation flowshops is considered with the objectives of minimizing the makespan and total flowtime of jobs. We present a multi-objective ant-colony algorithm (MOACA) for obtaining heuristically efficient or heuristically non-dominated solutions. Variants of the MOACA are proposed by varying the values of parameters in the MOACA and also by varying the concatenation of local search or improvement schemes that consider a compromise objective function. We make use of data set containing the benchmark flowshop problems of Taillard (1993) (up to 100 jobs), and generate the non-dominated solutions for every flowshop problem instance by using the different variants of the MOACA and the benchmark solutions consolidated by Framinan and Leisten (2007).

2 Formulation of the Multi-Objective Static Permutation Flowshop Scheduling Problem under Study

The static permutation flowshop scheduling problem consists in scheduling n jobs with given processing times on m machines, where the sequence of processing a job on all machines is identical and unidirectional for each job. In studying flowshop scheduling problems, it is a common assumption that the sequence in which each machine processes all jobs is identical on all machines (permutation flowshop). A schedule of this type is called a permutation schedule and is defined by a complete sequence of all jobs. We also consider only permutation sequences in the following.

Let

t_{ij} processing time of job i on machine j.

D_i due-date for job i.

n total number of jobs to be scheduled.

m total number of machines in the flowshop.

σ ordered set of jobs already scheduled, out of n jobs; partial sequence.

$q(\sigma, j)$ completion time of partial sequence σ on machine j (*i.e.* the release time of machine j after processing all jobs in partial sequence σ).

$q(\sigma i, j)$ completion time of job i on machine j, when the job is appended to partial sequence σ.

For calculating the start and completion times of jobs on machines in permutation flowshops, recursive equations are used as follows.

Initialize $q(\sigma i, 0)$, the completion time of job i on machine 0, equal to zero. This time indicates the time of availability of a job in the flowshop, and it is equal to 0 for all jobs in case of static flowshops.

For $j = 1$ to m do

$$q(\sigma i, j) := \max \{ q(\sigma, j) \,;\, q(\sigma i, j\text{-}1)\} + t_{ij}. \qquad (2.1)$$

The flowtime of job i, C_i, is given by

$$C_i = q(\sigma i, m). \qquad (2.2)$$

It is to be noted that $q(\phi, j)$ is equal to 0 for all j, where ϕ denotes a null schedule.

When all jobs are scheduled, the total flowtime F and the makespan C_{max} of jobs are obtained as follows:

$$F = \sum_{i=1}^{n} C_i \;, \qquad (2.3)$$

and

$$C_{max} = \max \{ C_i \,,\, i = 1, 2, \ldots, n \}. \qquad (2.4)$$

The objective is to simultaneously minimize F and C_{max}. Exceptions set aside, there exists no single solution minimizing both objectives simultaneously. An optimal solution then must have the property of non-dominance.

To present in brief the principle of non-dominance in the context of the problem under study, let us assume that the makespan and total flowtime of jobs yielded by sequence S are denoted by $C_{max}(S)$ and $F(S)$ respectively. For the sake of generality, we let $Z_1(S)$ and $Z_2(S)$ denote $C_{max}(S)$ and $F(S)$ respectively. Sequence S is said to dominate S' if $Z_r(S) \le Z_r(S') \;\forall\; r$, and $Z_r(S) < Z_r(S')$ for at least one r. Sequence S'' is efficient if there exists no other sequence S dominating S''. It is to be noted that the m-machine permutation flowshop scheduling problem with the consideration of a single objective, in most cases, was shown by Garey *et al.* (1976) to be NP-hard. It is therefore evident that researchers develop heuristic methods to obtain heuristically non-dominated solutions (without the guarantee of efficiency) in the case of multi-objective flowshop scheduling problems. A Sequence S'' is called heuristically non-dominated or heuristically efficient with respect to a given set of heuristic solutions if there exists no other known heuristic sequence S dominating S''. Suppose we have a set of heuristically non-dominated sequences, denoted by ψ. A new heuristic sequence S'' qualifies for entry into ψ if and only if for each sequence S in ψ there exists at least one r for which $Z_r(S'') < Z_r(S)$. Likewise, a sequence S' can be eliminated from the set ψ due to the inclusion of S'' if $Z_r(S'') \leq Z_r(S')$ $\forall$ r. Readers may see T'kindt and Billaut (2002) for a complete treatment on scheduling with multiple objectives. In the following, the a posteriori approach is considered, i.e., a set of heuristically efficient sequences with respect to the two objectives of minimizing total flowtime and minimizing makespan is to be determined.

3 Description of the Proposed Multi-Objective Ant-Colony Algorithm

3.1 General Structure of Ant-Colony Algorithms

ACO algorithms make use of simple agents, called ants, that iteratively construct solutions to combinatorial optimization problems. The solution generation or construction by ants is guided by (artificial) pheromone trails and problem-specific heuristic information. In the context of combinatorial optimization problems, pheromones indicate the intensity of ant-trails with respect to solution components, and such intensities are determined on the basis of the influence or contribution of each solution component with respect to the objective function. An individual ant constructs a complete solution by starting with a null solution and iteratively adding solution components until a complete solution is constructed. Typically, solution components which are part of better solutions used by ants over many iterations receive a higher amount of pheromone, and hence, such solution components are more likely to be used by the ants in future iterations of the ACO algorithm. This is enhanced by also making use of pheromone evaporation in updating trail intensities. In the context of application of ACO algorithms to scheduling problems, pheromone trail intensity (or desirability) of placing job i in position k of a sequence can be denoted by τ_{ik}. It is to be noted that for every job i for any possible position k, a pheromone value is stored and updated in each iteration of the ACO algorithm. An explanation on the structure of ACO algorithms is given in Stuetzle (1998), and Rajendran and Ziegler (2004).

3.2 Details of the Proposed Multi-Objective Ant-Colony Algorithm (MOACA)

We highlight the salient features of the proposed algorithm with respect to the search in the two-dimensional objective-function space enabled through the use of a compromise objective function incorporating relative weights for each objective function and the use of upper bounds on the makespan and total flowtime of jobs.

3.2.1 Characterization of the MOACA

In view of two objectives being considered, two seed sequences are used corresponding to every combination of the two relative weights related to the makespan and total flowtime of jobs, and these sequences are used to initialize the pheromone trail intensities τ_{ik}. A front that consists of non-dominated sequences obtained during the search process is maintained. The trail intensities and the best sequence obtained so far are used as the basis to construct multiple (in our study, two) ant sequences which are subsequently improved, with respect to the compromise objective function, by using different concatenations of two local search schemes, called JIS and JSS. We construct two ant sequences in view of the number of objectives being two; moreover, pilot runs with the construction of a greater number of multiple ant sequences have indicated the best performance of the proposed algorithm with two ant sequences, given our restriction on the total number of sequences enumerated in the MOACA. It is to be noted that every ant sequence that is generated (including every sequence generated in local search schemes) is checked for possible entry into the non-dominated front, so as to discover as many solutions lying on the multi-modal non-dominated front as possible.

In the MOACA, we define a compromise objective function for a given sequence S as follows:

$$Z(S) = w_1 \times (C_{max}(S) \,/\, up\_C_{max}) + w_2 \times (F(S) \,/\, up\_F), \tag{3.1}$$

where up_C_{max} refers to an upper bound on the makespan for a given problem, up_F refers to an upper bound on total flowtime of jobs, and $w_1 + w_2 = 1$ with w_1, $w_2 \geq 0$. This approach of using a compromise objective function with the incorporation of upper bounds on the makespan and total flowtime of jobs has been found to be effective in the case of multi-objective flowshop scheduling; the reason is that we basically normalize a heuristic solution's makespan and total flowtime of jobs, thereby avoiding the inconsistency in the magnitude of the makespan and total flowtime of jobs. In fact, similar approaches were also taken by Rajendran (1994) and (1995), and also by Sridhar and Rajendran (1996).

Note that to start with, for a given Taillard's problem instance, we use the upper bound on makespan that was reported by Taillard (1993) (denoted by *upmake* for a given problem instance), and we use the best upper bound on total flowtime that was reported by Rajendran and Ziegler (2004) (denoted by *upflow* for a given problem instance). Initialize up_C_{max} = *upmake*, and up_F = *upflow*. However, during the execution of the MOACA, better upper bounds, if obtained, are used to update up_C_{max} and up_F for their use in the compromise objective function. Note that the weights have to be appropriately chosen in order to discover many non-dominated solutions. We vary w_1 uniformly (and consequently w_2) in the range [0, 1]. In the MOACA, we initially set $w_1 = 0$, implying that we first seek to minimize total flowtime of jobs, and we increase w_1 in steps of 0.1, up to 1. This means that the basic MOACA is repeated 11 times, corresponding to different values of w_1 and w_2, and our experimental investigations have shown that the MOACA with these values for weights is able to discover many solutions lying on the non-dominated front (with no possible guarantee of Pareto optimality or efficiency). We now present the mechanics of the basic MOACA, for the given w_1 and w_2, in Sections 3.2.2, 3.2.3 and 3.2.4.

3.2.2 Generation of Two Initial Ant Sequences and Initialization of Trail Intensities

We generate one seed sequence by ordering jobs in the ascending order of the weighted sum of process times of jobs (i.e., in the non-decreasing order of

$\sum_{j=1}^{m}(m-j+1)t_{ij}$; see Rajendran (1993) for details), followed by the improvement scheme presented by Nawaz / Enscore / Ham (1983) if w_1 is less than or equal to 0.5, or by ordering jobs in the non-increasing order of the sum of process times of jobs, and then using the improvement scheme presented by Nawaz / Enscore / Ham (1983) if w_1 is greater than 0.5. Note that all partial and complete sequences (generated during these procedures) are evaluated by using Eq. (3.1), and the best partial (or complete) sequence is accordingly chosen. The second seed sequence is generated randomly by selecting the job to be placed in position k of the sequence with equal probability from the set of unscheduled jobs, $k = 1(1)n$. Check if each complete sequence can enter the existing non-dominated front. If so, enter it and update the front accordingly. Every seed sequence is subjected to the improvement schemes, namely, the job-index-based insertion scheme (called JIS), followed by the job-index-based swap scheme (JSS) in the given concatenation, with the consideration of $Z(S)$ for the given w_1 and w_2 (see Eq. (3.1)). The details of different concatenations of the JIS and JSS, namely, JIS-JSS-JIS-JSS, JIS-JIS-JSS-JIS and JIS-JIS-JIS-JSS, are presented later. The effectiveness of concatenation of the local search schemes is due to the fact that each of these schemes perturbs the seed sequence in different ways, thereby discovering many more local minima in the neighborhood than a single local search scheme applied more than once. These improvement schemes have been found to be effective in single-objective flowshop scheduling by earlier works as well (see Rajendran and Ziegler (2004) and (2005)). In fact, our computational experiments have also shown that the concatenation of such local search schemes has been found to perform better than the successive application of one single local scheme in terms of discovering many more solutions on the non-dominated front. The details of the JIS and JSS are given in the Appendix. Note that the JIS involves a relatively mild perturbation of the seed sequence, as opposed to the JSS. In fact, the JIS can be considered as an intensification of local search, while the JSS can be considered as a diversification of local search. It is also to be noted that each of the two local search schemes aims at improving the seed sequence with the consideration of the compromise objective function (as given in Eq. (3.1)) for the given w_1 and w_2, and that every sequence that is generated in a local search scheme is considered for possible entry in the non-dominated front. The two final sequences that are yielded by the application of concatenation of JIS and JSS on each of the two seed sequences are taken as the final ant sequences. These sequences, denoted by S^1 and S^2, are used to set the trail intensities for a given w_1 and w_2. Let these two sequences' compromise objective function values (computed by using Eq. (3.1)) be denoted by $Z(S^1)$ and $Z(S^2)$ respectively. Let the minimum of these two values be denoted by Z^*, and the corresponding sequence be denoted by S^*. We initialize pheromone trail intensities as follows:

$$\tau_{ik} = 1/(Z^*)^p, \ \forall\, k \text{ and } \forall i. \tag{3.2}$$

In the above, p (>= 1) denotes the index of power. Initialize *no_iter*, the number of iterations in respect of generation of ant sequences in the search process for the given w_1 and w_2, to 0. Subject S^1 and S^2 to an adjacent pairwise interchange of jobs (interchanging jobs found in positions k and k+1, for k = 1, 2, ..., n-1), thereby generating 2(n-1) sequences in the neighbourhood. Check every generated sequence for possible entry into the non-dominated front and also check for the consequent updating of the non-dominated front. Note that these $2(n\text{-}1)$ sequences do not have any impact on trail intensities, and that these sequences are generated to primarily explore the neighbourhood for non-dominated solutions.

3.2.3 Modification of Trail Intensities

We first modify the trail intensities as follows:

$$\tau_{ik} := \rho \times \tau_{ik}, \ \forall\, k \text{ and } \forall i, \tag{3.3}$$

where ρ denotes the persistence rate of pheromone trail intensities (or equivalently, 1-evaporation rate).

Then, we further modify the trail intensities τ_{ik} as follows, by taking into account the position occupied by a job.

For r = 1 and 2, do the following: /*corresponding to two sequences*/
 if $((Z(S^r) - Z^*) / Z^*) \le cut\_off$
 then
  for i = 1(1) n do the following: /*corresponding to n jobs*/
   for k = 1(1) n do the following: /*corresponding*/ /*to n positions*/
    if $|\, h^r(i) - k \,| \le \lfloor n/50 \rfloor$
    then

$$\text{set } \tau_{ik} := \tau_{ik} + 1 / (2\times(Z(S^r))^p). \tag{3.4}$$

In the above, *cut_off* refers to the threshold value with respect to the deviation of the compromise objective function value of a given sequence from the best value obtained so far in the MOACA, for the given w_1 and w_2. If the deviation is less than or equal to the *cut_off*, then we use the sequence to update the trail intensities. This is done so because we do not want to use an inferior sequence to be used in updating pheromone values, as otherwise, we would lose the good trail intensities that have been obtained in the search process. In the above, $h^r(i)$ refers to the position occupied by job i in sequence S^r. It is to be noted that we update the trail intensity for every job with respect to more than one position, depending upon the value of $\lfloor n/50 \rfloor$. This is done so because we believe that the trail intensities of such positions fairly close to the position of job i need to be updated in the same way as the position of job i, with the number of such positions being governed by the number of jobs in the given problem (also see the related observations by Rajendran and Ziegler (2004) and (2005) in the case of single-objective flowshop-scheduling problems). We have 2 in the denominator in Expression (3.4) because we use two ants in our MOACA. It is also to be noted that the value of *cut_off* is not static across all iterations (with each iteration involving the generation of two ant sequences) carried out for the given w_1 and w_2. After the generation of two ant sequences (to be presented in the following), we set *cut_off* equal to (*cut_off*×0.9), and we do the task of updating the pheromone values, as given in Expressions (3.3) and (3.4). In the current study, we initially set *cut_off* to 0.025.

In order to guide the MOACA towards discovering solutions possibly lying on the Pareto-optimal front by making use of S^*, the best sequence obtained so far (for the given w_1 and w_2, and with respect to Equation (3.1), we supplement the trail intensities as follows.

For i = 1(1) n do the following: /*corresponding to n jobs*/
  for k = 1(1) n do the following: /*corresponding to n positions*/
    compute $diff = (|\, h^*(i) - k \,| + 1)$
    and
    set $\tau_{ik} := \tau_{ik} + 1 \,/\, ((Z(S^*))^p \times (diff)^c)$. (3.5)

In the above, c denotes the power index for *diff*, and $h^*(i)$ refers to the position of job i in sequence S^*. It is evident that τ_{ik} is updated for job i with respect to position k depending upon the relative difference between this position and the position of job i in S^*. Power index c is introduced for enhancing the differentiation.

3.2.4 Construction of Two Ant Sequences and Their Improvement with Respect to the Compromise Objective Function by Local Search Schemes

In the MOACA, a complete sequence is built up, by starting from a null sequence and choosing a job by the following procedure in order to append it to the available partial sequence in position k, for k = 1, 2, …, n, and with the initial available partial sequence being a null set.

Set $T_{ik} = \sum_{q=1}^{k} \tau_{iq}$ and sample a uniform random number u in the range (0, 1).

If $u \leq 0.4$
then
  the first job in S^* that is not yet scheduled in the present partial sequence is chosen;
else
  if $u \leq 0.8$
  then
    among the set of the first $(4+\lfloor n/K \rfloor)$ jobs in S^* that are not yet scheduled in the present partial sequence, choose the job with the maximum value of T_{ik};
  else
    job i is selected from the same set of $(4+\lfloor n/K \rfloor)$ unscheduled jobs for position k as a result of sampling from the following probability distribution:

$$p_{ik} = (T_{ik} \,/\, \sum_{l} T_{lk}), \tag{3.6}$$

    where job l belongs to the same set of $(4+\lfloor n/K \rfloor)$ unscheduled jobs.

    Note that when there are unscheduled jobs less than this prescribed number, then all such unscheduled jobs are considered for possible selection.

In the above, K is a parameter that helps to decide on the number of unscheduled jobs to be considered while constructing an ant sequence. The rationale behind the selection of the job to be scheduled next is that the choice is governed between the best sequence and the best value of T_{ik} with equal probability, and the probabilistic choice of the job is done with half of the probability of going in for the first unscheduled job found in the best sequence. In addition, the number of unscheduled jobs considered for selection is not the same across all problem sizes, which is quite logical. Readers may see the related works by Rajendran and Ziegler (2004) and (2005) for single-objective flowshop-scheduling problems.

By performing the above procedure for $k = 1, 2, ..., n$, a complete ant sequence can be generated. Repeat the process for generating one more ant sequence. Check whether an ant sequence can enter the exiting non-dominated front; if so, enter it and accordingly update the front. Note that each of these two sequences is subjected to the JIS and JSS (in different combinations or concatenations) with the consideration of the compromise objective function for the given w_1 and w_2. Let us denote the two final sequences thus obtained by S^1 and S^2, with the values of the compromise objective functions denoted by $Z(S^1)$ and $Z(S^2)$ respectively. Update Z^* and S^*, if necessary, by comparing Z^* with $Z(S^1)$ and then with $Z(S^2)$. We wish to point out here that we use two different uniform random number streams, while developing an ant sequence: one for sampling u to decide on the choice between the best sequence and the set of unscheduled jobs; and another uniform random number stream for sampling from the probability distribution. Subject S^1 and S^2 to an adjacent pairwise interchange of jobs (interchanging jobs found in positions k and k+1, for k = 1, 2, ..., n-1), thereby generating 2(n-1) sequences in the neighbourhood; and check every generated sequence for possible entry into the non-dominated front and also check for the possible updating of the non-dominated front. Note that these 2(n-1) sequences do not have any impact on trail intensities and that these sequences are generated to primarily explore the neighbourhood for non-dominated solutions.

Set *cut_off* := (*cut_off*×0.9), and *no_iter* := *no_iter* + 1.

If *no_iter* is < 16, then repeat the steps given in Sections 3.2.3 and 3.2.4; else increase w_1 by 0.1, decrease w_2 accordingly, initialize *cut_off* to 0.025 and *no_iter* to 0, and go back to repeat the steps given in Sections 3.2.2, 3.2.3 and 3.2.4.

It is to be noted that we have opted to vary w_1 from 0 to 1, in steps of 0.1, and set the upper limit on the number of iterations to 16. This is done so in order to restrict the computational effort. Every variant of the proposed MOACA enumerates about $1500n^2$ sequences in the course of the entire search process. It is noteworthy that when each variant has been coded in FORTRAN (DOS version) and executed on a PC with Pentium 4 processor, 3 GHz, 512 MB RAM, a variant requires the execution time of about 10 hours to obtain the non-dominated solutions for all the 90 problem instances. It is also to be noted that the actual execution time is not large in view of the fact that the JIS and JSS are computationally fast schemes, unlike the relatively more-demanding crossover and mutation operations involved in genetic algorithms for permutation flowshop scheduling. Readers may see the related observations by Minella *et* al. (2008) in respect of the MOSA that also uses similar local search schemes involving job insertion or job swap. Of course, one can perform the MOACA with more enumeration of sequences by increasing the upper limit on the number of iterations so that an enhanced performance of the MOACA can possibly be achieved.

3.3 Step-by-Step Procedure of the MOACA

We now present the step-by-step procedure consolidating the salient features of the MOACA.

Step 1: Set $w_1 = -0.10$ and $w_2 = 1.10$. Obtain an upper bound on makespan and an upper bound on total flowtime for the given problem instance from the available literature, and hence obtain up_C_{max} and up_F.

Step 2: Set $w_1 := w_1 + 0.10$, and $w_2 := w_2 - 0.10$. Generate one seed sequence by ordering the jobs in the ascending order of $\sum_{j=1}^{m}(m-j+1)t_{ij}$ followed by the improvement scheme presented by Nawaz *et al.* (1983) if w_1 is less than or equal to 0.5, or by ordering jobs in the non-increasing order of the sum of process times of jobs, and then using the improvement scheme presented by Nawaz *et al.* if w_1 is greater than 0.5, with the consideration of the current w_1 and w_2. Generate the second seed sequence randomly. Update the non-dominated front by considering these two seed sequences. Every seed sequence is then subjected to the given concatenation of JIS and JSS. Call the final sequences S^1 and S^2. Update S^* and $Z(S^*)$ by using S^1 and S^2. In addition apply adjacent pairwise interchange of jobs to the sequences S^1 and S^2.

Note: Check each sequence generated for non-dominance and if necessary update the non-dominated front.

Initialize τ_{ik} as per Eq. (3.2). Set $cut\_off = 0.025$ and $no\_iter = 0$.

Step 3: Modify τ_{ik} as given by Eq. (3.3), and modify also by reckoning with S^1 and S^2, see Exp. (3.4), and thereafter reckoning with S^*, see Exp. (3.5).

Step 4: Construct two ant sequences by using the procedure given in Section 3.2.4, with each sequence thereafter improved by the given concatenation of JIS and JSS. Call the final sequences S^1 and S^2. Update S^* and $Z(S^*)$ by using S^1 and S^2. In addition apply adjacent pairwise interchange of jobs to the sequences S^1 and S^2.

Note: Check each sequence generated for non-dominance and if necessary update the non-dominated front.

Step 5: Set $cut\_off := cut\_off \times 0.9$ and $no\_iter := no\_iter + 1$. If $no\_iter < 16$ then go to Step 3; else return to Step 2 as long as $w_1 \leq 0.9$. Return the final set of heuristically non-dominated solutions for the given problem instance. STOP.

4 Performance Analysis of the MOACA

In line with previous researchers, we have considered the ninety benchmark flowshop scheduling problem instances by Taillard (1993), with the number of jobs being 20, 50 and 100, and with the number of machines being 5, 10 and 20. In order to evaluate the performance of a multi-objective flowshop scheduling algorithm, many researchers basically used the following metric in one form or another: the number of solutions contributed by an algorithm to the final or net non-dominated front (e.g., Ishibuchi and Murata (1998), Chang *et al.* (2002), Varadharajan and Rajendran

(2005), and Framinan and Leisten (2007)). We have also used a similar metric given as follows:

> number of solutions contributed by a given multi-objective algorithm to the net non-dominated front / total number of solutions in the net non-dominated front. (4.1)

This metric is relatively easy in terms of comprehending how a multi-objective algorithm performs in relation to other multi-objective algorithms.

As the first step, we have set $p = 2$, 1 and 1.5, with $K = 50$ and 20, $c = 1$ and $\rho = 0.75$ in the proposed MOACA. Note that for every given p, c and ρ, we experiment with $K = 50$ and 20. The reason is that the parameter K is involved in the generation of an ant sequence, and hence we would prefer to always experiment with these two values of K. The setting of ρ in the neighbourhood of 0.75 is found to work well by previous researchers as well (Stuetzle (1998), and Rajendran and Ziegler (2004) and (2005)). The corresponding variants are termed Variants 1-6 of the MOACA. From the performance analyses of these variants, we have observed that these variants perform not much differently on an average, and every variant does contribute to the final or net non-dominated front in a similar manner. Hence we have decided to freeze p at 1.5. We now set $c = 2$, and $\rho = 0.75$ and 0.7. We find that these variants, namely Variants 7-10, do perform well, especially in the case of the larger-sized problems. For the further two variants (namely Variants 11 and 12), we set $c = 1.5$ and $\rho = 0.725$. We find that these two variants also perform well, especially for the larger-sized problems. Note that in all these variants, the concatenation of JIS and JSS is done in the following manner: JIS-JSS-JIS-JSS.

As further analysis, we have decided to see the performance of the MOACA by changing the concatenation of the JIS and JSS. First picking up on Variants 11 and 12, we have experimented with the following concatenation of the JIS and JSS: JIS-JIS-JSS-JIS, followed by the concatenation of JIS-JIS-JIS-JSS. The corresponding variants are termed Variants 13 and 14 (derived from Variant 11), and Variants 15 and 16 (derived from Variant 12). Similarly, we have derived Variants 17-20 from Variants 9 and 10 respectively by implementing these two concatenations of JIS and JSS. Our performance analyses have shown that different concatenations of JIS and JSS have indeed served to discover additional non-dominated solutions, especially in the case of larger-sized problems. Table 1 presents the details of settings for different variants of the MOACA.

As mentioned earlier, we have made use of the benchmark solutions provided by Framinan and Leisten (2007), and consolidated the solutions yielded by all the variants of the MOACA with those obtained by Framinan and Leisten. The final sets of non-dominated solutions thus obtained for every problem instance are given in Tables 2a - 4c. We believe that these solutions can possibly serve as benchmarks for future research attempts as much the solutions obtained by Framinan and Leisten served for us as benchmarks. It is be noted again that the solutions obtained by Framinan and Leisten are through the implementation of their MOIGS with eight different values for d, and from the consolidation with the solutions reported by Varadharajan and Rajendran (2005).

We have also evaluated every variant of the MOACA and the set of solutions obtained by Framinan and Leisten (denoted by F&L), by using the metric given in

Table 1. Settings for MOACA variants

MOACA Variant	ρ	c	p	K	Concatenation of local search schemes
1	0.75	1	2	50	JIS-JSS-JIS-JSS
2	0.75	1	2	20	JIS-JSS-JIS-JSS
3	0.75	1	1	50	JIS-JSS-JIS-JSS
4	0.75	1	1	20	JIS-JSS-JIS-JSS
5	0.75	1	1.5	50	JIS-JSS-JIS-JSS
6	0.75	1	1.5	20	JIS-JSS-JIS-JSS
7	0.75	2	1.5	50	JIS-JSS-JIS-JSS
8	0.75	2	1.5	20	JIS-JSS-JIS-JSS
9	0.7	2	1.5	50	JIS-JSS-JIS-JSS
10	0.7	2	1.5	20	JIS-JSS-JIS-JSS
11	0.725	1.5	1.5	50	JIS-JSS-JIS-JSS
12	0.725	1.5	1.5	20	JIS-JSS-JIS-JSS
13	0.725	1.5	1.5	50	JIS-JIS-JSS-JIS
14	0.725	1.5	1.5	20	JIS-JIS-JSS-JIS
15	0.725	1.5	1.5	50	JIS-JIS-JIS-JSS
16	0.725	1.5	1.5	20	JIS-JIS-JIS-JSS
17	0.7	2	1.5	50	JIS-JIS-JSS-JIS
18	0.7	2	1.5	20	JIS-JIS-JSS-JIS
19	0.7	2	1.5	50	JIS-JIS-JIS-JSS
20	0.7	2	1.5	20	JIS-JIS-JIS-JSS

Exp. (4.1). The results of such an analysis are presented in Table 5. In Table 5, an entry in a given row under a given approach denotes the number of non-dominated solutions contributed by a given approach to the net non-dominated front for that problem instance. We then compute the metric given in Exp. (4.1), and sum it up with respect to that approach over ten problem instances. The average over these ten problem instances for that approach is then computed and reported, see the last row in a given problem set or size. Note that for the problem instances with jobs equal to 20,

Table 2a. Net set of non-dominated solutions obtained for the problem size (20×5)

Problem 1		Problem 2		Problem 3		Problem 4		Problem 5	
C_{max}	F	C_{max}	F	C_{max}	F	C_{max}	F	C_{max}	F
1278	14064	1359	16112	1081	13818	1293	16619	1235	14163
1313	14058	1360	16071	1085	13759	1299	16342	1239	14151
1315	14048	1361	15567	1086	13666	1301	15983	1243	14047
1324	14041	1364	15548	1095	13665	1304	15925	1244	14002
1339	14033	1368	15535	1096	13587	1306	15852	1250	13943
		1372	15525	1097	13560	1307	15844	1254	13927
		1377	15454	1099	13524	1309	15828	1264	13890
		1379	15450	1100	13505	1311	15819	1266	13885
		1383	15156	1107	13496	1313	15793	1278	13875
		1385	15151	1111	13418	1319	15758	1285	13872
				1122	13400	1320	15587	1289	13834
				1140	13358	1329	15484	1298	13827
				1183	13347	1354	15447	1301	13811
				1289	13301			1305	13763
								1311	13732
								1328	13668
								1338	13619
								1360	13552
								1387	13529

Table 2a. *(continued)*

Problem 6		Problem 7		Problem 8		Problem 9		Problem 10	
C_{max}	F	C_{max}	F	C_{max}	F	C_{max}	F	C_{max}	F
1195	14908	1242	14280	1206	14581	1230	14977	1108	13649
1198	14889	1245	14200	1211	14429	1232	14936	1112	13584
1200	14888	1252	13999	1212	14345	1245	14929	1113	13514
1202	14741	1254	13887	1213	14302	1246	14902	1115	13429
1203	14699	1255	13749	1214	14253	1247	14756	1133	13419
1207	14681	1264	13730	1217	14157	1248	14729	1134	13344
1210	14346	1265	13728	1222	14153	1249	14715	1138	13173
1213	14247	1266	13722	1226	14148	1253	14505	1150	13126
1217	14121	1267	13695	1229	14128	1255	14493	1151	13122
1218	14042	1274	13632	1233	14093	1256	14485	1153	13026
1224	13608	1278	13578	1234	14080	1259	14449	1163	12999
1233	13583	1283	13548	1240	14075	1271	14446	1179	12981
1241	13581			1245	14073	1272	14386	1184	12943
1245	13437			1246	14072	1281	14367		
1247	13412			1247	14059	1284	14329		
1248	13410			1252	14051	1336	14317		
1251	13391			1254	13994	1337	14295		
1252	13280			1320	13987				
1255	13274			1329	13948				
1256	13212								
1257	13171								
1260	13160								
1261	13139								
1266	13123								

Table 2b. Net set of non-dominated solutions obtained for the problem size (20×10)

Problem 1		Problem 2		Problem 3		Problem 4		Problem 5	
C_{max}	F	C_{max}	F	C_{max}	F	C_{max}	F	C_{max}	F
1582	22121	1664	23888	1496	20905	1377	19738	1419	19277
1583	21731	1666	23877	1501	20672	1380	19721	1420	19205
1590	21706	1667	23527	1508	20433	1385	19579	1422	19203
1592	21421	1668	23525	1515	20364	1386	19533	1432	18966
1595	21420	1671	23519	1521	20118	1387	19523	1435	18952
1608	21385	1672	23399	1534	20061	1392	19431	1446	18873
1629	21337	1676	23375	1546	20036	1393	19413	1463	18829
1640	21284	1683	23356	1547	20003	1394	19410	1466	18798
1641	21204	1684	23303	1577	19962	1397	19344	1473	18794
1656	21122	1690	23274	1589	19958	1399	19280	1476	18766
1685	21025	1692	23242	1615	19927	1403	19273	1485	18754
1686	21011	1694	23166	1624	19917	1406	19177	1486	18641
1698	21003	1699	23156	1650	19877	1409	19149		
1705	20957	1700	23112	1693	19861	1416	19094		
1707	20911	1701	22999	1703	19833	1424	19082		
		1706	22995			1425	19044		
		1708	22853			1432	19020		
		1728	22807			1437	18992		
		1737	22726			1443	18987		
		1744	22720			1445	18948		
		1764	22714			1451	18908		
		1779	22711			1473	18893		
		1781	22617			1476	18852		
		1782	22608			1493	18828		
		1818	22606			1494	18800		
		1827	22559			1509	18792		
		1831	22524			1525	18751		
		1841	22492			1558	18750		
		1847	22473						
		1872	22446						
		1893	22440						

Table 2b. (*continued*)

Problem 6		Problem 7		Problem 8		Problem 9		Problem 10	
C_{max}	F	C_{max}	F	C_{max}	F	C_{max}	F	C_{max}	F
1397	20725	1484	19232	1544	22075	1593	20779	1591	22719
1402	20612	1492	19166	1545	21827	1597	20765	1595	22575
1403	20512	1498	19159	1546	20927	1602	20763	1598	22334
1404	20374	1500	18894	1552	20845	1607	20761	1604	21945
1409	20278	1510	18846	1553	20756	1608	20725	1608	21930
1413	20127	1525	18765	1556	20674	1612	20651	1612	21882
1421	20102	1526	18658	1561	20489	1616	20592	1630	21872
1423	20049	1533	18598	1570	20480	1622	20591	1632	21675
1424	20027	1540	18584	1573	20471	1627	20564	1642	21662
1427	20024	1543	18526	1577	20466	1635	20538	1647	21659
1429	19896	1550	18476	1578	20381	1648	20487	1652	21632
1436	19862	1562	18445	1579	20374	1657	20454	1659	21581
1440	19856	1579	18409	1589	20358	1668	20421	1671	21497
1441	19838	1594	18377	1598	20347	1669	20419	1681	21462
1442	19775	1600	18376	1608	20288	1676	20412	1684	21459
1448	19774	1617	18363	1641	20241	1677	20374	1685	21453
1450	19761					1685	20356	1712	21418
1451	19721					1749	20347	1735	21405
1455	19711					1762	20330	1768	21402
1471	19666							1770	21359
1475	19641							1774	21352
1477	19543							1778	21320
1483	19522								
1498	19517								
1522	19495								
1523	19473								
1529	19422								
1551	19382								
1554	19357								
1568	19340								
1594	19331								
1635	19290								
1655	19283								
1659	19249								
1696	19245								

Table 2c. Net set of non-dominated solutions obtained for the problem size (20×20)

Problem 1		Problem 2		Problem 3		Problem 4		Problem 5	
C_{max}	F	C_{max}	F	C_{max}	F	C_{max}	F	C_{max}	F
2297	35831	2099	33261	2328	36809	2223	33282	2294	36054
2298	35764	2100	32912	2332	36578	2224	32841	2300	36040
2299	35724	2104	32874	2336	35985	2225	32546	2305	35992
2300	35665	2105	32786	2353	35829	2233	32516	2309	35834
2301	35623	2111	32769	2363	35821	2234	32231	2314	35608
2302	35384	2118	32762	2366	35739	2249	32124	2322	35528
2303	35358	2119	32734	2369	35363	2251	32121	2336	35451
2310	35322	2120	32684	2373	35251	2253	32025	2337	35440
2313	35274	2125	32681	2383	35243	2260	31993	2343	35365
2317	35237	2129	32647	2385	35217	2261	31928	2345	35215
2324	35195	2132	32489	2388	35120	2263	31855	2390	35214
2325	34965	2145	32482	2395	35094	2264	31826	2399	35154
2341	34961	2147	32462	2399	34991	2265	31804	2401	35131
2344	34954	2149	32360	2400	34959	2276	31753	2402	35076
2345	34738	2153	32339	2402	34917	2289	31726	2411	34942
2346	34581	2154	32316	2407	34840	2296	31714	2434	34805
2351	34533	2163	32205	2414	34783	2301	31708	2508	34782
2352	34467	2166	32089	2422	34732	2311	31690	2519	34710
2355	34374	2196	31906	2426	34707	2387	31677	2538	34667
2363	34220	2206	31826	2429	34703	2405	31661	2560	34659
2380	34139	2214	31777	2430	34679			2564	34649
2386	34126	2254	31716	2433	34614			2570	34645
2388	34026	2259	31713	2435	34480			2571	34616
2391	33998	2261	31612	2449	34400			2607	34605
2392	33901	2275	31597	2453	34388			2613	34602
2412	33827	2334	31587	2456	34385			2617	34590
2418	33799			2465	34377			2622	34557
2427	33742			2466	34364				
2434	33735			2474	34232				
2437	33623			2484	34127				
				2508	34125				
				2526	34110				
				2535	34107				
				2547	34101				
				2549	34084				
				2554	34082				
				2555	34072				
				2557	34055				
				2564	34051				
				2567	34016				
				2578	33977				
				2579	33932				
				2608	33920				

Table 2c. (*continued*)

Problem 6		Problem 7		Problem 8		Problem 9		Problem 10	
C_{max}	F	C_{max}	F	C_{max}	F	C_{max}	F	C_{max}	F
2230	34231	2276	34517	2200	34792	2237	34532	2180	33462
2232	33853	2278	33756	2202	34758	2243	34516	2183	33264
2234	33652	2282	33438	2205	34712	2248	34363	2191	33240
2239	33557	2292	33436	2209	34612	2253	34360	2196	33125
2242	33407	2299	33425	2210	34555	2258	34338	2202	32937
2252	33395	2305	33390	2212	34129	2260	34183	2229	32824
2253	33383	2307	33353	2221	34123	2281	34178	2231	32805
2257	33351	2320	33325	2222	33931	2289	34138	2238	32764
2258	33330	2324	33295	2224	33882	2292	34133	2242	32731
2260	33160	2334	33282	2234	33843	2297	34077	2245	32654
2263	32876	2335	33276	2237	33744	2308	34065	2246	32583
2270	32853	2336	33253	2238	33661	2310	34062	2249	32497
2281	32810	2340	33221	2242	33640	2319	34046	2250	32477
2284	32778	2343	33211	2243	33420	2320	34031	2270	32423
2292	32758	2350	33206	2257	33267	2336	34029	2287	32383
2304	32752	2353	33184	2266	33107	2337	34015	2308	32375
2307	32714	2356	33178	2273	33068	2343	33959	2309	32331
2318	32707	2359	33139	2284	33045	2356	33900	2329	32310
2320	32693	2368	33107	2294	32990	2360	33847	2338	32299
2324	32656	2407	32987	2297	32975	2372	33805	2339	32292
2334	32655	2415	32970	2299	32943	2379	33772	2345	32269
2358	32652	2453	32951	2311	32921	2418	33734	2365	32262
2359	32650	2466	32922	2312	32909	2419	33729		
2360	32625			2314	32897	2425	33727		
2365	32616			2318	32880	2427	33722		
2369	32604			2323	32865	2428	33641		
2372	32564			2330	32854	2448	33634		
				2331	32814	2455	33625		
				2341	32803	2458	33623		
				2351	32793	2486	33612		
				2360	32775				
				2373	32679				
				2380	32663				
				2391	32642				
				2393	32629				
				2394	32603				
				2396	32552				
				2408	32524				
				2415	32509				
				2433	32506				
				2470	32499				
				2476	32494				
				2478	32485				
				2492	32444				

Table 3a. Net set of non-dominated solutions obtained for the problem size (50×5)

Problem 1		Problem 2		Problem 3		Problem 4		Problem 5	
C_{max}	F	C_{max}	F	C_{max}	F	C_{max}	F	C_{max}	F
2724	67351	2838	76083	2621	65944	2753	72139	2864	70577
2728	67344	2841	76073	2622	65743	2757	70199	2865	70558
2729	67291	2843	75098	2630	65278	2758	70088	2886	70236
2731	67208	2848	69791	2641	65081	2764	70036	2887	70036
2735	65937	2849	69708	2642	65028	2766	69961	2904	69739
2743	65782	2853	69693	2645	64907	2767	69633		
2744	65776	2854	69656	2648	64851	2768	69613		
2745	65752	2857	69522	2660	64817	2775	69586		
2746	65726	2859	69173	2663	64550	2779	69549		
2747	65698	2860	69167	2665	64232	2782	69499		
2752	65218	2861	69088	2667	64108	2785	69490		
2774	65191	2862	69047	2671	64053	2788	69424		
2840	65168	2864	69011	2672	63930	2792	69408		
		2865	68920	2694	63879	2797	69297		
		2867	68894	2698	63861	2800	69256		
		2875	68840	2703	63859	2806	69080		
		2886	68836	2735	63856	2810	69067		
		2889	68811	2776	63838	2856	69000		
		2890	68798			2889	68968		
		2892	68685			2919	68958		
		2896	68683			2930	68864		
		2910	68641			2931	68814		
		2915	68631						
		2916	68622						
		2917	68617						
		2918	68610						
		2929	68599						
		2930	68589						
		2933	68580						
		2934	68575						
		2937	68540						
		2951	68507						
		2954	68491						
		2957	68457						
		2960	68415						
		2967	68413						

Table 3a. (*continued*)

Problem 6		Problem 7		Problem 8		Problem 9		Problem 10	
C_{max}	F	C_{max}	F	C_{max}	F	C_{max}	F	C_{max}	F
2829	72618	2725	72171	2683	70478	2554	72756	2782	71801
2832	69630	2732	70120	2686	69432	2560	72368	2783	70644
2839	68900	2736	69580	2694	68338	2561	67091	2784	70600
2841	68787	2737	69491	2697	68208	2564	66120	2785	70563
2845	68346	2741	68586	2703	68033	2565	65558	2789	70193
2846	68343	2743	68487	2704	67890	2566	65552	2791	70080
2847	68095	2745	67533	2705	65088	2569	65522	2792	69884
2882	67833	2746	67482	2706	65082	2570	65196	2793	69882
2886	67424	2758	67380	2707	65030	2571	64442	2794	69620
2888	67342	2760	67212	2710	64985	2572	64360	2796	69597
2894	67264	2767	66685	2713	64932	2573	64313	2803	69573
2978	67258	2768	66662	2718	64889	2577	64303	2833	69564
		2785	66545	2719	64883	2581	64190	2835	69546
		2811	66543	2727	64851	2584	64143	2836	69536
		2936	66508	2748	64835	2589	64122	2839	69516
				2810	64828	2590	64100	2841	69515
				2851	64804	2595	64072	2843	69508
						2596	63998	2844	69489
						2603	63966		
						2610	63929		
						2615	63862		
						2627	63854		
						2628	63846		
						2648	63788		
						2652	63721		
						2654	63686		
						2657	63627		
						2664	63561		
						2697	63559		
						2699	63382		
						2723	63371		
						2739	63350		

Table 3b. Net set of non-dominated solutions obtained for the problem size (50×10)

Problem 1		Problem 2		Problem 3		Problem 4		Problem 5	
C_{max}	F	C_{max}	F	C_{max}	F	C_{max}	F	C_{max}	F
3027	97529	2911	88604	2878	86459	3064	93360	3012	92036
3031	97447	2917	88575	2879	86207	3065	92283	3013	92013
3033	96868	2918	88213	2883	85762	3067	91415	3016	91982
3034	93859	2921	88028	2885	85740	3072	91243	3018	91375
3037	93457	2923	87720	2887	85112	3073	91236	3019	91112
3039	93395	2925	87170	2891	85075	3077	91104	3024	91040
3040	92537	2926	86816	2896	85050	3078	90722	3037	90983
3043	92513	2927	86753	2904	85027	3086	90696	3038	89182
3045	92332	2928	86603	2915	84984	3087	90582	3042	88694
3051	91407	2929	86399	2916	84722	3089	90476	3045	88455
3057	90587	2931	85928	2917	83316	3090	90139	3061	88131
3059	90415	2940	85913	2956	83034	3092	90046	3063	88124
3062	90363	2947	85904	2960	82822	3109	89960	3065	88044
3063	90171	2948	85781	2972	82487	3110	89915	3080	88025
3065	89894	2949	85716	2977	82399	3111	89756	3084	88010
3069	89312	2950	85285	2979	82154	3114	89403	3085	87668
3089	89273	2952	85240	2983	81943	3115	89260	3107	87667
3111	89060	2953	85189	2986	81900	3116	89053	3117	87638
3124	89045	2958	85140	2994	81897	3123	89005	3138	87616
3127	88853	2971	85087	2995	81889	3127	88736	3148	87606
3129	88570	2975	85017	2997	81838	3128	88720	3188	87583
3130	88523	2976	84916	3001	81742	3130	88483	3193	87510
3173	88467	2989	84885	3005	81545	3139	88434	3201	87462
3177	88461	2993	84842	3012	81458	3143	88209		
3202	88435	2994	84839	3013	81234	3150	88083		
3206	88387	2995	84835	3024	81231	3168	88066		
3215	88386	2996	84803	3028	80888	3216	88064		
3265	88299	3005	84673	3105	80828	3218	88028		
3273	88297	3015	84509			3239	87963		
		3020	84469			3262	87892		
		3034	84332			3264	87692		
		3059	84284			3274	87574		
		3083	84262			3287	87509		
		3085	84206			3289	87321		
		3090	84198						
		3095	84099						
		3100	84084						
		3106	83844						
		3108	83812						
		3116	83808						
		3136	83722						

Table 3b. (*continued*)

Problem 6		Problem 7		Problem 8		Problem 9		Problem 10	
C_{max}	F	C_{max}	F	C_{max}	F	C_{max}	F	C_{max}	F
3043	91681	3115	99295	3043	99269	2908	91310	3112	95762
3044	91470	3124	97762	3045	98936	2909	91160	3113	95376
3045	91268	3126	96113	3046	98444	2910	89958	3118	93950
3047	90923	3127	93536	3048	97405	2920	89740	3121	93676
3056	90902	3128	93535	3050	97236	2923	89384	3129	93632
3057	90901	3129	92846	3052	97222	2949	89152	3131	92599
3060	90720	3131	92808	3055	96960	2952	88649	3138	92289
3064	90563	3133	92305	3056	95518	2972	88346	3139	91716
3065	89754	3136	92229	3057	94768	2988	88278	3142	91666
3075	89376	3138	91984	3058	92976	2994	88126	3146	91342
3076	89361	3140	91586	3061	92738	2996	88095	3147	91275
3077	89335	3157	91198	3064	92537	3002	88018	3149	91256
3080	89174	3158	91167	3066	92496	3017	87488	3150	91148
3082	88786	3161	91150	3067	91428	3025	87269	3152	90902
3084	88749	3165	91082	3069	91420	3026	87250	3157	90839
3087	88704	3169	90859	3072	91389	3031	87063	3158	90081
3099	88698	3170	90814	3074	91327	3044	87031	3164	89992
3111	88679	3176	90802	3077	89908	3073	86984	3192	89946
3114	88603	3180	90770	3078	89746	3077	86977	3198	89804
3116	88579	3182	90678	3082	89709	3078	86974	3204	89709
3119	88552	3191	90509	3083	89595	3079	86894	3208	89535
3120	88128	3196	90485	3086	89553	3080	86866	3241	89231
3167	88113	3197	90446	3087	89541	3090	86862	3261	89209
3172	88066	3201	90402	3091	89504	3100	86631	3270	89082
3179	88019	3202	90391	3092	89474			3272	89075
3183	87996	3207	90349	3093	89416			3275	89054
3205	87873	3213	90337	3101	89336			3276	89051
3244	87850	3221	90300	3113	89322			3279	89033
3340	87826	3229	90261	3118	89316			3335	89019
		3230	90229	3128	89315			3449	88982
		3231	90196	3132	89169				
		3233	90132	3134	88906				
		3266	90095	3139	88863				
		3275	90067	3141	88790				
		3301	90046	3144	88742				
		3328	90042	3146	88737				
		3352	89989	3148	88673				
		3391	89929	3157	88608				
				3169	88593				
				3176	88585				
				3177	88527				
				3178	88334				
				3274	88237				
				3276	88224				
				3319	88185				
				3326	87993				

Table 3c. Net set of non-dominated solutions obtained for the problem size (50×20)

Problem 1		Problem 2		Problem 3		Problem 4		Problem 5	
C_{max}	F	C_{max}	F	C_{max}	F	C_{max}	F	C_{max}	F
3908	137314	3769	125037	3718	121229	3785	127750	3668	125806
3911	136711	3775	124846	3719	121194	3786	127731	3672	125107
3912	130906	3783	124831	3720	118930	3793	127710	3675	124900
3915	130783	3798	124094	3760	118911	3794	127700	3682	124472
3932	129913	3799	123802	3761	118906	3795	127600	3684	124444
3933	129685	3810	123778	3763	118893	3797	126628	3694	122330
3934	129376	3816	123464	3773	118850	3799	126470	3701	122102
3941	129338	3829	123201	3788	118817	3800	125246	3708	122094
3955	129170	3831	123103	3794	118769	3807	124900	3729	122082
3963	129169	3833	123078	3796	118739	3811	124869	3736	121771
3966	128600	3847	122626	3799	118647	3812	124863	3754	121673
3970	128544	3852	122559	3804	118593	3823	124146	3762	121393
3982	128483	3853	122208	3809	118525	3824	123909	3764	121339
3983	128371	3862	121942	3831	118517	3825	123838	3777	121152
3988	128362	3888	121916	3834	118464	3826	123718	3789	121066
3991	128357	3894	121903	3845	118269	3831	123648	3802	120956
4013	128306	3895	121898	3937	118086	3837	123640	3808	120898
4015	128219	3915	121895	3946	118083	3839	123568	3812	120857
4021	127837	3920	121881	4004	118036	3842	123315	3821	120773
4036	127770	3921	121345	4005	117926	3846	123314	3830	120723
4043	127661	3961	121169	4020	117636	3847	123286	3832	120545
4045	127655	3962	121050	4068	117619	3850	122646	3837	120516
4064	127603	3964	120562	4072	117600	3860	122611	3842	120342
4087	127518	4099	120486	4096	117556	3863	122583	3850	120329
4094	127338			4106	117500	3867	122390	3856	120256
4109	127308					3868	122389	3861	120213
4140	127302					3876	122306	3866	120199
4146	127040					3914	122297	3869	119810
4170	127037					3915	122152	3882	119775
4177	126861					3919	122103	3896	119628
4184	126846							3907	119502
4202	126713							3909	119348
								3917	119314
								3931	119313
								3954	119222
								3960	119183
								3969	119165
								3989	119156

Table 3c. *(continued)*

Problem 6		Problem 7		Problem 8		Problem 9		Problem 10	
C_{max}	F	C_{max}	F	C_{max}	F	C_{max}	F	C_{max}	F
3751	126155	3765	129180	3775	131886	3812	130514	3820	128480
3755	125986	3771	128924	3780	131831	3813	129684	3828	128410
3758	125764	3773	128327	3781	129295	3817	129331	3835	128407
3763	125663	3774	128285	3786	127910	3818	129020	3839	128345
3765	125601	3787	128235	3791	127519	3820	128129	3840	127939
3769	125590	3796	127847	3792	127448	3827	128126	3842	127933
3771	125569	3797	127466	3793	126788	3828	126339	3846	127769
3783	125558	3800	127450	3794	126593	3831	126321	3849	127624
3803	125379	3818	127410	3800	126546	3833	126188	3851	127047
3811	124897	3819	127388	3805	126515	3842	125873	3853	126829
3820	124219	3822	126995	3806	126469	3846	125577	3859	126820
3823	124052	3830	126474	3813	125832	3850	125575	3861	126796
3827	123906	3831	126472	3817	125813	3857	125573	3863	126772
3835	123822	3832	126419	3829	125782	3860	125552	3889	126617
3836	123818	3838	126397	3830	125760	3869	125522	3899	126139
3855	123791	3842	126032	3831	125421	3887	125484	3907	126124
3856	123423	3845	125914	3847	125397	3889	125429	3935	126103
3861	123339	3850	125905	3851	125381	3898	125026	3942	125775
3866	123325	3855	125841	3852	125289	3910	124723	3956	125712
3873	123274	3858	125715	3860	125282	3911	124359	4239	125702
3880	123034	3892	125700	3863	125259	3912	124324	4278	125542
3885	123001	3899	125553	3864	125222	3914	124200		
3889	122989	3901	125551	3865	125189	3921	124185		
3891	122671	3911	125550	3868	125094	3922	124180		
3904	122157	3912	125371	3884	124518	3929	124141		
3915	122114	3914	125366	3889	124512	3932	123818		
3921	122054	3924	125146	3910	124470	3934	123812		
4092	122032	3939	125101	3924	124453	3935	123809		
4098	122012	3948	125090	3944	124449	3936	123356		
4106	121895	3950	125058	3946	124445	3944	123292		
		3965	125043	3949	124381	3979	123081		
		3978	125033	3951	124369	3992	122879		
		3994	124972	3959	124322	3998	122779		
		4002	124959	3968	124199				
		4039	124937	3983	124141				
		4049	124894	3997	124088				
		4051	124816	4044	124072				
		4079	124725	4050	124033				
		4227	124706	4063	124019				
				4259	124007				
				4317	123883				

Table 4a. Net set of non-dominated solutions obtained for the problem size (100×5)

Problem 1		Problem 2		Problem 3		Problem 4		Problem 5	
C_{max}	F	C_{max}	F	C_{max}	F	C_{max}	F	C_{max}	F
5493	262040	5284	247380	5175	271196	5017	269625	5250	255617
5495	257719	5286	247261	5177	268438	5018	264631	5251	255583
5500	256992	5287	247127	5183	267517	5019	244402	5252	255567
5527	256068	5288	247056	5186	267344	5021	234041	5255	247072
5564	256056	5291	246957	5193	255989	5032	232875	5256	246971
5570	256010	5297	246937	5195	253303	5035	230917	5257	246765
5609	255943	5299	246879	5206	244864	5042	230812	5259	246663
		5301	246864	5208	243806	5043	230707	5260	245660
		5302	246245	5209	243748	5044	229982	5261	244612
		5311	245621	5212	242690	5082	229933	5263	244514
		5341	245585	5221	242187	5112	229881	5264	244450
		5345	245578	5239	241938	5182	229866	5267	244135
				5240	241564	5189	229857	5272	244104
				5244	240708	5195	229769	5276	244011
				5250	240634			5298	243919
				5251	240594			5303	243668
				5262	240509			5304	243475
				5267	240412			5305	243376
				5294	240378			5307	243301
				5297	240363			5320	243258
				5368	240282			5324	243155
				5369	240198			5333	242928
								5339	242822

Table 4a. (*continued*)

Problem 6		Problem 7		Problem 8		Problem 9		Problem 10	
C_{max}	F	C_{max}	F	C_{max}	F	C_{max}	F	C_{max}	F
5135	262520	5247	252470	5094	251687	5448	280016	5322	278222
5139	242167	5251	251231	5096	251658	5454	256252	5328	257564
5141	242024	5255	247623	5097	248092	5465	252622	5329	257174
5143	241709	5256	247599	5100	247674	5469	252421	5330	255535
5146	240991	5257	247270	5101	246505	5470	252258	5334	250817
5148	240787	5265	247167	5102	245524	5471	252173	5342	247238
5150	238910	5270	245856	5104	245488	5474	251859	5346	246752
5156	237401	5275	245558	5105	243726	5477	251854	5348	246746
5157	236947	5276	244786	5106	243411	5479	251584	5372	246241
5158	236466	5277	244483	5108	241799	5481	251560	5386	245651
5159	236098	5279	244321	5111	241677	5513	251541	5389	245545
5161	236039	5282	243988	5121	240670	5519	251448		
5162	236030	5296	243522	5127	240547	5523	251287		
5164	235841	5298	243281	5130	236569	5542	251277		
5172	235832	5305	242469	5133	236238				
5178	235607	5376	242417	5135	235532				
5179	235466			5140	235146				
5181	235425			5150	234557				
5183	235373			5152	234189				
5204	235347			5155	233910				
5220	235326			5159	233908				
5288	235270			5196	233754				
				5218	233738				

most algorithms are able to discover many of the non-dominated solutions on the net front. This is evident from comparing the elements in a given row with the number in the last column of the corresponding row. However, this is not the case when the number of jobs are equal to 50 and 100. It shows that as the number of jobs increases, it becomes increasingly difficult for any single algorithm to discover many non-dominated solutions. In addition, as stated earlier, the different concatenations of the JIS and JSS have served to discover many non-dominated solutions, especially in the case of relatively large-sized problems.

It appears that the proposed variants are able to discover many non-dominated solutions, especially for the larger-sized problems, as opposed to the benchmark solutions provided by Framinan and Leisten. It is interesting to note that we are not able to identify which variant is the best among all proposed ones. This is so because the permutation flowshop scheduling problems become harder to solve as the number of jobs increases, and it requires the implementation of many variants of the MOACA to discover as many non-dominated solutions possible. These observations point to the fact that it is indeed challenging to develop a single multi-objective flowshop scheduling algorithm that can discover many non-dominated solutions, all or most by itself.

Table 4b. Net set of non-dominated solutions obtained for the problem size (100×10)

Problem 1		Problem 2		Problem 3	
C_{max}	F	C_{max}	F	C_{max}	F
5781	339398	5362	299003	5691	300928
5782	339268	5364	297917	5692	300822
5785	337379	5365	297799	5695	299691
5787	317712	5367	297348	5696	299481
5789	317638	5370	297159	5698	299342
5792	315311	5372	297037	5700	298874
5799	315182	5373	296576	5701	298842
5800	314591	5375	296557	5702	298829
5801	314449	5377	289483	5703	298254
5802	313699	5380	289090	5704	298240
5807	313256	5386	286373	5705	296260
5810	312685	5387	283589	5720	295830
5811	312656	5391	283584	5724	295394
5812	311939	5394	283533	5726	295343
5814	311913	5395	283258	5731	295299
5834	311071	5403	282655	5732	295189
5863	309427	5407	282501	5736	295042
5865	309314	5410	282206	5737	295030
5866	309193	5414	281538	5738	294362
5869	309122	5418	281040	5748	294359
5873	308550	5422	280611	5750	294046
5874	306790	5426	280444	5753	293608
5875	306581	5434	279914	5755	293489
5878	306563	5437	279897		
5880	305547	5447	279259		
5884	305467	5449	278666		
5897	305449	5450	278656		
5902	305341	5452	278650		
5910	305076	5462	278591		
5914	304996	5464	278464		
5915	304846	5465	278418		
5920	304829	5575	278415		
5934	304500	5593	278229		
5935	304305	5648	278189		
6010	304148	5653	278142		
6022	304128	5661	278077		

Table 4b. (*continued*)

Problem 4				Problem 5			
C_{max}	F	C_{max}	F	C_{max}	F	C_{max}	F
5826	342759	6048	307189	5501	326804	5649	290309
5828	342586	6053	307102	5505	326777	5650	290256
5829	340868	6063	307047	5507	301808	5670	290138
5831	340587	6065	306547	5509	301785	5672	290086
5837	339158	6069	306527	5512	298541	5673	289995
5839	323104	6074	306430	5520	297036	5676	289959
5852	322128	6086	306249	5521	296912	5769	289957
5855	322122	6088	306165	5530	296881	5797	289915
5860	321668	6105	306153	5535	296620	5799	289827
5861	320315	6153	306137	5536	295339	5821	289714
5865	319834	6157	306034	5537	295293	5831	289682
5868	319585			5545	294866	5835	289667
5870	319446			5548	294856	5836	289589
5871	318276			5549	294668	5869	289498
5872	317697			5552	294610		
5874	317065			5554	293697		
5875	316522			5566	293475		
5880	316507			5569	293399		
5883	313202			5570	293396		
5884	311505			5571	293255		
5886	311480			5573	293120		
5891	310851			5576	293013		
5902	309761			5580	292715		
5907	309571			5584	292566		
5911	309244			5587	292561		
5922	309212			5588	292507		
5923	309156			5589	292506		
5933	309090			5595	292424		
5938	308370			5597	292346		
5943	308319			5600	292268		
5952	308301			5602	292165		
5957	308212			5605	292005		
5987	307969			5607	291647		
5994	307854			5621	291297		
6010	307675			5628	291114		
6037	307488			5631	290582		
6042	307235			5645	290567		
6047	307233			5646	290424		

Table 4b. (*continued*)

Problem 6		Problem 7		Problem 8		Problem 9		Problem 10	
C_{max}	F	C_{max}	F	C_{max}	F	C_{max}	F	C_{max}	F
5308	310735	5602	299370	5653	313225	5916	354188	5881	328588
5311	310461	5603	298022	5655	312447	5918	353627	5883	328000
5314	309833	5604	298018	5657	308086	5919	349544	5889	327950
5315	309763	5611	296590	5658	307545	5920	348927	5892	327706
5316	309576	5617	296255	5668	306254	5923	347760	5897	327126
5317	292932	5619	296149	5670	306097	5928	319642	5903	300553
5318	292018	5620	295646	5672	306018	5932	319589	5904	300541
5319	291083	5622	292096	5675	304991	5935	319029	5907	300520
5321	289478	5623	292018	5677	303990	5940	313635	5909	300223
5323	282469	5641	290126	5687	303534	5941	312810	5910	299016
5326	281203	5643	288820	5689	303284	5942	312755	5911	298938
5331	280751	5662	288389	5690	302373	5955	312314	5912	298285
5332	280337	5673	287045	5694	301378	5958	312309	5914	298114
5334	280324	5676	286988	5695	299976	5966	311647	5916	298045
5345	279544	5696	286907	5697	299968	5969	311282	5918	297909
5347	279248	5701	285714	5698	299909	5971	310409	5921	297824
5348	278753	5702	284972	5699	299598	5975	310374	5932	297333
5350	278714	5794	284844	5709	299062	5979	310081	5933	297131
5358	278438	5832	284811	5717	298993	5982	310012	5938	296920
5359	278282	5839	284795	5721	298957	5988	309728	5951	296736
5361	277590	5902	284792	5723	298119	5989	309149	5961	296729
5404	277316	5931	284765	5726	298072	6010	308630	5962	296575
5405	277299	5933	284700	5730	298027	6025	308627	5978	296442
5406	277240	5945	284678	5732	297881	6027	308614	6001	296406
5413	277205	5959	284671	5745	297642	6028	308121	6002	296398
5414	277084			5753	297475	6032	308099	6015	296343
5422	276743			5761	296902	6035	307375	6023	296140
5425	276612			5770	296787	6045	307314		
5429	276591			5777	296511	6065	307269		
5432	276275			5825	296203	6098	307260		
5434	276227			5864	296191	6122	307083		
5435	276196			5871	296094	6144	307034		
5440	275644			5881	296083	6163	306982		
5449	275639			5896	296031	6178	306794		
5452	275164			5915	295714				
5454	274997			5988	295638				
5462	274813								
5465	274749								
5510	274701								
5611	274700								

Table 4c. Net set of non-dominated solutions obtained for the problem size (100×20)

Problem 1				Problem 2		Problem 3			
C_{max}	F	C_{max}	F	C_{max}	F	C_{max}	F	C_{max}	F
6350	394882	6521	375831	6306	396288	6383	400928	6682	379809
6361	391134	6571	375818	6307	395091	6389	400907	6710	379412
6370	390878	6581	375537	6309	394851	6390	400820	6711	379373
6373	390861	6585	375381	6312	394722	6391	400743	6821	379168
6374	390802	6588	375000	6314	394441	6394	400720	6822	379119
6375	390755	6590	374806	6315	393407	6395	392250	6823	379111
6377	389788	6593	374671	6317	393346	6425	391946	6830	379110
6380	389685	6651	374663	6328	393284	6426	387453	6832	379103
6386	386479	6736	374010	6332	391493	6452	386915	6833	379079
6387	386441	6786	373687	6338	391188	6471	386879	6838	379026
6389	386403	6790	373563	6357	391020	6475	386455	6842	378955
6392	386402	6799	373534	6367	390464	6477	384609	6868	378783
6395	386283	6800	373462	6382	390404	6489	384063		
6402	385681	6809	373250	6383	390189	6503	384058		
6403	385482	6814	373218	6385	390149	6504	384040		
6405	385028	6838	373193	6386	390119	6506	383913		
6415	384986	6840	373148	6397	389714	6513	383880		
6418	384977	6902	373140	6398	389576	6514	383868		
6420	384015	6915	373051	6403	387929	6515	383813		
6423	383628			6405	387610	6522	383017		
6424	383613			6408	387375	6523	382953		
6426	383595			6411	387010	6525	382821		
6440	383142			6412	386878	6529	382789		
6444	382900			6413	386726	6530	382774		
6446	382899			6420	385568	6536	382178		
6447	382390			6423	384944	6546	382078		
6451	381468			6426	384892	6547	381796		
6452	381426			6429	383777	6548	381783		
6455	381416			6431	383224	6567	381432		
6457	381372			6433	383211	6575	381220		
6464	381365			6538	382977	6579	381125		
6475	380330			6543	382973	6580	381074		
6476	380215			6549	382925	6581	381052		
6481	380196			6557	382658	6582	380687		
6487	379899			6562	382625	6599	380225		
6492	379866			6572	382540	6600	380222		
6493	379847			6594	382249	6601	380152		
6497	379795			6599	381994	6610	380146		
6506	378838			6625	380997	6651	380101		
6510	378524			6816	380868	6653	379948		
6514	375922			6818	380840	6666	379813		
6520	375833			6835	380664	6667	379812		

Table 4c. (*continued*)

Problem 4				Problem 5		Problem 6	
C_{max}	F	C_{max}	F	C_{max}	F	C_{max}	F
6363	403871	6690	382236	6433	394215	6488	394212
6364	401043	6692	382176	6439	388296	6490	394184
6369	399804	6699	381909	6442	387483	6491	394131
6370	399801	6711	381861	6443	386821	6495	394128
6372	399748	6716	381532	6445	386813	6501	394056
6375	398295	6727	381290	6447	386485	6506	392865
6377	398225	6820	381012	6448	386464	6508	391704
6380	397772	6823	380950	6450	386448	6513	391193
6383	397764	6825	380885	6451	386210	6514	391152
6385	396809	6924	380771	6452	386194	6533	389728
6388	396808	6928	380594	6458	385961	6536	389501
6392	396788			6461	385900	6547	389401
6393	396571			6468	385883	6555	388696
6395	396169			6474	384655	6557	387309
6398	395924			6475	384458	6558	386700
6400	395858			6488	384267	6563	386689
6405	394698			6489	384114	6564	386365
6407	392664			6491	383889	6569	386332
6409	392493			6499	383632	6583	385920
6414	392489			6504	381632	6587	385885
6417	392201			6510	381470	6591	385066
6423	390799			6516	381031	6592	384814
6447	389527			6519	381000	6593	384692
6450	388340			6538	380632	6602	384684
6456	387805			6546	380283	6618	383656
6462	387791			6565	380246	6625	383452
6464	387637			6571	380112	6675	383147
6496	387134			6572	379488	6679	383093
6503	387084			6601	379214	6680	383014
6512	386838			6602	378929	6681	382480
6516	386682			6615	378847	6693	382430
6526	386593			6619	378771	6695	382385
6530	386104			6621	378717	6699	382189
6538	386012			6631	378533	6705	381665
6541	385885			6639	378509	6722	381548
6543	385628			6642	377717	6731	381444
6545	385608			6643	377579	6757	380832
6548	383630			6656	377482	6786	380640
6571	383263			6658	377466	6789	380601
6573	383056			6667	377336	6790	380479
6582	383033			6670	377302	6804	380466
6616	382980			6673	377257	6816	380253
6619	382857			6682	376806	6827	380235
6644	382855			6699	376432	6829	379890
6660	382854			6701	376421	6864	379836
6682	382414			6709	375943		
6686	382393			6732	375857		
6688	382306			6733	375346		

Table 4c. (*continued*)

Problem 7				Problem 8		Problem 9		Problem 10	
C_{max}	F	C_{max}	F	C_{max}	F	C_{max}	F	C_{max}	F
6394	403384	6654	383872	6541	411389	6413	397716	6528	409325
6395	403352	6658	383581	6543	411309	6415	397701	6529	409286
6397	398870	6663	383551	6549	411152	6419	397305	6544	407571
6398	398859	6665	383232	6557	410992	6425	396813	6545	401166
6400	398775	6669	383185	6566	410727	6426	388701	6551	400822
6419	398477	6677	382680	6567	410658	6430	388686	6556	400049
6421	398436	6679	382665	6568	409888	6433	388652	6559	399751
6431	398154	6688	382360	6570	409723	6435	388620	6564	398717
6432	396999	6698	381782	6573	405352	6451	388522	6569	398087
6435	396899	6699	381671	6575	405308	6456	388273	6570	396745
6442	396896	6701	381579	6579	405213	6468	387912	6573	396017
6445	396365	6824	381504	6585	402822	6470	387370	6583	395563
6450	393029			6592	402554	6479	387006	6585	395375
6456	392973			6599	401899	6480	386888	6588	394639
6458	392923			6600	400210	6483	386846	6603	393528
6461	392831			6614	400202	6494	386821	6606	393255
6466	392755			6615	400093	6497	386799	6624	392822
6470	392704			6638	400061	6507	386701	6625	392802
6471	392650			6639	399929	6523	386308	6626	392699
6473	391149			6641	399572	6532	385566	6628	392668
6477	391045			6643	398987	6533	385554	6637	391442
6478	390977			6644	398953	6540	385510	6644	391163
6485	390884			6652	398752	6543	385057	6654	390946
6490	390566			6653	398713	6545	384624	6663	389961
6499	389357			6654	398649	6568	383717	6667	389957
6509	389329			6662	398398	6571	383642	6672	389951
6510	388908			6677	398346	6689	383329	6674	389815
6512	388044			6678	397274	6695	383090	6678	389796
6513	387837			6679	396833	6701	382870	6688	389734
6515	387810			6680	396752	6774	382841	6693	389711
6516	387804			6694	396682	6786	382622	6708	389661
6549	387666			6706	395013	6795	382492	6714	389348
6554	385481			6711	394922	6803	382326	6721	388859
6563	385460			6721	394850	6812	382320	6725	388567
6577	385174			6731	394669	6834	382269	6734	388468
6578	385009			6733	393458	6863	382215	6742	388048
6581	384358			6847	393310	6871	382200	6764	387826
6627	384326			6849	393077	6874	382151	6769	387817
6628	384311			6874	392935	6907	382089	6813	387803
6635	384299			6891	392906	6919	381969	6845	387564
6636	384046			6982	392592	6944	381918	6852	387169
6639	384008			7018	392477			6858	387096

Table 5. Contributions of the MOACA variants and the solutions from Framinan and Leisten to the net non-dominated front

(n×m) (20×5)	**MOACA variants**																				**F&L**	**Number of solu-tions in the net front**
Problem Number	**1**	**2**	**3**	**4**	**5**	**6**	**7**	**8**	**9**	**10**	**11**	**12**	**13**	**14**	**15**	**16**	**17**	**18**	**19**	**20**		
1	4	3	4	3	4	3	3	2	3	1	2	4	4	4	4	4	4	4	4	4	1	5
2	3	2	3	0	3	2	0	2	2	1	2	2	0	0	0	1	0	0	0	1	6	10
3	4	6	4	5	4	5	3	3	2	2	2	5	0	1	2	4	0	0	0	4	6	14
4	7	4	7	4	7	4	5	3	6	2	7	2	8	5	7	7	7	5	7	6	0	13
5	12	12	12	12	12	12	13	12	13	12	13	13	14	13	14	13	12	14	13	13	13	19
6	10	8	10	8	10	8	9	9	9	7	10	12	9	8	8	8	9	8	8	7	8	24
7	0	0	0	0	0	0	1	3	1	5	1	0	0	0	0	0	0	4	0	0	2	12
8	9	9	9	9	9	9	6	7	7	7	7	8	4	7	5	5	6	7	5	5	2	19
9	2	11	2	11	2	11	5	5	5	7	5	7	5	10	4	8	4	8	5	4	6	17
10	4	5	4	5	4	5	5	5	4	3	5	8	8	6	4	5	7	3	5	3	2	13
Average	0.39	0.40	0.39	0.37	0.39	0.39	0.33	0.33	0.35	0.30	0.35	0.41	0.37	0.37	0.34	0.39	0.34	0.37	0.33	0.34	0.30	

Absolute numbers, except Average; Average is relative contribution considering all problems in the respective set

Table 5. *(continued)*

(*n*×*m*) (20×10)	MOACA variants																				F&L	Number of solutions in the net front
Problem Number	**1**	**2**	**3**	**4**	**5**	**6**	**7**	**8**	**9**	**10**	**11**	**12**	**13**	**14**	**15**	**16**	**17**	**18**	**19**	**20**		
1	8	4	8	6	8	6	2	4	4	4	3	4	6	4	4	5	3	3	2	6	6	15
2	13	3	13	3	13	3	10	9	10	11	9	8	10	12	6	12	11	5	9	12	20	31
3	7	4	8	4	8	4	3	2	1	6	6	3	5	3	6	8	7	5	1	2	6	15
4	8	8	8	8	8	8	8	11	4	4	11	4	9	11	7	6	5	7	7	9	2	28
5	5	2	5	2	5	2	5	5	2	3	1	2	0	1	4	3	0	2	2	2	1	12
6	4	7	4	7	4	7	5	7	7	6	7	8	14	1	5	9	10	3	4	6	0	35
7	7	9	7	9	7	9	5	8	5	8	5	10	12	13	13	15	13	14	13	14	12	16
8	10	2	10	2	10	2	5	1	4	0	5	0	2	5	1	0	3	3	0	1	2	16
9	9	12	8	13	9	12	9	7	11	12	11	10	2	7	5	6	4	6	8	2	8	19
10	5	5	5	5	5	5	5	4	5	8	8	10	5	3	10	3	4	5	4	1	5	22
Average	0.40	0.28	0.40	0.30	0.41	0.30	0.28	0.28	0.25	0.31	0.31	0.29	0.30	0.30	0.32	0.34	0.29	0.28	0.24	0.27	0.31	

Table 5. *(continued)*

($n \times m$) (20×20)	**MOACA variants**																				**F&L**	**Number of solutions in the net front**
Problem Number	**1**	**2**	**3**	**4**	**5**	**6**	**7**	**8**	**9**	**10**	**11**	**12**	**13**	**14**	**15**	**16**	**17**	**18**	**19**	**20**		
1	10	8	7	8	7	8	10	9	6	8	14	9	8	12	11	11	6	8	12	7	23	30
2	4	6	4	6	4	6	4	1	4	2	7	2	3	7	3	7	1	3	3	4	9	26
3	20	20	20	20	20	20	13	14	12	13	20	21	14	16	6	11	11	9	7	15	23	43
4	10	6	10	6	10	6	10	10	10	6	10	8	4	6	7	8	7	8	4	8	6	20
5	7	2	9	2	7	2	2	1	8	0	5	1	7	1	5	1	0	6	2	4	10	27
6	7	9	7	9	7	9	4	8	6	10	4	7	6	6	8	9	6	7	4	8	6	27
7	14	7	14	7	14	7	13	7	10	9	13	11	8	14	5	16	9	13	5	10	17	23
8	12	12	6	12	6	12	3	13	0	6	2	10	10	5	8	5	8	5	9	11	10	44
9	8	6	8	6	8	6	5	10	5	12	10	4	9	10	9	12	7	10	3	10	2	30
10	6	7	6	10	6	7	4	9	5	10	3	10	8	10	11	5	6	7	8	11	6	22
Average	0.34	0.28	0.32	0.29	0.32	0.28	0.25	0.28	0.25	0.27	0.31	0.29	0.26	0.31	0.27	0.31	0.22	0.28	0.20	0.31	0.39	

Table 5. *(continued)*

($n \times m$) (50×5)	MOACA variants																				F&L	Number of solutions in the net front
Problem Number	**1**	**2**	**3**	**4**	**5**	**6**	**7**	**8**	**9**	**10**	**11**	**12**	**13**	**14**	**15**	**16**	**17**	**18**	**19**	**20**		
1	0	0	0	0	0	0	1	3	1	2	1	0	2	0	1	1	1	0	0	0	0	13
2	4	0	5	0	5	0	3	0	7	1	4	0	1	0	0	1	1	0	0	10	0	36
3	1	1	0	1	0	2	2	0	3	3	2	2	0	1	0	0	1	0	2	0	0	18
4	0	4	0	0	0	4	2	0	6	1	0	4	3	0	1	0	1	0	0	0	0	22
5	0	1	0	0	0	0	1	1	0	0	0	0	0	0	0	0	0	1	1	0	0	5
6	0	0	0	1	0	0	0	2	0	3	0	1	1	1	0	0	0	0	2	1	0	12
7	2	0	2	0	2	0	2	1	0	0	0	1	0	1	0	0	3	2	2	0	1	15
8	0	1	0	1	0	1	1	1	0	1	0	5	0	0	1	0	6	0	0	1	0	17
9	0	3	0	1	0	1	5	0	2	0	3	2	5	0	0	1	3	3	6	0	0	32
10	0	3	0	0	0	2	1	6	0	0	7	1	0	0	0	0	0	0	0	0	0	18
Average	0.03	0.08	0.03	0.02	0.03	0.05	010	0.11	0.08	0.07	0.08	0.09	0.06	0.02	0.02	0.01	0.09	0.04	0.08	0.04	0.01	

Table 5. *(continued)*

$(n \times m)$ (50×10)	MOACA variants																				F&L	Number of solutions in the net front
Problem Number	**1**	**2**	**3**	**4**	**5**	**6**	**7**	**8**	**9**	**10**	**11**	**12**	**13**	**14**	**15**	**16**	**17**	**18**	**19**	**20**		
1	3	0	3	1	3	1	4	2	1	0	4	4	2	1	0	2	2	3	0	0	0	29
2	3	0	3	1	3	1	0	0	1	6	4	0	1	7	0	3	7	1	2	5	0	41
3	3	8	3	1	3	8	1	0	1	0	5	2	3	0	2	0	0	1	0	1	0	28
4	0	1	1	1	0	0	0	0	0	2	5	0	0	1	0	8	7	3	0	2	3	34
5	0	0	0	0	0	0	0	0	0	0	2	0	0	5	0	0	0	0	11	5	0	23
6	0	0	0	1	0	1	0	0	0	2	6	8	8	0	0	1	1	0	1	1	0	29
7	0	1	0	2	0	1	7	2	0	4	4	6	0	1	0	3	2	3	3	1	0	38
8	1	0	1	0	4	0	0	0	0	1	1	5	2	12	4	2	2	3	7	2	0	46
9	0	3	6	0	0	0	3	1	0	1	0	0	2	0	3	2	0	0	3	0	0	24
10	0	4	0	3	0	2	0	0	0	5	0	6	3	8	0	0	2	1	0	0	0	30
Average	0.03	0.06	0.06	0.03	0.04	0.05	0.05	0.02	0.01	0.06	0.10	0.10	0.07	0.10	0.03	0.06	0.06	0.04	0.09	0.05	0.01	

Table 5. *(continued)*

(*n*×*m*) (50×20)	MOACA variants																				F&L	Number of solutions in the net front
Problem Number	1	2	3	4	5	6	7	8	9	10	11	12	13	14	15	16	17	18	19	20		
1	2	0	4	2	5	2	0	0	8	1	1	0	3	0	0	2	1	0	3	0	0	32
2	0	4	0	0	0	1	4	0	0	1	0	0	0	0	4	3	0	5	3	0	0	24
3	0	0	1	0	0	0	0	0	4	0	0	0	0	0	5	1	6	0	3	3	2	25
4	1	0	0	0	0	0	0	3	0	10	0	0	0	0	1	0	5	7	0	3	0	30
5	0	0	0	0	0	0	0	2	9	0	0	0	0	0	1	2	9	0	15	0	0	38
6	0	0	0	0	0	0	4	3	0	2	0	0	0	0	2	7	0	1	7	0	4	30
7	0	0	0	0	0	1	3	11	5	0	3	5	1	4	3	0	0	0	3	0	0	39
8	0	0	0	8	0	6	0	8	1	0	1	2	1	5	0	4	0	8	2	1	0	41
9	0	0	0	0	0	0	3	0	0	3	0	0	6	0	6	0	11	2	0	1	1	33
10	0	0	0	0	0	0	0	0	0	1	0	2	1	4	1	0	10	1	1	0	0	21
Average	0.01	0.02	0.02	0.03	0.02	0.03	0.05	0.07	0.08	0.06	0.01	0.03	0.04	0.04	0.08	0.06	0.15	0.08	0.11	0.03	0.02	

Table 5. *(continued)*

($n \times m$) (100×5)	MOACA variants																				F&L	Number of solutions in the net front
Problem Number	1	2	3	4	5	6	7	8	9	10	11	12	13	14	15	16	17	18	19	20		
1	0	0	0	0	0	0	0	0	0	0	2	0	1	0	0	0	4	0	0	0	0	7
2	0	2	0	0	0	0	3	1	1	0	0	0	1	0	0	0	0	0	4	0	0	12
3	2	0	1	0	0	0	1	1	1	7	1	2	3	3	1	0	1	0	0	0	0	22
4	4	0	4	0	4	1	3	0	0	0	1	3	0	0	0	0	0	0	0	0	0	14
5	3	3	2	0	2	0	0	2	0	1	6	1	0	3	0	0	0	0	0	2	0	23
6	0	2	6	0	1	0	0	1	7	1	1	0	0	0	0	2	0	0	1	0	0	22
7	0	2	0	2	0	2	3	2	4	1	2	2	0	1	0	0	1	1	0	0	0	16
8	1	0	7	1	1	1	0	0	0	0	3	5	1	1	0	2	1	1	0	0	0	23
9	2	1	1	0	0	0	0	0	1	1	0	0	0	0	0	4	0	0	3	2	0	14
10	1	2	1	0	0	0	1	0	0	0	2	0	0	1	0	0	2	1	0	0	0	11
Average	0.08	0.08	0.12	0.02	0.05	0.02	0.08	0.04	0.08	0.05	0.12	0.07	0.04	0.05	0.01	0.05	0.09	0.02	0.06	0.02	0.00	

Table 5. *(continued)*

($n \times m$) (100×10)	MOACA variants																				F&L	Number of solutions in the net front
Problem Number	**1**	**2**	**3**	**4**	**5**	**6**	**7**	**8**	**9**	**10**	**11**	**12**	**13**	**14**	**15**	**16**	**17**	**18**	**19**	**20**		
1	2	0	1	2	3	1	3	0	3	8	2	0	0	1	0	0	7	0	4	0	0	36
2	2	0	2	0	0	0	0	2	3	0	3	0	4	1	1	0	10	0	10	0	0	36
3	8	3	0	0	2	0	0	5	0	0	2	1	0	0	0	0	2	0	0	0	0	23
4	1	0	5	3	8	0	2	3	5	8	1	0	0	4	0	1	8	0	4	1	0	49
5	2	0	0	0	3	2	0	0	4	0	5	2	7	2	0	6	0	11	2	6	0	52
6	0	0	3	0	0	0	4	4	3	1	12	0	2	3	0	0	0	0	7	1	0	40
7	0	2	0	5	0	1	4	0	1	0	4	0	1	0	1	1	2	3	2	0	0	25
8	2	1	4	0	3	2	0	0	2	0	0	6	1	6	1	0	0	7	2	3	0	36
9	3	3	2	0	0	0	0	1	0	0	0	0	1	8	3	1	6	2	1	3	0	34
10	0	2	5	0	0	0	6	0	0	2	3	0	4	1	0	2	1	0	1	0	0	27
Average	0.07	0.04	0.06	0.03	0.05	0.02	0.06	0.05	0.05	0.05	0.09	0.03	0.05	0.07	0.02	0.03	0.10	0.06	0.09	0.03	0.00	

Table 5. *(continued)*

(*n*×*m*) (100×20)	MOACA variants																				F&L	Number of solutions in the net front
Problem Number	1	2	3	4	5	6	7	8	9	10	11	12	13	14	15	16	17	18	19	20		
1	3	0	0	0	0	0	7	0	19	0	0	0	0	0	11	4	5	0	8	3	1	61
2	4	0	0	0	0	0	0	0	7	2	9	1	6	0	2	0	2	0	2	7	0	42
3	2	3	9	0	1	0	5	2	2	7	7	3	0	1	0	0	1	4	7	0	0	54
4	1	0	0	0	3	0	0	1	2	3	0	0	8	4	15	5	8	7	1	0	1	59
5	0	2	3	0	2	0	0	0	0	0	0	9	1	2	0	0	19	2	2	8	0	48
6	3	0	5	0	0	0	0	0	2	0	2	5	7	6	0	0	6	5	4	0	0	45
7	0	2	2	0	0	0	0	0	0	5	2	7	4	3	2	0	8	15	0	4	0	54
8	0	0	1	0	0	3	8	3	7	0	2	0	0	6	0	0	2	0	10	0	0	42
9	6	0	0	0	0	0	4	0	0	0	1	0	0	2	4	0	8	8	0	8	0	41
10	0	1	1	0	0	0	4	0	3	0	0	1	9	2	6	3	2	5	3	2	0	42
Average	0.04	0.02	0.04	0.00	0.01	0.01	0.06	0.01	0.08	0.03	0.05	0.05	0.07	0.06	0.08	0.02	0.13	0.09	0.08	0.07	0.00	

5 Conclusions

The problem of scheduling in permutation flowshops with the objectives of minimizing the makespan and total flowtime of jobs was investigated. A new multi-objective ant-colony algorithm, called MOACA, has been developed with many unique features. Twenty variants of MOACA have been proposed. Benchmark flowshop scheduling problems have been solved by using these variants of the MOACA, and a non-dominated solution front is obtained by consolidating the solutions obtained from these variants and the benchmark solutions available in the literature. It is evident from the computational evaluation that the proposed variants of the MOACA are quite effective in discovering many non-dominated solutions. We believe that the non-dominated solutions obtained by us could serve as possible benchmarks for future researchers as much as we have benefited from the earlier researchers. The complete set of non-dominated solutions for every problem instance is given to serve an as easy reference for future researchers.

Acknowledgments. The first author thanks the Alexander-von-Humboldt Foundation for supporting him through the Fellowship in 2003, 2004 and 2006. Thanks are also due to Varadharajan, Madhushini and Christian Petri for their help in consolidating the results. Special thanks are due to Jose Framinan and Rainer Leisten for sharing their benchmark solutions with us.

Appendix

The step-by-step procedure of the job-index-based insertion scheme (JIS) is presented below.

Step 1: Let the input sequence to the JIS be denoted, in general, by S. Let $Z(S)$ denote its compromise objective function value for the given w_1 and w_2. Let $[k]$ denote the job found in position k of S. Initialize $i = 0$.
Step 2: Set $i := i + 1$.
Step 3: For $k = 1$ to n do the following:

if $i \neq [k]$
then
remove job i from its current position in S, insert job i in position k of S and adjust the sequence accordingly by not changing the relative positions of other jobs in S. Let the resultant sequence be denoted by Σ^k; calculate its makespan and total flowtime denoted respectively by $C_{max}(\Sigma^k)$ and $F(\Sigma^k)$; let its compromise objective function value be denoted by $Z(\Sigma^k)$; check if Σ^k enters the non-dominated front, and if so, accordingly update the front; also, if $C_{max}(\Sigma^k) < up\_C_{max}$, set $up\_C_{max} = C_{max}(\Sigma^k)$; and likewise, if $F(\Sigma^k) < up\_F$, set $up\_F = F(\Sigma^k)$.
else
set $k' = k$.

Step 4: Determine sequence Σ^l such that
$Z(\Sigma^l) = \min\{ Z(\Sigma^k) \text{ for } k = 1, 2, \ldots, n, \text{ and } k \neq k' \}$.
If $Z(\Sigma^l) < Z(S)$ then set $S = \Sigma^l$ and $Z(S) = Z(\Sigma^l)$.

Step 5: Go back to Step 2 if $i < n$; else stop. Sequence S is the output sequence from the JIS.

The step-by-step procedure of the job-index-based swap scheme (JSS) is presented below.

Step 1: Let the input sequence to the JSS be denoted, in general, by S. Let $Z(S)$ denote its compromise objective function value for the given w_1 and w_2. Let $[k]$ denote the job found in position k of S. Initialize $i = 0$.
Step 2: Set $i := i + 1$.
Step 3: For $k = 1$ to n do the following:
if $i \neq [k]$
then
generate sequence Σ^k which differs from S only by having swapped jobs i and $[k]$; calculate its makespan and total flowtime; let its compromise objective function value be denoted by $Z(\Sigma^k)$; check if Σ^k enters the non-dominated front, and if so, accordingly update the front; also, if $C_{max}(\Sigma^k) < up\_C_{max}$, set $up\_C_{max} = C_{max}(\Sigma^k)$; and likewise, if $F(\Sigma^k) < up\_F$, set $up\_F = F(\Sigma^k)$
else
set $k' = k$.
Step 4: Determine sequence Σ^l such that
$Z(\Sigma^l) = \min\{ Z(\Sigma^k) \text{ for } k = 1, 2, \ldots, n, \text{ and } k \neq k' \}$.
If $Z(\Sigma^l) < Z(S)$ then set $S = \Sigma^l$ and $Z(S) = Z(\Sigma^l)$.
Step 5: Go back to Step 2 if $i < n$; else stop. Sequence S is the output sequence from the JSS.

References

Allahverdi, A.: A new heuristic for m-machine flowshop scheduling problem with bicriteria of makespan and maximum tardiness. Computers & Operations Research 31, 157–180 (2004)
Allahverdi, A., Aldowaisan, T.: New heuristics to minimize total completion time in m-machine flowshops. International Journal of Production Economics 77, 71–83 (2002)
Armentano, V.A., Arroyo, J.E.C.: An application of a multi-objective tabu search algorithm to a bicriteria flowshop problem. Journal of Heuristics 10, 463–481 (2004)
Arroyo, J.E.C., Armentano, V.A.: Genetic local search for multi-objective flowshop scheduling problems. European Journal of Operational Research 167, 717–738 (2005)
Bagchi, T.P.: Multiobjective scheduling by genetic algorithms. Kluwer Academic Publishers, Boston (1999)
Ben-Daya, M., Al-Fawzan, M.: A tabu search approach for the flow shop scheduling problem. European Journal of Operational Research 109, 88–95 (1998)
Campbell, H.G., Dudek, R.A., Smith, M.L.: A heuristic algorithm for the n-job, m-machine sequencing problem. Management Science 16, B630–B637 (1970)
Chakravarthy, K., Rajendran, C.: A heuristic for scheduling in a flowshop with the bicriteria of makespan and maximum tardiness minimization. Production Planning and Control 10, 707–714 (1999)

Chang, P.-C., Hsieh, J.-C., Lin, S.G.: The development of gradual priority weighting approach for the multi-objective flowshop scheduling prob¬lem. International Journal of Production Economics 79, 171–183 (2002)

Chung, C.-S., Flynn, J., Kirca, O.: A branch and bound algorithm to minimize the total flow time for m-machine permutation flowshop problems. International Journal of Production Economics 79, 185–196 (2002)

Corne, D.W., Knowles, J.D., Oates, M.J.: The pareto envelope-based selection algorithm for multiobjective optimization. In: Schoenauer, M., Deb, K., Rudolph, G., Yao, X., Lutton, E., Guervos, J.J.M., Schwefel, H.P. (eds.) PPSN 2000. LNCS, vol. 1917, pp. 839–848. Springer, Heidelberg (2000)

Corne, D.W., Jerram, N.R., Knowles, J.D., Oates, M.J.: PESA-II: Region-based selection in evolutionary multiobjective optimization. In: Spector, L., Goodman, E.D., Wu, A., Langdon, W.B., Voigt, H.M., Gen, M., Sen, S., Dorigo, M., Pezeshk, S., Garzon, M.H., Burke, E. (eds.) Proceedings of the Genetic and Evolutionary Computation Conference (GECCO-2001), pp. 283–290. Morgan Kaufmann, San Francisco (2001)

Daniels, R.L., Chambers, R.J.: Multiobjective flow-shop scheduling. Naval Research Logistics 37, 981–995 (1990)

Deb, K., Pratap, A., Agarwal, S., Meyarivan, T.: A fast and elitist multiobjective genetic algorithm: NSGA-II. IEEE Transactions on Evolutionary Computation 6, 182–197 (2002)

Dong, X., Huang, H., Chen, P.: An improved NEH-based heuristic for the permutation flowshop problem. Computers & Operations Research 35, 3962–3968 (2008)

Dorigo, M.: Optimization, Learning and Natural Algorithms (in Italian). PhD thesis, Dipartimento di Elettronica, Politecnico di Milano, Italy (1992)

Dorigo, M., Maniezzo, V., Colorni, A.: The ant system: Optimization by a colony of cooperating agents. IEEE Transactions on Systems, Man and Cybernetics – Part B 26, 29–41 (1996)

Framinan, J.M., Leisten, R.: An efficient constructive heuristic for flowtime minimisation in permutation flow shops. OMEGA 31, 311–317 (2003)

Framinan, J.M., Leisten, R.: A heuristic for scheduling a permutation flowshop with makespan objective subject to maximum tardiness. International Journal of Production Economics 99, 28–40 (2006)

Framinan, J.M., Leisten, R.: A multi-objective iterated greedy search for flowshop scheduling with makespan and flowtime criteria. OR Spectrum, published online before print, August 4 (2007)

Framinan, J.M., Leisten, R., Ruiz-Usano, R.: Efficient heuristics for flowshop sequencing with the objectives of makespan and flowtime minimisation. European Journal of Operational Research 141, 559–569 (2002)

Framinan, J.M., Ruiz-Usano, R., Leisten, R.: Comparison of heuristics for flowtime minimisation in permutation flowshops. Computers & Operations Research 32, 1237–1254 (2005)

Garey, M.R., Johnson, D.S., Sethi, R.: The complexity of flowshop and jobshop scheduling. Mathematics of Operations Research 1, 117–129 (1976)

Geiger, M.J.: On operators and search space topology in multi-objective flow shop scheduling. European Journal of Operational Research 181, 195–206 (2007)

Gelders, L.F., Sambandam, N.: Four simple heuristics for scheduling a flow-shop. International Journal of Production Research 16, 221–231 (1978)

Ho, J.C.: Flowshop sequencing with mean flow time objective. European Journal of Operational Research 81, 571–578 (1995)

Ignall, E., Schrage, L.: Application of the branch-and-bound technique to some flowshop scheduling problems. Operations Research 13, 400–412 (1965)

Ishibuchi, H., Murata, T.: A multi-objective genetic local search algorithm and its application to flowshop scheduling. IEEE Transactions on Sys¬tems, Man and Cybernetics – Part C: Applications and Reviews 28, 392–403 (1998)
Johnson, S.M.: Optimal two- and three-stage production schedules with setup times included. Naval Research Logistics Quarterly 1, 61–68 (1954)
Kalczynski, P.J., Kamburowski, J.: On the NEH heuristic for minimizing the makespan in permutation flow shops. OMEGA 35, 53–60 (2007)
Kalczynski, P.J., Kamburowski, J.: An improved NEH heuristic to minimize makespan in permutation flow shops. Computers & Operations Research 35, 3001–3008 (2008)
Laha, D., Chakraborty, U.K.: A constructive heuristic for minimizing makespan in no-wait flow shop scheduling. International Journal of Advanced Manufacturing Technology (2008) (DOI: 10.1007/s00170-008-1454-0)
Liao, C.-J., Tseng, C.-T., Luarn, P.: A discrete version of particle swarm optimization for flowshop scheduling problems. Computers & Operations Research 34, 3099–3111 (2007)
Liu, J., Reeves, C.R.: Constructive and composite heuristic solutions to the $P//\sum C_i$ scheduling problem. European Journal of Operational Research 132, 439–452 (2001)
Merkle, D., Middendorf, M.: An ant algorithm with a new pheromone evaluation rule for total tardiness problems. In: Oates, M.J., Lanzi, P.L., Li, Y., Cagnoni, S., Corne, D.W., Fogarty, T.C., Poli, R., Smith, G.D. (eds.) EvoIASP 2000, EvoWorkshops 2000, EvoFlight 2000, EvoSCONDI 2000, EvoSTIM 2000, EvoTEL 2000, and EvoROB/EvoRobot 2000. LNCS, vol. 1803, pp. 287–296. Springer, Heidelberg (2000)
Minella, G., Ruiz, R., Ciavotta, M.: A review and evaluation of multi-objective algorithms for the flowshop scheduling problem. INFORMS Journal on Computing published online before print, April 2 (2008)
Miyazaki, S., Nishiyama, N.: Analysis for minimizing weighted mean flowtime in flowshop scheduling. Journal of the Operations Research Society of Japan 23, 118–132 (1980)
Miyazaki, S., Nishiyama, N., Hashimoto, F.: An adjacent pairwise approach to the mean flowtime scheduling problem. Journal of Operations Research Society of Japan 21, 287–299 (1978)
Murata, T., Ishibuchi, H., Tanaka, H.: Multi-objective genetic algorithm and its applications to flowshop scheduling. Computers & Industrial Engineering 30, 957–968 (1996)
Nawaz, M., Enscore Jr, E.E., Ham, I.: A heuristic algorithm for the m-machine, n-job flowshop sequencing problem. OMEGA 11, 91–95 (1983)
Pasupathy, T., Rajendran, C., Suresh, R.K.: A multi-objective genetic algorithm for scheduling in flow shops to minimize the makespan and total flow time of jobs. International Journal of Advanced Manufacturing Technology 27, 804–815 (2006)
Rajendran, C.: Two-stage flowshop scheduling problem with bicriteria. Journal of the Operational Research Society 43, 871–884 (1992)
Rajendran, C.: Heuristic algorithm for scheduling in a flowshop to minimize total flowtime. International Journal of Production Economics 29, 65–73 (1993)
Rajendran, C.: A heuristic for scheduling in flowshop and flowline-based manufacturing cell with multi-criteria. International Journal of Production Research 32, 2541–2558 (1994)
Rajendran, C.: Heuristics for scheduling in flowshop with multiple objectives. European Journal of Operational Research 82, 540–555 (1995)
Rajendran, C., Ziegler, H.: Ant-colony algorithms for permutation flowshop scheduling to minimize makespan / total flowtime of jobs. European Journal of Operational Research 155, 426–438 (2004)
Rajendran, C., Ziegler, H.: Two ant-colony algorithms for minimizing total flowtime in permutation flowshops. Computers & Industrial Engineering 48, 789–797 (2005)
Ruiz, R., Stuetzle, T.: A simple and iterated greedy algorithm for the permutation flowshop scheduling problem. European Journal of Operational Research 177, 2033–2049 (2007)

Ruiz, R., Maroto, C., Alcaraz, J.: Two new robust genetic algorithms for the flowshop scheduling problem. OMEGA 34, 461–476 (2006)

Sridhar, J., Rajendran, C.: Scheduling in flowshop and cellular manufacturing systems with multiple objectives – a genetic algorithmic approach. Production Planning & Control 7, 374–382 (1996)

Srinivas, N., Deb, K.: Multiobjective optimization using nondominated sorting in genetic algorithms. Evolutionary Computation 2, 221–248 (1994)

Stuetzle, T.: An ant approach to the flow shop problem. In: Proceedings of the 6th European Congress on Intelligent Techniques & Soft Computing (EUFIT 1998), Verlag Mainz, Aachen, pp. 1560–1564 (1998)

Taillard, E.: Benchmarks for basic scheduling problems. European Journal of Operational Research 64, 278–285 (1993)

Tasgetiren, M.F., Liang, Y.-C., Sevkli, M., Gencyilmaz, G.: A particle-swarm optimization algorithm for makespan and total flowtime minimization in the permutation flowshop sequencing problem. European Journal of Operational Research 177, 1930–1947 (2007)

T'kindt, V., Billaut, J.-C.: Multicriteria scheduling: Theory, models and algorithms. Springer, Berlin (2002)

T'kindt, V., Monmarche, N., Tercinet, F., Lauegt, D.: An ant colony optimization algorithm to solve a two-machine bicriteria flowshop scheduling problem. European Journal of Operational Research 142, 250–257 (2002)

Varadharajan, T.K., Rajendran, C.: A multi-objective simulated-annealing algorithm for scheduling in flowshops to minimize the makespan and total flowtime of jobs. European Journal of Operational Research 167, 772–795 (2005)

Widmer, M., Hertz, A.: A new heuristic method for the flowshop sequencing problem. European Journal of Operational Research 41, 186–193 (1989)

Wang, C., Chu, C., Proth, J.-M.: Heuristic approaches for n/m/F/$\sum C_i$ scheduling roblems. European Journal of Operational Research 96, 636–644 (1997)

Woo, H.S., Yim, D.S.: A heuristic algorithm for mean flowtime objective in flowshop scheduling. Computers and Operations Research 25, 175–182 (1998)

Zitzler, E., Thiele, L.: Multiobjective evolutionary algorithms: A comparative case study and the strength pareto approach. IEEE Transactions on Evolutionary Computation 3, 257–271 (1999)

Multi-objective Simulated Annealing for Permutation Flow Shop Problems

E. Mokotoff

Universidad de Alcalá
Department of Economics
Plaza Victoria 3, 28802 Alcalá de Henares, Spain
ethel.mokotoff@uah.es

Summary. In this chapter we present a Multi-Objective Simulated Annealing algorithm to deal with the Permutation Flow Shop Scheduling Problem in a real context. We have designed the models taking into account results obtained from a study conducted in the Spanish Ceramic Tile Sector. The proposed methods consist in obtaining a good approximation of the efficient frontier. Starting with a set of initial sequences, the algorithm samples a point in its neighbourhood. If this generated sequence is dominated, we still accept it with a certain probability. Different heuristics and constructive algorithms are used to compute initial good sequences and lower bounds for the different criteria. Makespan and flow time are considered. The procedure is good enough to give efficient solutions with little computational effort. A computational experiment has been carried out to check the performance of the proposed algorithms. Different metrics for comparing algorithms have been computed, and have been analyzed together with the CPU time. We have studied how the number of initial solutions, the neighbouring procedure, and other parameters, affect the results. For all the tested instances a net set of potentially efficient schedules has been obtained.

1 Introduction

In this chapter we consider a classic permutation flow shop scheduling problem for which, after more than 50 years of scientific research, there is an important gap between theory and practice. This problem results in the context where multi-purpose machines are used to manufacture different jobs and, for every one, the operations are carried out in the same order among the machines. So, to find the schedule that optimizes a certain performance measure simply means finding the optimal job sequencing, that is to say the order in which those jobs should be processed, as in the production of textiles and ceramic tiles. Ceramic tiles are produced in processing lines composed of several stages: molding press, dryer, glazing line, kiln, quality control, finally, packing and delivery [4, 96]. To the complexity that naturally arises in this problem, considering only one criterion [33], we have to add the additional complexity that comes from the multivariant condition of corresponding alternative schedules. In fact the description and valuation of alternative decisions are not naturally accomplished by only one criterion, but by several (*e.g.* makespan, flow-time, completion-time, tardiness, inventory, utilization, etc.). This is certainly the natural framework of the Multicriterion Decision Making discipline (MDM). A solution which is optimal

U.K. Chakraborty (Ed.): Comput. Intel. in Flow Shop and Job Shop Sched., SCI 230, pp. 101–150.
springerlink.com

with respect to a given criterion might be a poor candidate for another. The trade-offs involved in considering several different criteria provide useful insights for the decision-maker. Thus considering Combinatorial Optimization (CO) problems with more than one criterion is more relevant in the context of real-life scheduling problems. Research in this important field has been scarce when compared to research in single-criterion scheduling. Until the late 1980's, only one criterion was considered in scheduling problems. Furthermore, until the 1990's, most work in the area of multiple criteria scheduling consists of bi-criteria studies of the single machine case [45].

Of course, to expect to find the "Optimum" schedule must usually be discarded. We would be satisfied to find the Pareto optimal alternatives. At this point we have to let some subjective considerations intervene, such as the decision-maker preferences. It is actually an MDM Problem, and at the present time, there is no other rational tool to apply to discard alternatives. Only with the breakthrough of metaheursitcs in solving CO problems, did researchers begin to adapt metaheuristics to solve Multi-Objective Combinatorial Optimization problems. Then, the acronym MOCO started to appear in the scientific literature to refer to Multi-Objective Combinatorial Optimization problems and the techniques specially developed to deal with them. Multi-Objective Simulated Annealing (MOSA) methods are metaheuristics based on Simulated Annealing (SA) to tackle MOCO problems. SA has demonstrated its ability to solve combinatorial problems such as vehicle routing, production scheduling, timetabling, etc. Based on this MOSA scheme, we have developed our models to provide the decision-maker with efficient solutions for the scheduling problem we are dealing with.

The aim of this chapter is to present the proposed MOSA techniques and their performance analysis, after a review regarding the permutation flow shop scheduling problem, the MOCO theory, including recent developments considering more than one optimization criterion (the detailed theorems and proofs have been omitted to avoid a huge chapter). The main proposed procedures find a good approximation of the set of non-dominated solutions in a relatively short time. We carried out an intensive computational experiment by making use of the 90 benchmark problems given by Taillard [113]. The performance analysis includes a set of metrics specific for evaluating Multi-Objective Optimization algorithms (MOO). The influence on the number of potential efficient solutions, the neighborhood search procedure and SA parameters have been analyzed together with the CPU time. With all these experiments we have obtained a net set of potentially efficient schedules and we have updated some published net set, for the same instances.

In the next section, the classical permutation flow shop problem statement is presented. Since we are facing the multi-objective nature of the problem, we will briefly introduce multi-objective theory and notations (section 3), followed by a brief survey on MOCO algorithms devoted to scheduling problems (section 4). In section 5 we present the proposed approaches based on the MOSA scheme. Section 6 reports on the computational experiment. We conclude, in section 7, with a summary discussion on research directions.

2 Permutation Flow Shop Scheduling Problem

In the classical permutation flow shop scheduling problem, there are n jobs and m machines, or stages. Each job needs to complete one operation on each of the machines during a fixed processing time. So, the aim is to find the schedule, or job sequence, that optimizes certain performance measures. In this chapter we focus attention on the permutation flow shop situation, where all jobs must pass through all machines in the same order ([87] presents a comparative study of permutation versus non-permutation flow shop scheduling problems).

The scheduling process involves just finding the optimal job sequencing. Nevertheless, the computational complexity usually grows exponentially with the number of machines, m, making the problem intractable. This problem, like almost all deterministic scheduling problems, belongs to the wide class of CO problems, many of which are known to be NP-hard [33]. What it means is that it is unlikely that efficient optimization algorithms exist to solve them. Only a few scheduling problems have been shown to be tractable, in the sense that they are solvable in polynomial time. For the remaining ones, the only way to secure optimal solutions is usually by enumerative methods, requiring exponential time. The investigation has focused on two approaches: developing approximation algorithms, and optimally solving restricted, more tractable, cases. Thus, heuristic methods have been developed, some of them showing an acceptable performance.

Many real life problems can be modeled as permutation flow shop scheduling ones. On production lines, it is common to find multi-purpose machines carrying out different products. We are working with the ceramic tile manufacturing sector, however many problems could be mentioned when we speak about scarce resources, or machines, dedicated to the production of some goods, or jobs.

2.1 Notation

We will use the notation that follows:

J: set of n jobs J_i (i=1,...,n)
M: set of m machines M_j (j=1,...,m)
p_{ij}: processing time of job J_i on machine M_j
d_i: due date of job J_i, time limit by which J_i should be completed
r_i: time at which the job J_i is ready to be processed
w_i: priority or weight of job J_i
C_i: completion time of job J_i
$C_{\max}$: the maximum completion time of all jobs J_i (this is the schedule length, which is also called the makespan)
F_i: flow time of job J_i, $F_i = C_i - r_i$, if $r_i = 0$, then $F_i = C_i$
L_i: lateness of job J_i, $L_i = C_i - d_i$
T_i: tardiness of job J_i, $T_{\max} = \max\{L_i, 0\}$
E_i: earliness of job J_i, $E_{\max} = \max\{-L_i, 0\}$

The optimal value of any criterion is denoted with an asterisk, *e.g.* $C^*_{\max}$ denotes the optimal makespan value.

We will use the three-parameter notation, $\alpha / \beta / \gamma$, introduced by Graham et al. [38] and, extended for T'kindt and Billaut [109] to MultiCriteria scheduling problems. The first field specifies the machine environment (F represents general permutation flow shop); the second, job characteristics; and the third refers to the chosen optimality criterion for single criteria models, and it extends to cover multicriteria as well as methodology.

2.2 Definitions

Consider a set of n independent jobs J_i (i=1,...,n) to be processed, each of them on a set of m machines M_j (j=1,...,m), that represent the m stages of the production process. Every job requires a known, deterministic and non-negative processing time, denoted as p_{ij}, for completion at each machine. Each machine processes the jobs in the same order, thus knowing the order of jobs the resulting schedule is entirely fixed. Any feasible solution is then called a *permutation schedule* or a *sequence*. In a single-criterion problem we look for the permutation of jobs from set J that would optimize the performance criterion, while for more than one criterion the objective is to find out the set of Pareto optimal solutions. The most used criterion is the minimization of the total completion time of the schedule, often referred to as makespan (C_{max}). But there are many performance criteria to be considered when solving scheduling problems.

2.3 Criteria

French [31] presents the following classification:

Criteria based upon completion time measures

- F_{max}= max{F_1, F_2,..., F_n}, the maximum flow time
- C_{max}= max{C_1, C_2,..., C_n}, the maximum completion time
- $\sum F_i/n$ or $\sum F_i$, mean flow time or total flow time, respectively
- $\sum C_i/n$ or $\sum C_i$, mean completion time or total completion time, respectively
- $\sum w_i C_i$, weighted completion time
- $\sum w_i F_i$, weighted flow time

Flow time is applied as a criterion when the cost function is related to the job standing time. Completion time reflects a criterion where the cost depends on the finish time. In the event of all ready times being zero, r_i=0, $\forall i$, completion time and flow time functions are identical. Maximum criteria should be used when interest is focused on the whole system. When some jobs are more important than others, weighted measures could be considered.

Criteria based upon due date measures

- L_{max}= max{L_1, L_2,..., L_n}, maximum lateness
- T_{max}= max{T_1, T_2,..., T_n}, maximum tardiness
- $\sum L_i/n$ or $\sum L_i$, mean lateness or total lateness, respectively
- $\sum T_i/n$ or $\sum T_i$, mean tardiness or total tardiness, respectively

- $\sum w_i L_i$, weighted lateness
- $\sum w_i T_i$, weighted tardiness
- $\sum U_i$, total tardy jobs. The indicator function U_i denotes whether the job J_i is tardy, then $U_i = 1$, or on time, then $U_i = 0$.

When maintaining customer satisfaction by observing due dates, or any other *just in time* concept has to be considered, measures related to the notion of how much is lost by not meeting the due dates are applied. If the penalty is applied only to the delays, tardiness measures are used. When there is a positive reward, or penalization, for completing a job early and that reward/penalization is larger the earlier a job is completed, lateness measures are appropriate. In the case where all the due dates are zero, d_i=0, $\forall i$, tardiness or lateness are identical to completion time functions.

All of the above mentioned criteria are regular in the sense that they are non-decreasing functions of job completion times. French's classification includes some non-regular criteria, such as measures based upon the inventory and utilization costs. For example, to measure the idle time of a machine, the following criterion is used.

- $I_j = C_{max} - \sum_{i=1}^{n} p_{ij}$, total time during which machine M_j is waiting for a job or has finished processing jobs, but the total process of jobs has not finished jet.

In this chapter we focus on the maximum completion time criterion (makespan) and the total flow time, even though we shall also refer to other measures. In the literature, the most common criterion is the makespan. Only a relative few published works are devoted to flow time and tardiness measures.

2.4 Assumptions

Unless explicitly indicated, in the text that follows we assume that:

- Each job is an entity, composed of m operations, which cannot be processed on more than one machine simultaneously.
- At every machine, there are no precedence constraints among operations of different jobs.
- No preemption is allowed. That is to say, once an operation has started, it must be completed before another operation may initiate on the same machine.
- No cancellation. Each job must be finished.
- Processing times are independent of sequencing.
- Job accumulation is allowed. Jobs can be waiting for a free machine.
- Machine idle time is allowed. The machines can be waiting for jobs or for the end of the total process.
- No machine can process more than one job simultaneously.
- Machines never break down and are available throughout the scheduling period.
- Ready times are zero for all jobs.

- There is no randomness:
  - the number of jobs, n, is known and fixed;
  - the number of machines, m, is known and fixed;
  - the processing times, p_{ij} (i=1,...,n; j=1,...,m), are known and fixed;
  - all other specifications, needed to define a particular problem, are known and fixed.

The assumptions listed above characterize the classical permutation flow shop models. However, it is possible to find in the literature variants of permutation flow shop problems which do not accomplish these features.

Computational Complexity

Since the early Johnson Algorithm [54] that solves $F2//C_{max}$ in polynomial time, only a few restricted cases have been shown to be efficiently solvable. Minimizing the sum of completion times is still NP-complete for two machines [33].

The following cases have been shown to be polynomially solvable:

- $F/p_{ij}=1, intree, r_i/C_{max}$
- $F/p_{ij}=1, prec/C_{max}$
- $F2/chains/C_{max}$
- $F2/chains, pmtn/C_{max}$
- $F2/r_i/C_{max}$
- $F2/r_i, pmtn/C_{max}$
- $F3//C_{max}$
- $F3/pmtn/C_{max}$
- $F/p_{ij}=1, outtree/L_{max}$
- $F2//L_{max}$
- $F2/pmtn/L_{max}$
- $F2//\sum C_i$
- $F2/pmtn/\sum C_i$
- $Fm/p_{ij}=1, chains/\sum w_i C_i$
- $Fm/p_{ij}=1, chains/\sum U_i$, for each $m \geq 2$
- $Fm/ p_{ij}=1, chains/\sum T_i$, for each $m \geq 2$

Review of permutation flow shop scheduling algorithms, considering only a single-criterion

Despite the large amount of papers dealing with flow shop problems, most of the research has been devoted to the permutation problem. From the pioneer paper by Johnson [54] until the present day, a lot of papers devoted to permutation flow shop problem have been published. The majority of them consider the problem of minimizing the makespan.

Johnson's rule, states that job i must precede job k in a sequence if $\min\{p_{i1}, p_{k2}\} < \min\{p_{k1}, p_{i2}\}$. Thus, jobs with shorter processing time in the first machine are set to be processed before, and jobs with shorter processing time in the second machine are set to be processed after. The algorithm that applies this rule is optimal for m=2, and can approximate solutions for m>2 [13].

For the problem restricted to n=2, and general m, the graphical method due to Akers [3] gets the minimum makespan [9].

[48, 69] propose the earliest branch and bound algorithms applied to permutation flow shop. [60] presents a general bounding scheme for permutation flow shop problem, considering makespan. Though the original intention was to improve branch and bound techniques (in vogue at the time of its publication), their contributions are still useful in saving computational effort when looking for non-dominated solutions. [86] presents a branching rule. [102] proposes a Goal Programming formulation. [113] presents, besides very useful benchmarks, a lower bound for the makespan. [14] presents two branch and bound algorithms.

Heuristics and metaheuristics have been mainly developed for CO problems. In contrast to exact methods that guarantee optimality, heuristic methods seek near optimal solutions in a reasonably bounded time. Metaheuristics are more general than heuristics, in the sense that they are applicable to different problems, while heuristics are usually problem-dependent.

A constructive algorithm builds a solution, starting from the input data (without it being necessary to know a previous feasible solution), following a set of rules. There is a class of algorithms which share a similar way of making a schedule: a sorting list with all the jobs is made. The accuracy of any list scheduling algorithm is intimately related to the priority rule applied. There are more than one hundred dispatching rules, as can be seen in [81, 42].

The most important constructive algorithms dedicated to the F//Cmax problem can be classified by their design as a list scheduling algorithm. In order to minimize the makespan, the list of jobs must be made in such a way as to give higher priority to the jobs consuming more total processing time. That is to say, the jobs with the longest total processing time should not be placed at the last positions of the list. Based on this premise, a simple algorithm is presented by Nawaz, Enscore and Ham (NEH algorithm, in the following) in [75]. NEH algorithm produces very good sequences in comparison with heuristics existing even up to the present. The results of the proposed algorithm show that it performs especially well on large flow shop problems, in both the static and dynamic sequencing environments. [112] presents an important improvement in saving computational effort for the NEH algorithm.

[39] presents three algorithms to deal with total flow time and maximum flow time (not simultaneously). [63, 89] present constructive algorithms for the $F//\sum C_i$ problem. The first one is based on the principle of job insertion, and the second one could be thought as an extension of the NEH algorithm and performs very well.

Improvement algorithms need a feasible solution as a starting-point and are intended to improve it by iterative small changes. This iterative improvement can be achieved by means of many different processes.

Threshold Algorithms are designed according to three techniques: Iterative Improvement, Threshold Accepting and Simulated Annealing (SA), the most popular one.

Considering $F//C_{max}$, [80] presents four SA algorithms varying the neighbor generating method. Their results show that insertion performs better than swapping. The SA algorithms presented by [78] have similar performance than [80]. Only the algorithm presented in [50] seems to perform better for large instances. [62] introduces SA in the NEH algorithm and [119] presents a parallel SA.

[83] presents an application of SA to the $F//\sum w_i T_i$. In this paper the authors introduced the Random Insertion Perturbation Scheme that is employed in some later papers (One of our proposed neighbouring generating procedure is based on this technique).

In [61] SA is applied to solve the $F//\sum C_i$ problem. [68 and 92] consider also this problem, [68] using pair-wise exchange and [92] Ant Colony techniques. [91] presents heuristics dealing with the total weighted flow time.

Based on Johnson's rule, [58] proposes an improvement heuristic which uses job passing.

Tabu Search is probably the most tested local search concerned with scheduling problems. Some applications to the flow shop scheduling problem have been presented in: [112, 93] and, more recently, in [37].

Unlike the previously-mentioned techniques, Genetic and Evolutionary Algorithms (GA and EA, in the following) start with a set of solutions instead of only one: [94] applies GA to the flow shop scheduling problem. Differential evolutionary optimization is applied to permutation flow shop scheduling problem for minimizing makespan, mean flow time and total tardiness, individually considered, in [79].

Research on metaheuristics is quite extensive. Ruiz and Maroto [97] and Dorn et al. [22] survey this field.

Real-life scheduling problems require more than one criterion. Nevertheless, the complex nature of flow shop scheduling has prevented the development of models with multiple criteria. In the following, we will consider the Multi-Objective Flow Shop Scheduling problems.

For further information about deterministic scheduling and flow shop, considering only single-criterion problems, we refer the reader to the books and PhD thesis of: Blazewicz et al. [8]; Brucker [10]; Ruiz [96]; Pinedo [85]; Andrés [4]; Schulz [101] and Parker [82]; or the survey papers of: Lawler et al. [65]; Dudek et al. [23]; Monma and Rinnooy Kan [72] and the earliest Baker [7].

3 Multi-objective Combinatorial Optimization Problem

Quality is, in real-life, a multidimensional notion. A schedule is valued on the basis of a number of criteria, for example: makespan, work-in-process inventories, idle times, observance of due dates, etc. If only one criterion is taken into account, no matter what criterion is considered, some aspect of the quality of the schedule will result regardless. An appropriate schedule can not be obtained unless one observes the whole set of important criteria. The multidimensional nature of the problem at hand leads us to the area of MultiCriteria Optimization (see Ehrgott and Wiecek [28], for a state of the art).

When a problem appears as a multicriteria case, it is necessary to take into account different objective functions. The solution may vary according to the criterion considered individually. If the criteria are not conflicting, it is possible to obtain a global optimal solution. In the vast majority of cases, they are conflicting and thus the knowledge of the decision-maker preferences is necessary to solve the problem.

Considering only one regular criterion, the general permutation flow shop scheduling has been shown to be NP-hard, and to belong to the CO field (except for the restricted special cases already mentioned).

Even though MDM, as well as CO, have been intensively studied by many researchers for many years, it is surprising that a combination of both, *i.e.* Multi-Objective Combinatorial Optimization (MOCO), was not widely studied until the last decade, as it is not long since interest in this field has been shown [27]. The proliferation of metaheuristic techniques has encouraged researchers to apply them to this highly complex problem.

In this section we will present a brief introduction to MOCO problems, including a general problem formulation, the most important theoretical properties, and the existing methods for dealing with this kind of problem.

3.1 Formulation of a MOCO Problem

A MOCO problem is a discrete optimization problem, where each feasible solution X has n variables, x_i, constrained by a specific structure, and there are K objective functions, z_k, to be optimized. Without loss of generality we can formulate the problem as follows:

$$\underset{X \in D}{Min}\, z_k(X),\ k = 1,\ldots,K \tag{1}$$

where functions z_k are the objectives, X is the vector that represents a feasible solution (a sequence for the flow shop scheduling problem), and D is the set of feasible solutions: a discrete set.

The criteria (reviewed in the previous section) are of two different kinds:

- sum function: $\sum f_i$
- bottleneck function: $f_{max} = \max\{f_1, f_2, \ldots, f_n\}$

We call a feasible solution, $X^{(e)} \in D$, efficient, non-dominated, or Pareto optimal, if there is no other feasible solution $X \in D$ such that,

$$z_k(X) \leq z_k\left(X^{(e)}\right), \forall k \tag{2}$$

with at least one strict inequality.

The corresponding vector of objective values,

$$z\left(X^{(e)}\right) = \left(z_1\left(X^{(e)}\right), z_2\left(X^{(e)}\right), \ldots, z_K\left(X^{(e)}\right)\right) \tag{3}$$

is called non dominated vector.

The set of feasible Efficient solutions, $X^{(e)}$, is denoted by E, and the set of non-dominated vectors by ND.

3.2 Some Theoretical Concepts

A general result for Multi-Objective Linear Programming (MLP) problems is that the set of efficient solutions for the MLP problem,

$$\min\{\mathbf{c}X : \mathrm{A}X = \mathbf{b},\ X \geq 0\} \tag{4}$$

is exactly the set of solutions of

$$\min\{\sum_{j=1,\ldots K}\lambda_j c_j X : AX=\mathbf{b}, X\geq 0\}, \tag{5}$$

where $\sum_{j=1,\ldots K}\lambda_j = 1$, $\lambda_j > 0, j=1, \ldots K$.

It is important to point out that we are dealing with a CO problem, which means that the transformation of the objective functions into a linear function (aggregating into weighted sums) does not transform the problem into a Linear Programming one. Except in some special cases, *e.g.* preemption allowance, or where idle time insertion is advantageous, for which Linear Programming can be applied, the discrete structure of a MOCO problem persists. An important consequence is the fact that the previous result for MLP is not valid, so there could be some efficient solutions not optimal for any weighted sum of the objectives. The set of these solutions are named Non-supported Efficient solutions (NE), whereas the set of the remaining ones are called Supported Efficient solutions (SE) [27].

The cardinality of the NE set depends on the number of sum objective functions. For a problem with more than one sum objective function, NE has many more solutions than SE.

Despite these results which constitute the essence of the difficulty of MOCO problems, many published works ignore the existence of NE.

Concerning computational complexity, in obtaining the set of efficient solutions MOCO problems are in general NP-complete. Results are presented by Ehrgott [25]. The cardinality of E for a MOCO problem may be exponential in the problem size [29], therefore algorithms could determine just an approximation of E in many cases. Thus, methods may be exact or approximate, and metaheuristics are nowadays being applied intensively to MOCO problems.

3.3 MultiCriteria Optimization Methods

The "minimization" concept in the above formulation is not restricted to one meaning. At this point we have to point out that MOO was originally conceived to find a set of Pareto optimal alternative solutions, because hoping to find the minimum schedule must usually be discarded. The MDM always assumes that subjective considerations, such as the decision-maker preferences, have to intervene. Besides the classic classification for optimization methods between exact or approximation, it is usual to distinguish the MOO methods according to when the decision-maker intervenes in the resolution process, as follows:

- *a priori*: All the preferences are known at the beginning of the decision-making process. The search for the solution is carried out on the basis of the known information.
- *interactive*: The decision-maker intervenes during the search process. Computing steps alternate with dialogue steps. At each step a satisfying compromise determination is achieved. It requires the intensive participation of the decision-maker.
- *a posteriori*: The set of efficient solutions (the complete set or an approximation of it) is generated. This set can be analyzed according to the decision-maker preferences. The choice of a solution from the set of efficient solutions is an *a posteriori* approach.

If the problem criteria show a hierarchical structure, more important criteria should be minimized before less important ones. Thus, optimization methods can be classified as hierarchical or simultaneous.

In bicriteria models, if z_1 is more important than z_2, then it seem to be natural to minimize with respect to z_1 first, and choose, from among these optimal solutions, the optimum with respect to z_2. This hierarchical approach is called lexicographic optimization, and is denoted by $\alpha/\beta/\text{Lex}(z_1, z_2)$.

In a general case, lexicographic minimization consists in comparing the objective values of a feasible solution X, with respect to another Y, in a lexicographical order, denoted by $<_{\text{lex}}$. Objective functions are ranked according to their importance. We say $X <_{\text{lex}} Y$, if, and only if, there is a j such that $z_j(X) <_j z_j(Y)$, and there is not any $h<j$, such that $z_h(Y) <_h z_h(X)$. This means that the first objective function index, $i \in 1,...,K$, for which $z_i(X)$, is not equal to $z_i(Y)$, $z_i(X) < z_i(Y)$.

Simultaneous optimization has to be applied when there is no dominant relation among the criteria. Optimizing with respect to one criterion at a time leads to unbalanced results. It is common, in a case such as this, to use a composite objective function with the original criteria. It gives rise to another classification, because we can generate solutions by means of scalarization and non-scalarizing methods.

Scalarization is made by means of a real-valued scalarizing on the objective functions of the original problem [117]. Well known examples of scalarization methods are the following.

The Weighted Sum approach consists in building a new objective criterion with the original ones [49]. This composite function can be linear (in the majority of cases), where the scalar coefficients represent the relative importance of every criterion, or it may present a more complex composition. Despite the apparent simplicity of the methods, it conceals two difficulties:

i) the difficulty of expressing the decision-maker preferences by means of a function (interactive approaches overcome this drawback, *e.g.* AHP procedures could be useful, [99]);
ii) the computational complexity of minimizing the function in a direct manner.

The set of all supported efficient solutions can be found considering a wide diversified set of weights (Parametric Programming may be used to solve this problem). [102, 118] apply this technique, considering a linear combination of makespan and flow time. [104] proposes a linear combination of the makespan and a total cost function, for unrelated parallel machine models.

The distance to the ideal point approach [46] consists in minimizing the distance to an ideal solution. The ideal point is settled according to the optimum of each individual single-criterion. It is also known as the compromise solution method.

The ε-constraint [17] and the Target-Vector approaches are scalarization as well as hierarchical methods. A constraint system representing levels ε_i of satisfaction, for some criteria, is established, and the objective is to find a solution which provides a value, as close as possible, to the pre-defined goal for each objective. A single-objective minimization subject to constraints of levels ε_i for the other objective functions is formulated. The formulation is solved for different levels ε_i, to generate the entire Pareto optimal set. Some authors consider that the main criteria must be fixed by constraints, others put the main criteria in the objective of the formulation by turn. It would depend on the mathematical programs to solve. [66, 24] present algorithms to minimize the makespan, subject to a determined flow time level (the first one is devoted to preemptive job models). [35] proposes minimizing the makespan, subject to a bound on the number of preemptions. [105] considers the problem of minimizing the makespan and the number of preemptions, for a set of jobs, constrained to due dates.

When a set of goals for each criterion is known, the target vector approaches are appropriate. The most popular is Goal Programming (introduced by [18]), for which the minimization of the deviation from the specified goals is the aim.

Non-scalarizing approaches do not explicitly use this kind of scalarizing function. For example, Lexicographic and Max-ordering are non-scalarizing approaches.

Max-ordering chooses the alternative with the minimum value of the worst values. After a normalization process, z_j is the worst value of X, if and only if,

$$z_j(X) = \max(z_1(X), z_2(X), \ldots, z_K(X)) \tag{6}$$

Then, X is the best alternative, if, and only if, there is not Y such that $z_{j(y)}(Y) < z_{j(x)}(X)$.

Only a few algorithms have been developed based on branch and bound techniques for MOCO problems [26].

The *two phases* method [114] consists in determining the set of supported efficient solutions by means of a weighted sum scalarization algorithm, and then, in the second phase, searching for the non-supported ones, following a specific problem-dependent method.

Approximation for MOO is a research area which has gained increasing interest in recent years. Multi-Objective Metaheuristics seek an approximate set of Pareto optimal solutions. The main question is how to ensure that the obtained non-dominated set covers the Pareto front as widely as possible. In the beginning, methods were adaptations of single-objective optimization. Nowadays they have their own entity. They are initially inspired by EA or neighborhood search. Furthermore, recent developments are more hybridized, given rise to Multi-Objective Hyperheuristic methods. A hyperheuristic can be thought as a heuristic method, which iteratively selects the most suitable heuristic amongst many [12].

The problem of obtaining a uniformly distributed set of non-dominated solutions is of great concern in Pareto optimization. The specification of the search direction, by tuning weights, is the method that directly attempts to drive the current solution towards the desired region of the trade-off frontier. Hyperheuristic approaches attempt to do it by applying the neighbourhood search heuristic that is more likely to drive the solution in the desired direction. This technique can be applied to single-solution and population-based algorithms.

Most of the published works in MOO are a priori methods since they assume that the decision-maker preferences can be expressed. The hierarchical approach penalizes too much the less important criteria, while setting a criterion as the most important one. In reality, the decision-maker preferences are usually smooth, giving less importance to the main criterion and more to the less important criteria. Considering a composed function of the criteria involved in the problem, it is implicitly assumed that the decision-maker preferences are accurately reflected in this objective function. The decision-maker knows the preferable schedule, but it is not easy to express this preference in a function. In general, a priori approaches give a solution to the problem, which cannot usually be trusted to be the most preferred solution.

To be confident with a particular solution to a problem with multiple objectives, the decision-maker active involvement is required. In interactive methods, she indicates their preferences during the process of solution, guiding the search direction. [1] proposes an interactive particle-swarm metaheuristic for MOO. The approach presented by [53] can be placed between the a priori and interactive procedures. The method that this paper presents includes some interaction with the decision-maker, but is based on the assumption that decision-maker preferences are already relatively well-defined at the beginning of the solution process.

For methods that should offer the complete set of efficient solutions (a posteriori approaches), it is guaranteed that no potential preferable solution has been eliminated, but the number of efficient solutions can be overwhelmingly high to warrant proper examination by the decision-maker.

In the following we are going to focus on scheduling and flow shop applications of the MOO.

We refer to [115, 27] for further information on MOCO theory. [64, 55] present overviews to the metaheuristics applied to solve MOCO problems. [52] compares metaheuristics for bicriteria optimization problems. For each particular metaheuristics, we refer the reader to the following references:

- For Multi-Objective Genetic Algorithms (MOGA), to [2, 52]. For general Evolutionary Multi-Objective Algorithms, to [19, 34].
- For MOSA, to [103, 114, 41, 70].
- For Multi-Objective Tabu Search, to [32].

4 Multicriteria Scheduling Review

Starting with the *just-in-time* philosophy, the earliness–tardiness problem becomes one of the most appealing bicriteria in Scheduling Theory. Early completion time results in the need to store the product until it can be shipped. [44] presents an extensive review for the case where the due dates have been determined already, which is contrary to the due date assignment model (one has the freedom to determine the optimal due date, at a certain cost), for which we refer to the survey by [36].

We refer to [44, 77] for a survey of the field of scheduling with controllable processing times, in which the processing times can be compressed at the expense of some extra cost, which is called the compression cost. Hoogeveen [44] also presents an overview of bi-criteria worst-case analysis.

In this section we will focus on Multi-Objective flow shop scheduling problems. For further information on general Multi-Objective Scheduling we refer to the following surveys or books:

[98] provides the earliest survey of papers on multiple-objective scheduling. Subsequently, [74, 108, 44] have been published, and they present exhaustive surveys of MultiCriteria Scheduling problems. [64] reviews metaheuristics for general Multi-Objective

problems and presents the application of these techniques to some Multi-Objective Scheduling problems.

The book of T'kindt and Billaut [109] can be useful as a good reference work, and also an introduction to any field of Multicriteria Scheduling.

4.1 Multicriteria Flow Shop Scheduling Problem Review

Permutation flow shop scheduling research has been mostly restricted to the treatment of one objective at a time. Furthermore, attention focused on the makespan criterion. However, the total flow time performance measure has also received some attention. These two measures, each of which is a regular performance measure, constitute a conflicting pair of objectives [95]. Specifically, the total flow time criterion is a work in process inventory performance measure, and the makespan criterion is equivalent to the mean utilization performance measure. While total flow time is a customer-oriented performance measure, the makespan criterion is a firm-oriented performance measure. Therefore, the set of efficient solutions to a bicriteria model that seeks to optimize both measures simultaneously would contain valuable trade-off information crucial to the decision-maker, who has to identify the most preferable solution, according to her preferences.

Solving a bi-criteria model for a general number of machines implies heavy computational requirements, since both criteria makespan and total flow time, lead to NP-hard problems even when they are treated individually. Due to the fact that only the $F2//C_{max}$ problem can be solved in polynomial time (the rest of flow shop scheduling problems are *NP-complete*), research production concentrates on heuristics and enumerative approaches. The majority of research on bicriteria flow shop problems concerns the two-machine case, in which some combination of $\sum C_i$ and C_{max} has to be minimized.

Since $F2//\sum C_i$ is NP-hard in the strong sense, any lexicographic approach including $\sum C_i$ will be NP-hard too. [88, 76, 40, 110], present heuristics for the two-machine flow shop problem, where total flow time has to be minimized among the schedules that minimize makespan (lexicographical approach). Local search algorithms based on Ant Colony Optimization have been proposed by [111]. [47] presents a technique named Local Dynamic Programming. [110] presents a branch and bound algorithm, which can solve problem instances of up to 35 jobs to optimality.

[74, 106] present heuristics and branch and bound algorithms for the $F2// f(\sum C_i, C_{\max})$ problem.

For the two-machine flow shop scheduling problem of minimizing makespan and sum of completion times simultaneously, [100] presents an *a posteriori* approach based on branch and bound.

[21] presents a branch and bound algorithm for the $F2// f(C_{\max}, T_{\max})$, and a heuristic to approximate the set of non-dominated solutions for the more general $F// f(C_{\max}, T_{\max})$ problem.

[67] presents branch and bound algorithms for the $F2// f(C_{\max}, \sum U_i)$ and $F2// f(C_{\max}, \sum T_i)$ problems.

[102, 118] consider a linear combination of makespan and flow time. [22] presents a comparison of four iterative improvement techniques for flow shop scheduling problems that differ in local search methodology. These techniques are iterative deepening, random search, tabu search and GA. The evaluation function is defined according to the gradual satisfaction of explicitly represented domain constraints and optimization functions. The problem is constrained by a greater variety of antagonistic criteria that are partly contradictory.

[43, 90] propose heuristic procedures for the general m machine case, considering $\sum C_i, C_{\max}, \sum I_j$. They are based on the idea of minimizing the gaps between the completion times of jobs on adjacent machines (one of our proposed improvement techniques was inspired by this paper). [120] applies Ant Colony Optimization to the same problem.

[6] presents a MOGA that improves the previous MOGA presented by [107]. [73] presents a MOGA considering the makespan, total flow time and total tardiness, based on a weighted sum of objective functions with variable weights. This algorithm belongs to the class of evolutionary multi-objective optimization algorithms and [51] shows that this algorithm can be improved by adding a local search procedure to the offspring. [15] applies subpopulation GA to the same problems. Artificial chromosomes are created and introduced into the evolution process to improve the efficiency and the quality of the solution.

[5] proposes a MOGA algorithm with preservation of dispersion in the population, elitism, and use of a parallel bi-objective local search so as intensify the search in distinct regions. The algorithm is applied to the makespan-maximum tardiness and makespan-total tardiness problems.

[30] investigates *a priori* and *a posteriori* heuristics. The *a posteriori* heuristic does not require a decision-maker preference structure and uncovers non-dominated solutions by varying the weight criteria in an effective way.

[16] proposes a GA algorithm for the $F// f(\sum C_i, C_{\max})$ problem based on the concept of *gradual priority weighting* (the search process starts along the direction of the first selected objective function, and progresses such that the weight for the first objective function decreases gradually, and the weight for the second objective function increases gradually). [116] applies a similar idea for a MOSA. [84] presents a Pareto GA with Local Search, based on ranks that are computed by means of crowding distances. Both papers apply the same initial population and improvement schemes.

[11] applies Dantzig-Wolfe reformulation and Lagrangian relaxation to an Integer Programming formulation to minimize a total cost of job function that includes: earliness, tardiness and work in process inventory costs.

[34] presents a study of the problem structure and the effectiveness of local search neighborhoods within an evolutionary search framework on Multi-Objective flow shop scheduling problems.

5 Proposed Algorithms

We present a new approximation algorithm for the Pareto solution set of the MOCO problem defined by minimizing makespan and total flow time in the classical permutation flow shop scheduling problem. The most promising practical approach to MOCO consists in generating efficient solutions with metaheuristic procedures. Different approaches are applicable to tackle MOCO problems, each of them having their own advantages and drawbacks. The chosen approach depends essentially on the aim of the study. SA (introduced by Kirkpatrick et al. [56]), has demonstrated their ability in solving combinatorial intractable problems considering just one criterion [59]. [103] presents a broad study of the application of SA to MOO. (A brief survey of published papers in this field has already been presented in the previous section).

SA is a generic technique (based on an analogy to physical cooling studied by statistical mechanics), and has to be adapted in the context of the specific problem being studied. It is basically an improvement technique, by which an initial solution is improved by means of local perturbations. All MOSA methods have in common:

- An acceptance rule for new solutions, with some probability that depends on the temperature level.
- A scheme of cooling.
- A mechanism for browsing the efficient frontier.
- Information is obtained from the set of solutions.

The proposed method is based on the MOSA scheme that follows.

5.1 MOSA Scheme

The procedure begins with an initial iterate solution, X_0, that belongs to a set S of initial points (feasible solutions of $F//(C_{\max}, \sum C_i)$, which are good solutions for one of the two simplified single-criterion versions of the problem). X_0 is then sampled with a point Y in its neighbourhood. But instead of accepting Y if it is better than the current iterate regarding an objective function, we now accept it, if it is not dominated by the current solution. In this case, we make Y the current iterate, add it to the Potentially Efficient solution set (PE), and throw out any point in PE that is dominated by Y.

On the other hand, if Y is dominated by X_0, we still make it the current iterate with some probability. This randomization is introduced in the procedure to attempt to reduce the probability of getting stuck in a poor locally optimal solution.

The solutions that are generated, during the optimization process, make iterative updates to the PE point set, to get closer to the Pareto optimal set (E). The only complicated aspect of this algorithm is the necessity of generating solutions in several directions of the bi-objective space search. So, to be able to cover the entire efficient frontier, a diversified set of points must be generated. Neighbourhood search procedures play a crucial role in the performance of the algorithm.

At each time during the search, the selection of the next heuristic to be used is based on the quality of the current seed. A set of simple neighbourhood exploration heuristics has been developed. Then, the approach proposed here selects the most appropriate neighbourhood heuristic at certain points during the search, in order to uncover the solution in the Pareto optimal front.

The objective function of the MOCO problem plays here the role of acceptance rule. The discrete nature of the problem at hand makes it possible that some efficient solutions do not minimize any aggregated function of the criteria. Only supported efficient solutions will be admitted for entrance into the PE. In order to avoid the non-supported efficient solutions to enable entrance into the PE, we have developed a bi-objective model where, simultaneously, both criteria are minimized. For just bi-criteria models, checking whether a solution is dominated by another, is not computationally costly, and besides, updating the non-dominated solution set have to be face up in any case (for more than two criteria models, the use of aggregated functions may be absolutely justified).

A set of feasible initial solutions, S, is constituted. For each initial solution $X_0 \in S$, the following procedure is applied.

- Initialization ($X_n = X_0$, $N_{count}=n=0$)
- Iteration n
  - Sample a neighbor Y
  - Evaluate Y
  - If Y is acceptable: $X_n = Y$, $N_{count}=0$. Else, we accept the solution with probability

$$p = \exp\left(-\frac{\Delta\phi}{T_n}\right)$$

$$X_{n+1}\begin{cases} \xleftarrow{p} & Y, \qquad N_{count} = 0 \\ \xleftarrow{1-p} & X_n, N_{count} = N_{count} + 1 \end{cases}$$

  - Update PE.
  - $n=n+1$. If $n(\text{mod}N_{step})=0$, then $T_n=\alpha T_{n-1}$, else $T_n=T_{n-1}$. If $N_{count} = N_{stop}$ or $T_n < T_{stop}$, then stop. Else iterate.

This generic scheme is completed with the different particularities that are described in the following sections.

5.2 Set of Initial Solutions

The quality of seed solutions helps to reduce the search space. In this model we propose using constructive techniques to compute a set, S, of initial feasible sequences, which are good for one of the criteria at a time. The size of S, may take values from 2 to N, N being a parameter of the algorithm.

The first two solutions are obtained by means of the two simple but effective constructive algorithms: NEH [75], looking for the minimum C_{max}, and the algorithm for the $F // \sum C_i$ problem presented by Rajendran [89]. We recall both of them here.

NEH algorithm (X_1)
The steps for generating the NEH seed sequence can be fully described as follows:

Step 1: For each job i calculate the total processing time $p_i = \sum_{j=1}^{m} p_{ij}$.

Step 2: List the jobs according to descending order of p_i.

Step 3: Schedule the first two jobs (from the list) in order to minimize the partial makespan (as if there were only these two jobs).
Step 4: For k=3 to n, insert the job k at the position which minimizes the partial makespan, among the k possible places.

X_1 is considered as a good solution for the makespan criterion. The computation of the minimum partial makespan in Step 4 is made by the algorithm presented by Taillard [112].

Rajendran algorithm (X_2)
The steps for generating the Rajendran seed sequence are analogous with the NEH algorithm. The difference is in the way of making the list of jobs. Here the schedule is made as follows:

Step 1: For each job i calculate the index $w_i = \sum_{j=1}^{m}(m-j+1)p_{ij}$.
Step 2: List the jobs according to ascending order of w_i.
Step 3: Schedule the first two jobs (from the list) in order to minimize the partial total flow time (as if there were only these two jobs).
Step 4: For k=3 to n, insert the job k at the position which minimizes the partial total flow time, among the k possible places.

X_2 is considered as a good solution for the total flow time criterion.

Sequence X_1 and X_2, obtained in Step 4 of the corresponding algorithms, become the seed sequences to be given as input to the Improvement Schemes presented in the following section.

This common list scheduling procedure is also applicable re-combining *making list* (ordering by p_i or w_i), and *criterion to be minimized* (makespan or total flow time). X_3 is obtained following the NEH algorithm, only altering in Steps 3 and 4 the minimization criterion. Now the Steps 3 and 4 will read:

Step 3: Schedule the first two jobs (from the list) in order to minimize the partial total flow time (as if there were only these two jobs).
Step 4: For k=3 to n, insert the job k at the position which minimizes the partial total flow time, among the k possible places.

So, X_3 is considered a *good* solution for the total flow time criterion.

By analogy, Rajendran algorithm is applied to obtain X_4, a *good* solution for the makespan criterion. X_4 is obtained following the Rajendran algorithm, only altering in Steps 3 and 4 the minimization criterion. Now the Steps 3 and 4 will read:

Step 3: Schedule the first two jobs (from the list) in order to minimize the partial makespan (as if there were only these two jobs).
Step 4: For k=3 to n, insert the job k at the position which minimizes the partial makespan, among the k possible places.

All of these four generation algorithms share the same four-step structure. (These four initial solutions are used by the MOGA algorithm presented in [84]).

To obtain eight solutions for the set S, we have followed this strategy: At Step 3, in each of the four algorithms we keep both partial schedules, and proceed to Step 4 for each of both seeds. Thus, we obtain eight, instead of four, initial solutions.

For a larger *S*, at Step 4 (of each of the eight partial schedules) we conserve the *k* generated partial schedules, and proceed with every partial schedule until the permutation is complete. For $k = 3$, we will count on twenty-four feasible solutions. Continuing with this strategy we can generate as many initial solutions as desired. Therefore, [*S*] becomes a parameter for the algorithm (where [] denotes cardinality).

In the computation of the initial solutions, the procedure keeps the useful data in order to save computational effort (*e.g.* job lists, best partial schedules, etc.). With this technique it is possible to obtain a selective list of efficient solutions as seeds, instead of just randomly-generated ones.

When sampling solutions, only those who pass the domination control are taken into account for listing in the PE solutions set. The rest of the generated solutions are only used as input (for improvement or neighbouring search) and discarded later.

5.3 Improvement Techniques

Improvement of the initial solutions and neighbouring generation are carried out by simple neighbourhood exploration heuristics. The objective of these procedures is to approximate the trade-off surface in a more efficient way by using those movements that are more promising according to the quality of the current solution.

One can set a relation between the optimization criterion for which the iterate solution presents the least deviation (which coincides, in general, with the minimizing criterion for which it has been calculated) and the criterion taken into account for the improvement technique, thereby distinguishing three kinds of movement strategies:

- Direct search: if the best criterion value corresponds to, or if the seed was calculated considering, makespan/total flow time, then, the improvement technique looks for solutions with less makespan/total flow time.
- Cross search: if the best criterion value corresponds to, or if the seed was calculated considering, makespan/total flow time, then the improvement technique looks for solutions with less total flow time/makespan.
- Combining search: one of the two criteria is chosen for applying the improvement technique at random.

These procedures induce a privileged direction of search to the efficient frontier. So, to be able to cover the entire efficient frontier, a diversified combination of initial solutions and neighbouring generation heuristics must be considered.

In the MOSA scheme described previously, a neighbouring solution of the current permutation must be chosen. The most important neighborhoods based on a single permutation as an input are:

- Exchange, swapping the positions of two jobs at *i* and *k*, with $i \neq k$. The remaining jobs in the sequence conserve their positions.
- Insertion, forward or backward shift, removing a job at *i* and reinserting it at a different position k, with $k > i$ in forward case, and with $k < i$ in backward case. The remaining jobs in the sequence must be re-arranged in order to keep their relative positions.

In our development we have implemented the improvement and perturbation schemes, which are described in the following section.

Improvement Scheme

Instead of inducing the search direction by tuning weights, to improve the distribution of non-dominated solutions we apply different neighbouring search heuristics based on the features of the current solution. Heuristics are selected in order to achieve improvements on the objective with relative worse value, while keeping the quality of better value on the other objective.

While *insertion* have been shown to lead to superior results compared with *exchange*, for flow shop scheduling problems with C_{max} objective [112], it seems not to be possible to derive a similar general rule when considering total flow time criterion. So, for improving makespan, we have just implemented insertion. Instead, for flow time, we try with insertion and exchange.

In the valuation of a neighbor, it is very important to save computational effort in order to check a larger neighbourhood. When inserting or exchanging jobs in a schedule, it should be possible to discard some potentially dominated candidate permutation with small computational requirements just considering the corresponding partial schedules. With this in mind, we have developed two neighbouring generating heuristics: one devoted to search for neighbouring solution superior than the current one regarding makespan; another sampling better solution according to total flow time measure.

Improving Makespan

In order to reduce the search space, we have developed a technique based on elimination by domination conditions. Furthermore, we compute the lower bound for the makespan introduced by [113]. Thus, if we find out a permutation having this makespan value, we stop searching on decreasing value on the C_{max} axis, and concentrate effort in exploring the direction of reducing $\sum C_i$, in the neighborhood of the permutation with $C^*_{\max}$. The lower bound is computed as:

$$LB(C_{\max}) = \max_j(\min_i \sum_{k=1}^{j-1} p_{ik} + \sum_{i=1}^{n} p_{ij} + \min_i \sum_{k=j+1}^{m} p_{ik} \tag{7}$$

The algorithm for the $F // C_{\max}$ problem (improvement over NEH), by Taillard [112], is actually a procedure to compute the value of the partial makespan when a job *i* is added in a partial schedule at position *k*. We employ this algorithm embedded in our neighbourhood search heuristic as a shortcut to evaluate a partial permutation. So, we do not need to compute the objective function for the complete schedule. Based on domination criteria for partial schedules [48, 71], we have developed our elimination neighbouring search.

Any partial schedule of *t* jobs, $J^p_{(t)}=\{J_1, J_2, \ldots J_t\}$, where $t=1, 2, \ldots, n$, is a sequence of the indexes corresponding to the jobs in $J^p_{(t)}$, and it could be named as $\sigma_I(J^p_{(t)})$. The completion time for a partial schedule $\sigma_I(J^p_{(t)})$ on machine *k*, where $k=1, 2, \ldots, m$, is denoted by $C(\sigma_I(J^p_{(t)}), k)$. It was proved that:

Elimination Criterion 1. If $C(\sigma_{II}(J^p_{(t)}), k) \leq C(\sigma_I(J^p_{(t)}), k)$ for k=1, 2, …, *m*, then $\sigma_{II}(J^p_{(t)})$ dominates $\sigma_I(J^p_{(t)})$.

For the case where $\sigma_{II}(J^{p}_{(t)})$ and $\sigma_{I}(J^{p}_{(t-1)})$ are partial schedules of $J^{p}_{(t)} \supset J^{p}_{(t-1)}$, being $J^{p}_{(t)}$-$J^{p}_{(t-1)}=\{J_j\}$, $\Delta_k=C(\sigma_{II}(J^{p}_{(t)}), k) - C(\sigma_{I}(J^{p}_{(t-1)}), k)$ is defined. It was proved that:

Elimination Criterion 2. If $\Delta_{k-1} \leq \Delta_k \leq p_{jk}$ for k=2, 3, …, m, then $\sigma_{II}(J^{p}_{(t)})$ dominates $\sigma_{I}(J^{p}_{(t-1)})$.

Both theorems allow us to discard any completion of a partial schedule $\sigma_{I}(J^{p}_{(t)})$ or $\sigma_{I}(J^{p}_{(t-1)})$, because a schedule at least as good exists among the completion of another partial schedule $\sigma_{II}(J^{p}_{(t)})$.

The improvement scheme proposed in this section is based on the sequential insertion of a job in the current sequence at each possible different position. Since jobs with larger total processing time at the beginning of the schedule bring, in general, schedules with less makespan value, the proposed scheme selects, for insertion, a subset of jobs which are located at the first β% of the total positions in the current sequence. Hence, the set of t jobs scheduled at {1, 2, …, t} positions, in the current permutation, σ_{Xk}, where $t=\beta\%n$, is selected for exploration consisting in checking whenever a better partial permutation, involving these t jobs, could be built.

Theoretically, we have to check and compare, for each job placed at i on σ_{Xk}, with i=1 to t, the makespan that results when this job is placed at a different position j, with j=1 to t. Nevertheless, the elimination criteria described above leads to efficiency gains. The Elimination Criterion 2 will filter any potential permutation generated by moving a job for which, to be placed at a different position with respect to its position in the current schedule, will not yield a sequence with less total completion time. Only for a potential permutation that passes this control, specified for a job to be moved, we check then for the different positions. By the Elimination Criterion 1, any potential schedule σ_Y, for which the current schedule σ_{Xk} is at least as good, will be discarded. If one partial schedule is not eliminated, then the corresponding complete schedule, σ_Y, becomes the generated neighbouring solution, $\sigma_{Xk}=\sigma_Y$, and the lower bounds used for computations are updated.

Improving Total Flow Time

Following the ideas of [43, 90], we have developed an improvement heuristic looking for permutations with less total flow time values, but attempting not to loss the level obtained in makespan. The original idea was to minimize gaps between successive operations that would lead to a better quality solution. The pair of jobs with the most positive gaps has to be placed at the beginning, while the pair of jobs with the most negative gaps has to be placed at the end of the schedule. During the total processing of the whole set of jobs, the larger gaps would have more chance of being compensated with the negative gaps corresponding to the pair of jobs scheduled at the end of the sequence.

In order to improve the quality of solutions in total flow time measure the following heuristic is implemented.

The improvement scheme proposed in this section is based on the interchange of adjacent pairs of jobs with positive gaps in the current permutation. Since the objective is to minimize gaps, the jobs are listed in descending order of gaps. Exchanging adjacent jobs with larger gaps is more likely result in a permutation which yields less flow time value. The procedure selects, for exchanging, a subset of jobs, $J^{p}_{(t)}$, which are located at the later β% of the total positions in the current sequence σ_{Xk}. The exploration consists in checking whenever a new permutation obtained by exchanging an adjacent pair of jobs of $J^{p}_{(t)}$, yields a schedule with less flow time value.

Step 1: The subset of jobs placed at the last t positions of σ_{Xk}, where t=β%n, is selected to constitute the set $J^{p}_{(t)}$.
Step 2: For the jobs of $J^{p}_{(t)}$, the gaps between every pair of adjacent jobs in σ_{Xk}, is then computed as

$$G_i = \sum_{j=1}^{m} p_{ij} - \sum_{j=1}^{m} p_{i+1j} \quad \text{, for } i=n-t,\ n-t+1,\ldots,\ n-1 \tag{8}$$

Step 3: Jobs in $J^{p}_{(t)}$ are listed in descending order of gaps G_i. Jobs with negative gaps are not included in the list, and ties are broken in descending order of this similar gap:

$$G_i^{'} = \sum_{j=1}^{m} (m-j+1)p_{ij} - \sum_{j=1}^{m} (m-j+1)p_{i+1j} \quad \text{, for } i=n-t,\ n-t+1,\ldots,\ n-1 \tag{9}$$

that is computed only in the case of a tie.
Step 4: The first job in the list, $J_{\sigma Xk(i)}$, scheduled at position i in the current permutation σ_{Xk}, will be set at i+1, in a new permutation σ_{II}, and its counterpart, the job placed at i+1 in σ_{Xk}, will be set at position i in σ_{II}.
Step 5: If $\exists j / C_{i,j}(\sigma_{II})+C_{i+1,j}(\sigma_{II})<C_{i,j}(\sigma_{Xk})+C_{i+1,j}(\sigma_{Xk})$, or $C_{i,m}(\sigma_{II})<C_{i+1,m}(\sigma_{Xk})$, then σ_{II} is accepted as a new permutation, $\sigma_{Xk} = \sigma_{II}$, then return to Step 1. Otherwise proceed to Step 6.
Step 6: Remove $J_{\sigma Xk(i)}$ from the list of jobs. If the list is not exhausted, then return to Step 4.

Perturbation of X_k
In our algorithm, we have implemented the following two perturbation schemes:

Scheme A
In this simple procedure one of these three different procedures is randomly chosen.

Swapping
Two integer numbers, i and k, in the range (1, 2, … n), are chosen. The job at i will be set at position k, and the job at k will be set at position i.

Insertion
Two integer numbers, i and k, in the range (1, 2, … n), are chosen. Job at i will be inserted at position k. If $i<k$, then the job at k will be set at position k-1. However, if $i>k$, then the job at k will be set at position k+1.

At random
Randomly, swapping or insertion is chosen.

Scheme B

Based on the Random Insertion Perturbation Scheme, introduced by [83], we have developed a perturbation scheme that explores the neighbourhood of the current permutation, X_k, and yields a neighbor Y with a good objective value in conformance with a preferable criterion. According to the three kinds of movement strategies defined previously, this criterion will be determined. Let z_i be the preferable criterion. For the permutation $\sigma_{Xk}=\{I_1, I_2,\ldots I_n\}$, where I_i, with i=1, 2,...n, is the index of the job scheduled at the position i in σ_{Xk}, we will check its neighbourhood for finding out a good neighbor with respect to z_i. As it is known that insertion brings better improvement than exchange, this procedure generates potential permutations by inserting, forward and backward, removing each job J_{Ii}, and reinserting it in a different position at random. For each job, J_{Ii}, where $I_i \neq I_1$, and $I_i \neq I_n$, the procedure will choose randomly two positions for insertion. One position to its right, choosing randomly a number between i+1 and n, for forward insertion, and another position to its left, choosing randomly a number between 1 and i-1, for backward insertion. For jobs in extreme positions, I_1 and I_n, only one direction of insertion can be chosen. For I_1 only forward insertion is possible to apply, hence, to select a new position, a random number between 2 and n must be generated. In a similar way, I_n can only be inserted at positions to its left, so a number between 1 and n-1 has to be chosen. Thus, the z_i value of the 2(n-1) potential permutations has to be evaluated and the permutation with minimum z_i, is then selected as the neighbouring solution Y.

5.4 Updating Potential Efficient Set

When a neighbouring solution Y is accepted and made the current solution, $X_k=Y$, the set of PE solutions should be updated. If X_k is a new non-dominated solution, it should be added to the archive set and the archive set should be updated. Any solution dominated by the added one will be removed from the set. For the efficiency of this algorithm the updating PE process is crucial.

In order to save computational effort in updating PE, non-dominated solutions are always stored in ascending order of one of the criterion values, thus their other criterion values will be in descending order.

This arrangement constitutes a fast method of finding out dominated instances with respect to the new solution, and of updating the PE.

5.5 Simulated Annealing Parameters

The acceptance rule is essential in an SA algorithm. p is the probability, for a dominated permutation, of being admitted to the PE set. This probability is computed by $p = \exp\left(-\frac{\Delta\phi}{T_n}\right)$, where the numerator of the exponent evaluates the candidate solution, and the denominator is the temperature at any iteration. In the SA technique, temperature is reduced at every step of iterations. This cooling process makes the possibility of admitting a dominated solution to be decreased during the search process. By means of this high probability, at the beginning of the process, one attempts to avoid being trapped in a local optimum.

The deviation function for computing this probability is normalized as follows:

$$\Delta\phi = \sum_k \left(\frac{z_k(y)-z_k(x_n)}{z_k(x_n)}\right)^2 \times 100 \qquad (10)$$

With this normalization we diminish the influence of the different dimensions of the criteria, hence we have a dimensionless quantity which indicates the relative deviation of the quality of the generated solution, Y, from that of the current one, X_n. Since $\Delta\phi$ is not dependent upon the instance size, the initial and final temperature values can be fixed more reasonably and accurately to minimize the computational effort without sacrificing the quality of the final solution.

Similar SA parameters have been employed by previous SA applications. Particularly the single objective algorithm presented by [83] and the MOSA procedures introduced by [70] and [116], in which the present work has found inspiration.

After a study carried out by varying SA parameters we have determined the setting values. Here we point out some aspects of them.

Initial temperature should permit the acceptance of inferior quality solutions. The algorithm starts with a temperature value of 475, and finishes when the temperature is below 20. This value is set to limit the inferior quality of acceptance of a generated permutation by 50%. This means that the probability of accepting a solution with deviation of performance criteria of 50% is 0.9 at the beginning of the iterations and 0.08 at the later iterations. The temperature will be reduced by the factor α=0.968. This reduction takes place at every *length of step* iterations (with or without improvements). Thus, the temperature will be at 100 different steps ($T_f=\alpha^{100}T_0$). In order to control the computational effort a stopping criterion must be fixed, thus the number of iterations without improvement, N_{stop} has been fixed. Furthermore, the length of the temperature step, N_{step}, is essential in driving the cooling process. After the mentioned analysis, we have fixed the following values: N_{stop}=2500 and N_{step} =500.

6 Evaluation of MOCO Approaches

6.1 Metrics

For the MOO algorithms, the analysis of performance is more complex than for single-objective ones. The goal of multiple objective metaheuristic procedures is to find a good approximation of the set of efficient solutions. It is unlikely that the whole set of efficient solutions (E) is fully known. While the outcomes from compared algorithms are different, they can still be all equally Pareto efficient.

Usually, the three following conditions are considered as desirable for a good multi-objective algorithm:

1. The distance of the obtained PE solutions to the E should be minimized.
2. The distribution of the solutions found should be uniform.
3. The larger the number of obtained solutions, *i.e.* the cardinality of PE, the better the algorithm.

The last two conditions present more weaknesses than strengths. If E does not present a uniform distribution, or $[E]=1$, the algorithm that obtains the proper E will not fulfill conditions 2 and 3. Furthermore, an algorithm that just reports a huge number of solutions does not ensure their quality (in terms of efficiency). To have an idea of quality, a reference set of E (R in the following) should be considered. The ideal R is the set E. However, for MOCO problems it is unlikely that the whole E is known (except for small size instances, with non-practical application). A useful practice is having a set R as close to E as possible, then filtering the PE output with R. The obtained net set of non-dominated solutions in the net set is $N=\{X$ is Pareto efficient in $(PE \cup R)\}$, and it will be at least as good as R. One can measure the quality of the output as the percentage of solutions in PE that survive the filtering process with the R set:

$$Q_1(PE) = \frac{[PE \cap N]}{[PE]} 100\% \qquad (11)$$

[20] presents a quality measure of the percentage of reference solutions found by the algorithm:

$$Q_2(PE) = \frac{[PE \cap R]}{[R]} 100\% \qquad (12)$$

Both of the above metrics are cardinal. However, in the case of real-life MOCO problems it may be impossible to obtain, in a reasonable time, a significant percentage of efficient solutions. Obtaining near-efficient solutions would also be highly appreciated. Following [57], a more general and economic criterion may be to concentrate on evaluating the distance of solutions to the efficient frontier. The C metric by [121], and the *Dist1R* and *Dist2R* metrics by [20], can serve this purpose. We have chosen them because they are not difficult to compute, and they seem to be complementary (to each other) with respect to the properties analyzed by [57].

The C metric, also a cardinal measure, compares two sets of PE, A and B. A reference set, R, is not required and it is really easy to compute as:

$$C(A,B) = \frac{[b \in B / \exists a \in A : a \prec b]}{[B]} \qquad (13)$$

The following statements can aid the understanding of $C(A,B)$:

- If $C(A,B)=1$, all solutions in B are weakly dominated by A.
- If $C(A,B)=0$, none of the solutions in B are weakly dominated by A.

When two algorithms are compared, $C(A,B)$ and $C(B,A)$ must be computed, because they are not necessary complementary. Unless $C(A,B)=1$ and $C(B,A)<1$, it is not possible to establish that A weakly outperfoms B.

As non-cardinal measures we have the *Dist1R* and *Dist2R*, but in obtaining them, R is required. Their computations, although more complicated than C, do not imply a high complexity. They are based on an achievement scalarizing function:

$$d(X,Y) = \max_{k=1,\dots K} \{0, \lambda_k (z_k(Y) - z_k(X))\} \tag{14}$$

where $X \in R$, $Y \in$ PE, and $\lambda_k = \dfrac{1}{\left(\max\limits_{X \in R} z_k(X) - \min\limits_{X \in R} z_k(X) \right)}$

Dist1R is defined as:

$$Dist1_R(PE) = \frac{1}{[R]} \sum_{x \in R} \left\{ \min_{Y \in PE} \{d(X,Y)\} \right\} \tag{15}$$

While *Dist1R* measures the mean distance, over the points in *R*, of the nearest point in PE, *Dist2R* gives the worst case distance, thus is defined as:

$$Dist2_R(PE) = \max_{X \in R} \left\{ \min_{Y \in PE} \{d(X,Y)\} \right\} \tag{16}$$

The lower the values the better PE approximates *R*. Moreover, the lower the ratio *Dist2R*/ *Dist1R* the more uniform the distribution of solutions from PE over *R*. *Dist1R* induces a complete ordering and let to weak outperformance relations.

Combining PE yielded by different algorithms, a net set of non-dominated solutions, *N*, for an instance problem is obtained. The *N* set is very useful as a reference for many evaluations of new developments. An important contribution is updating the published *N* set obtained for benchmark instances.

6.2 Computational Experiments

The proposed methods have been investigated with respect to their effectiveness in solving 90 test instances presented in [113], with the number of jobs varying from 20 to 100, and the number of machines varying from 5 to 20. Each setting of the algorithm has been tested in each of these instances. The quality of the obtained approximations is analyzed regarding the Q_1(PE), Q_2(PE), *C*(*A*,*B*), *Dist1R* and *Dist2R* measures described in the previous section.

To compare the performance of the proposed improvement approaches we have also implemented the Job-Index-Based Insertion Scheme (JIBIS), Overall-Seed Sequence-Based Insertion Scheme (OSSBIS), and Job-Index-Based Swap Scheme (JIBSS), employed in MOSAI and MOSAII [116], and PGA-ALS [84]. Thus, we have 5 variants considering the improvement technique: none improvement (*N*), direct search (D), cross search (I), combining search (C), JIBIS-OSSBIS-JIBSS (J). For perturbation, we have tested the two schemes: A and B, described previously. In our experiment we have tried with initial solution sets of 2, 4 and 8 points. In order to confirm the hypothesis of the superiority of simultaneous optimization (S) versus aggregated function (A), we have also tested a model where the objective function is the weighted sum of the makespan and the flow time, generating different weight vectors (λ_1, λ_2), with λ_1>1, λ_2>1, and $\lambda_1+\lambda_2=1$. The scheme presented in Table 1 describes how the proposed algorithms are coded. In the following we refer to them with their corresponding code.

With the PE for the 90 instances obtained by means of all these algorithms, we have built a net set to be used as reference (*R*) for this computational experiment.

Table 1. Code for the proposed algorithms

Improvement	Perturbation	Initial Solutions	O.F.	CODE
N	B	4	S	NB4S
I	B	4	S	IB4S
D	B	4	S	DB4S
C	B	4	S	CB4S
J	B	4	S	JB4S
I	B	2	S	IB2S
I	B	8	S	IB8S
J	B	2	A	JB2A
D	B	2	A	DB2A
J	A	4	S	JA4S
D	A	4	S	DA4S
N	A	4	S	NA4S

In column Improvement: *N* means no-improvement; I means cross search; D means direct search; C means combined search; and J means JIBIS-OSSBIS-JIBSS. In column Perturbation: A/B means that scheme A/B has been applied. In column Initial Solutions the number of the initial seeds is indicated. In column O.F.: S means simultaneous optimization, and A means aggregated function. The final column indicates the acronym of the algorithm in each row.

We have also updated the net sets for the cases published in [116, 84]. We have made a net based on results from:

1. Net of MOSA I, MOSA II, GPWGA [16], *a posteriori* [30], MOGLS [51], ENGA [6], published in [116] (size instances: 20x20, 50x20, and 100x20).
2. Net of PGA-ALS, MOGLS, ENGA, GPWGA, published in [84] (size instances: 50x5, 50x10, 50x20, 100x5, and 100x10).
3. PE of the proposed algorithms (size instances: 20x5, 20x10, 20x20, 50x5, 50x10, 50x20, 100x5, 100x10, and 100x20).

Tables 2 shows the net sets corresponding to the size instance problems 50x20 and the proportion of the solution on the final net contributed by every algorithm is reported in Table 3. Only as an example we comment on an experiment with Problem 1, instance size 50x20. The net set published in [116] has been updated with the net set published in [84]. Then, the resulting net set has been updated just with the output of one of our proposed algorithm, CB4S (details are showed in Table 4). The interesting significance of this test is that, in spite of the robustness of MOSAI, MOSAII and PGA-ALS, the proposed algorithm is superior to them in the sense that it does not give a large percentage of dominated solutions in the resulting PE.

In results presented by [116], after comparing with the PE sets obtained from different algorithms (updating net sets), solutions from MOSA I, GPWGA, *a posteriori* and ENGA, are null. The number of reported solutions of MOSA II, after filtering (one can suppose, before it was even superior), is 23 and just 7 of them persisted the

Table 2. Net set of non-dominated solutions obtained from various multi-objective flow shop scheduling algorithms for the benchmark problems given by [113], size (50×20)

Problem 1		Problem 2		Problem 3		Problem 4		Problem 5	
C_{max}	$\sum C_i$	C_{max}	$\sum C_i$	C_{max}	$\sum C_i$	C_{max}	$\sum C_i$	C_{max}	$\sum C_i$
3900	136088 03	3742	128765 09	3697	123829 08	3769	128724 09	3664	126889 09
3901	135335 09	3744	128585 09	3700	122933 13	3773	128259 09	3667	125957 02
3905	133503 13	3745	127960 09	3724	122919 02	3775	128253 09	3670	125939 02
3912	133422 13	3747	127168 09	3746	120940 04	3792	127464 02	3674	125909 02
3913	133178 13	3754	126954 09	3766	120826 04	3807	127401 11	3679	125842 09
3915	133172 13	3765	126754 09	3768	120822 04	3810	127165 13	3684	125228 09
3953	131506 06	3772	126710 13	3782	120684 04	3811	127062 13	3700	125195 09
3954	131481 06	3775	126332 13	3792	120654 04	3816	127055 13	3701	125190 09
3958	131153 06	3777	126223 13	3795	120321 06	3850	126443 06	3704	125099 07
3997	130965 01	3800	125940 04	3804	119927 01	3851	126357 06	3713	124617 07
3998	130351 01	3840	125783 04	3810	119884 01	3853	126311 06	3714	124558 04
4018	130311 01	3847	125781 04	3812	119853 01	3860	125953 06	3735	124062 01
4022	130283 01	3854	124417 04	3814	119842 01	3863	125008 01	3746	124061 01
4030	130076 01	3860	124336 04	3817	119150 01	3874	124431 01	3747	123698 01
4031	129835 01	3864	123367 06	3818	119050 01	3886	124197 01	3761	123222 01
4036	129807 01	3869	123282 06	3822	118978 01	3888	124149 01	3767	121800 01
4049	129451 01	3872	123049 06	3824	118925 01	3900	123599 06	3786	121568 01
4068	129436 01	3875	123025 06	3825	118894 01	3910	123524 06	3862	121357 01
4164	129332 06	3925	122972 04	3842	118880 01	3982	123506 06	3864	121241 01

Table 2. *(continued)*

4173	129309	06	3930	122878	04	3853	118777	01	3984	123247	06	3868	121203	01
4175	129295	06	3950	122766	01	3859	118761	01	3985	123211	06	3875	121186	01
4181	129293	06	3954	122663	01	3879	118731	01	3993	123102	06	3878	121172	01
4189	129212	05	3964	122477	01	3882	118628	01	4019	123010	06	3887	121079	01
4203	129137	05	3968	122430	01	-	-	-	4072	122959	06	3902	121047	06
4216	129088	05	3983	122374	01	-	-	-	4076	122904	06	3905	121014	06
4218	129040	05	3986	122366	01	-	-	-	4127	122886	06	3912	120901	06
4220	129034	05	4020	122309	01	-	-	-	4150	122847	06	3927	120624	01
4221	128982	05	4060	121870	06	-	-	-	-	-	-	3964	120780	01
4257	128888	05	4072	121708	06	-	-	-	-	-	-	4022	120475	02
-	-	-	4079	121556	06	-	-	-	-	-	-	-	-	-
-	-	-	4088	121443	06	-	-	-	-	-	-	-	-	-
-	-	-	4099	121383	06	-	-	-	-	-	-	-	-	-
-	-	-	4104	121275	06	-	-	-	-	-	-	-	-	-
-	-	-	4109	121175	06	-	-	-	-	-	-	continued on next page		

For each problem: In the 1st column makespan is indicated. In the 2nd column total flow time is indicated. In the 3rd column the algorithm that yielded the corresponding solution is indicated as follows: 1 is PGA-ALS; 2 is IB8S ; 3 is CB4S; 4 is MOSAI; 5 is MOGLS; 6 is MOSAII; 7 is DB4S; 8 is IB4S; 9 is JB4S; 10 is NB4S;11 is DB2A; 12 is GPWGA; and 13 is JB2A.

Table 2. *(continued)*

Problem 6		Problem 7		Problem 8		Problem 9		Problem 10	
C_{max}	$\sum C_i$	C_{max}	$\sum C_i$	C_{max}	$\sum C_i$	C_{max}	$\sum C_i$	C_{max}	$\sum C_i$
3724	130563 10	3763	131057 10	3785	129403 03	3826	133374 02	3815	131662 07
3725	130556 10	3767	131054 10	3787	129401 03	3829	133277 02	3820	131439 07
3727	130497 10	3770	131049 10	3796	129312 03	3830	130069 06	3822	131148 07
3731	130480 10	3773	130828 10	3803	129301 08,10	3833	129452 06	3826	131146 07
3734	130231 10	3778	130658 10	3805	129283 03	3848	129432 06	3828	131031 07
3736	130080 07	3781	130635 10	3807	128778 06	3862	129262 06	3838	130964 07
3739	129647 07	3791	130614 10	3808	128629 06	3864	128145 01	3840	130821 07
3740	129176 08	3792	130605 10	3816	128484 06	3872	128139 01	3851	130743 10
3743	128707 08	3796	130143 10	3817	128451 06	3879	127913 01	3854	130469 04
3745	128480 08	3804	129612 04	3819	127629 06	3880	127654 01	3855	128334 01
3746	128471 07	3817	129582 01	3823	127583 06	3883	127306 01	3869	128130 01
3754	128010 08	3818	128781 01	3836	127515 06	3896	127209 01	3896	127440 04
3782	127931 01	3823	128778 01	3854	127135 04	3905	127141 01	3907	127398 04
3790	127698 01	3829	128396 01	3860	126727 04	3916	126976 01	3914	126755 06
3791	126759 01	3833	128313 01	3865	125871 01	3917	126582 06	3915	126697 06
3807	126439 01	3837	128277 01	3869	125840 01	3921	126454 06	3918	126694 06
3810	126346 01	3840	127916 01	3878	125759 01	3924	126150 01	3922	126618 06
3814	126314 01	3841	127912 01	3884	125579 01	3933	125771 06	3936	126462 06
3821	125839 01	3854	127910 01	3889	125331 01	3954	125619 01	3952	126423 06
3827	125682 01	3867	126695 01	3891	125236 01	3965	125355 06	4046	126388 06
3834	125579 01	3892	126540 01	3894	125203 01	3974	125334 06	-	- -
3836	125567 01	3894	126455 01	3901	125079 01	3981	125192 06	-	- -

Table 2. (*continued*)

3840	125468	01	3902	126336	01	3905	125063	01	3982	125187	06	-	-	-
3841	125462	01	3927	126311	01	3910	125024	01	3983	125142	06	-	-	-
3843	125438	01	3953	126063	04	3918	124859	01	3984	125137	06	-	-	-
3845	125422	01	3954	126045	04	3929	124794	01	3988	125111	06	-	-	-
3848	124824	06	4032	125949	04	3931	124760	01	3990	125031	06	-	-	-
3862	124633	06	4098	125884	01	3935	124742	01	3991	125026	06	-	-	-
3878	124553	06	4145	125853	07	4076	124696	01	4004	124874	06	-	-	-
3886	124538	06	4147	125852	07	4127	124677	01	4006	124857	06	-	-	-
3891	123973	06	4269	125835	11	4131	124539	07	4013	124701	06	-	-	-
3933	123824	06	4273	125827	11	4161	124529	07	4040	124618	06	-	-	-
3944	123793	06	-	-	-	4176	124518	07	4043	124467	06	-	-	-
3946	123766	06	-	-	-	-	-	-	4044	124385	06	-	-	-
3956	123683	06	-	-	-	-	-	-	4052	124326	06	-	-	-
3959	123601	06	-	-	-	-	-	-	4056	124317	06	-	-	-
3964	123592	06	-	-	-	-	-	-	4143	124299	06	-	-	-
3979	123535	06	-	-	-	-	-	-	-	-	-	-	-	-
4008	123163	06	-	-	-	-	-	-	-	-	-	-	-	-
4011	123121	06	-	-	-	-	-	-	-	-	-	-	-	-
4052	123093	06	-	-	-	-	-	-	-	-	-	-	-	-
4053	123057	06	-	-	-	-	-	-	-	-	-	-	-	-
4116	122998	06	-	-	-	-	-	-	-	-	-	-	-	-
4156	122910	06	-	-	-	-	-	-	-	-	-	-	-	-
4177	122884	06	-	-	-	-	-	-	-	-	-	-	-	-

For each problem: In the 1st column makespan is indicated. In the 2nd column total flow time is indicated. In the 3rd column the algorithm that yielded the corresponding solution is indicated as follows: 1 is PGA-ALS; 2 is IB8S ; 3 is CB4S; 4 is MOSAI; 5 is MOGLS; 6 is MOSAII; 7 is DB4S; 8 is IB4S; 9 is JB4S; 10 is NB4S;11 is DB2A; 12 is GPWGA; and 13 is JB2A.

Table 3. Proportion of the solution on the final net contributed by every algorithm for the benchmark problems given by [113], size (50×20)

Algorithm	Problem 1	Problem 2	Problem 3	Problem 4	Problem 5	Problem 6	Problem 7	Problem 8	Problem 9	Problem 10
1-PGA-ALS	31,03	20,59	60,87	14,81	48,28	31,11	46,88	47,06	27,03	10,00
2-IB8S	0,00	0,00	4,35	3,70	13,79	0,00	0,00	0,00	5,41	0,00
3-CB4S	3,45	0,00	0,00	0,00	0,00	0,00	0,00	14,71	0,00	0,00
4-MOSA I	0,00	20,59	21,74	0,00	3,45	0,00	12,50	5,88	0,00	15,00
5-MOGLS	24,14	0,00	0,00	0,00	0,00	0,00	0,00	0,00	0,00	0,00
6-MOSA II	24,14	32,35	4,35	55,56	10,34	42,22	0,00	20,59	67,57	35,00
7-DB4S	0,00	0,00	0,00	0,00	6,90	6,67	6,25	8,82	0,00	35,00
8-IB4S	0,00	0,00	4,35	0,00	0,00	8,89	0,00	2,94	0,00	0,00
9-JB4S	3,45	17,65	0,00	11,11	17,24	0,00	0,00	0,00	0,00	0,00
10-NB4S	0,00	0,00	0,00	0,00	0,00	11,11	28,13	0,00	0,00	5,00
11-DB2A	0,00	0,00	0,00	3,70	0,00	0,00	6,25	0,00	0,00	0,00
12-GPWGA	0,00	0,00	0,00	0,00	0,00	0,00	0,00	0,00	0,00	0,00
13-JB2A	13,79	8,82	4,35	11,11	0,00	0,00	0,00	0,00	0,00	0,00

For each problem, the figures indicate the percentage of solutions on the net obtained by each algorithm.

Table 4. The process of updating Net set published by [84] and [116], with the output of one of the proposed algorithm: CB4S, for the instance: Problem 1, size 50x20

NET			NET 1			NET 2			PE(DB4S)		
C_{max}	$\sum C_i$		C_{max}	$\sum C_i$		C_{max}	$\sum C_i$		C_{max}	$\sum C_i$	
4036	129807	PGAALS	3928	138212	MOSAII	4182	129314	PGAALS	4267	129205	-
4031	129835	PGAALS	3929	138137	MOSAII	4036	129807	PGAALS	4233	131799	-
4018	130311	PGAALS	3932	138095	MOSAII	4031	129835	PGAALS	3976	133707	-
4049	129451	PGAALS	3936	138078	MOSAII	4018	130311	PGAALS	3966	133753	-
4068	129436	PGAALS	3938	138030	MOSAII	4049	129451	PGAALS	3957	133855	-
4030	130076	PGAALS	3953	131506	MOSAII	4068	129436	PGAALS	3956	133867	-
4022	130283	PGAALS	3954	131481	MOSAII	4030	130076	PGAALS	3955	133914	-
3998	130351	PGAALS	3958	131153	MOSAII	4022	130283	PGAALS	3921	134022	*
3997	130965	PGAALS	4009	130558	MOSAII	3965	133658	PGAALS	3920	134074	*
3953	131506	MOSAII	4037	130317	MOSAII	3984	131387	PGAALS	3919	134107	*
3954	131481	MOSAII	4060	130217	MOSAII	3973	131538	PGAALS	3916	134423	*
3958	131153	MOSAII	4067	130153	MOSAII	3971	131728	PGAALS	3912	134424	*
4164	129332	MOSAII	4071	130110	MOSAII	3969	131922	PGAALS	3911	134438	*
4173	129309	MOSAII	4084	130099	MOSAII	3967	132131	PGAALS	3907	134612	*
4175	129295	MOSAII	4098	130083	MOSAII	3965	132133	PGAALS	3905	134995	*
4181	129293	MOSAII	4100	129685	MOSAII	3962	132782	PGAALS	3901	135529	*
4189	129212	MOGLS	4105	129632	MOSAII	3998	130351	PGAALS	3900	136088	*
4203	129137	MOGLS	4108	129572	MOSAII	3996	131378	PGAALS	-	-	-
4216	129088	MOGLS	4137	129525	MOSAII	3997	130965	PGAALS	-	-	-
4218	129040	MOGLS	4164	129332	MOSAII	-	-	-	-	-	-
4220	129034	MOGLS	4173	129309	MOSAII	-	-	-	-	-	-
4221	128982	MOGLS	4175	129295	MOSAII	-	-	-	-	-	-
4257	128888	MOGLS	4181	129293	MOSAII	-	-	-	-	-	-
3921	134022	PE(CB4S)	4189	129212	MOGLS	-	-	-	-	-	-
3920	134074	PE(CB4S)	4203	129137	MOGLS	-	-	-	-	-	-
3919	134107	PE(CB4S)	4216	129088	MOGLS	-	-	-	-	-	-
3916	134423	PE(CB4S)	4218	129040	MOGLS	-	-	-	-	-	-
3912	134424	PE(CB4S)	4220	129034	MOGLS	-	-	-	-	-	-
3911	134438	PE(CB4S)	4221	128982	MOGLS	-	-	-	-	-	-
3907	134612	PE(CB4S)	4257	128888	MOGLS	-	-	-	-	-	-
3905	134995	PE(CB4S)	-	-	-	-	-	-	-	-	-
3901	135529	PE(CB4S)	-	-	-	-	-	-	-	-	-
3900	136088	PE(CB4S)	-	-	-	-	-	-	-	-	-

Each solution is followed by the acronym of the algorithm that has yielded it. Exception is made in column corresponding to CB4S, because it shows the output of CB4S, not a Net set. Asterisks in this column indicate a solution that is still in the final net set.

Table 5. Averages values of distance measures *Dist1R*, *Dist2R*, *Dist1R/Dist2R* for IB4S, DB4S, CB4S, JB4S, IB2S and IB8S

n	*m*	IB4S	-	-	DB4S	-	-	CB4S	-	-	JB4S	-	-	IB2S	-	-	IB8S	-	-
-	-	Dist1	Dist2	Dist1/Dist2	Dist1	Dist2	Dist1/Dist2	Dist1	Dist2	Dist1/Dist2	Dist1	Dist2	Dist1/Dist2	Dist1	Dist2	Dist1/Dist2	Dist1	Dist2	Dist1/Dist2
20	5	37,51	37,83	0,63	0,22	0,49	0,53	37,41	37,77	0,54	37,76	38,07	0,69	37,69	37,96	0,74	36,35	36,66	0,49
20	10	0,64	0,80	0,58	50,03	50,21	0,55	0,27	0,41	0,52	14,92	15,08	0,63	0,61	0,85	0,60	0,22	0,38	0,53
20	20	0,19	0,38	0,42	0,26	0,51	0,46	0,20	0,42	0,44	0,26	0,49	0,46	0,19	0,35	0,55	0,06	0,16	0,31
50	5	0,28	0,50	0,53	0,17	0,38	0,42	0,25	0,40	0,53	0,54	0,70	0,74	0,35	0,60	0,57	0,23	0,43	0,51
50	10	0,27	0,44	0,53	0,10	0,29	0,33	0,26	0,41	0,54	0,34	0,51	0,63	0,37	0,60	0,59	0,17	0,40	0,36
50	20	0,21	0,36	0,57	0,13	0,32	0,39	0,13	0,30	0,44	0,22	0,36	0,55	0,30	0,46	0,60	0,07	0,19	0,33
100	5	0,15	0,41	0,34	0,19	0,40	0,40	0,23	0,45	0,44	0,36	0,69	0,44	0,48	0,73	0,46	0,11	0,28	0,48
100	10	0,07	0,22	0,22	0,04	0,21	0,18	0,02	0,09	0,48	0,08	0,24	0,34	0,08	0,25	0,32	0,06	0,16	0,65
100	20	0,12	0,26	0,45	0,16	0,36	0,54	0,14	0,35	0,32	0,17	0,31	0,50	0,20	0,40	0,39	0,05	0,14	0,37

Table 6. Average values of distance measures *Dist1R*, *Dist2R*, and *Dist1R/Dist2R* for DA4S, JA4S, NA4S, NB4S, JB2A and DB2A

n	*m*	DA4S -		-	JA4S -		-	NA4S-		-	NB4S-		-	JB2A -		-	DB2A-		-
-	-	Dist1	Dist2	Dist1/Dist2	Dist1	Dist2	Dist1/Dist2	Dist1	Dist2	Dist1/Dist2	Dist1	Dist2	Dist1/Dist2	Dist1	Dist2	Dist1/Dist2	Dist1	Dist2	Dist1/Dist2
20	5	43,39	43,65	1,27	41,91	42,22	0,79	43,55	43,84	0,84	37,77	38,00	0,74	38,40	38,72	0,70	40,87	41,13	0,78
20	10	60,64	60,86	1,32	56,62	56,92	0,68	60,67	60,88	0,80	45,69	45,82	0,70	105,27	105,64	0,58	36,48	36,81	0,59
20	20	0,87	1,07	1,25	0,67	0,89	0,74	0,88	1,08	0,81	0,21	0,40	0,50	0,44	0,80	0,54	0,48	0,79	0,57
50	5	0,59	0,86	1,52	0,84	0,97	0,86	0,88	1,05	0,84	0,52	0,73	0,68	0,84	1,02	0,81	0,80	0,98	0,79
50	10	0,75	1,08	1,49	0,77	0,99	0,76	0,95	1,20	0,78	0,31	0,55	0,54	0,58	0,89	0,64	0,51	0,79	0,62
50	20	0,77	1,01	1,33	0,72	0,95	0,76	0,87	1,07	0,80	0,21	0,41	0,49	0,58	0,87	0,67	0,57	0,85	0,65
100	5	0,44	0,82	3,33	0,55	0,99	0,47	0,72	1,18	0,54	0,37	0,72	0,50	0,80	1,24	0,52	0,82	1,28	0,52
100	10	0,21	0,49	3,34	0,25	0,54	0,43	0,32	0,66	0,46	0,06	0,18	0,50	0,25	0,59	0,40	0,25	0,59	0,40
100	20	0,53	0,80	2,01	0,50	0,82	0,56	0,61	0,91	0,61	0,18	0,35	0,39	0,40	0,78	0,49	0,40	0,78	0,49

updating process. On the other hand, MOGLS presents 7 solutions in results published in [116], and all of them are still in our net set.

In a similar way, results reported by [84], after comparing PGA-ALS with the PE of MOGLS, ENGA and GPWGA, only PGA-ALS survived, and the number of PGA-ALS solutions in the published results for the net set is 19, of which, 9 are still in the net set after updating. (The different results for MOGLS, ENGA and GPWGA, in both papers, is surely due to the different parameter setting).

To have a numerical idea of these comments, the percentage of the number of solutions in the final net set over the number of solutions in the set before filtering are the following:

MOSA I=0%, MOSA II=30%, GPWGA=0%, *a posteriori*=0%, MOGLS=100%, ENGA=0%, PGA-ALS =47%, CB4S=59%. (Considering other of the proposed algorithms this tendency is similar, *e.g.* the corresponding percentage for IB4S is 53, and for DB4S is 75).

It is important to note that the PE considered for CB4S (IB4S or DB4S) is just the output of the algorithm, and even though, when comparing with results from net sets, it is only outperformed by MOGLS. (Only for CB4S, IB4S, and DB4S algorithms the computed percentage values coincide with Q_1(PE)).

The advantage of yielding a large PE set (as by MOSAI, MOSAII and PGA-ALS algorithms) is the possibility of covering the efficient frontier with a more diversified set of solutions, even though they are not efficient. To clarify this idea, we present, in Fig. 1, the efficient frontier obtained for the Problem 10, size 50x20, from the Taillard's benchmarks. In spite of having a high percentage of non-efficient solutions, PGA-ALS gives a wide set of near-efficient solutions.

In order to evaluate the diversification, we use the DistR_1, DistR_2 and $\text{DistR}_1/\text{DistR}_2$ metrics. To compute these metrics, the complete set of PE is required. In Tables 5 and 6 we present the average results for each size of the 90 benchmark instances obtained for the variants of the proposed algorithms.

We have also computed Q_1 and Q_2 metrics and for pair-wise comparison between different algorithms, $C(A,B)$ have been calculated. Because of limited space only average figures of the obtained results are presented (Table 7 and 8 present the average values of Q_1 and Q_2, respectively). However, we comment on the most important results. It could be observed that the rules direct search (DB4S) and combined search (CB4S) yield more solutions which are kept in the final net set (efficient). On the other hand, cross search (IB4S) and JIBIS-OSSBIS-JIBSS (JB4S) give similar results. JB4S was implemented, following [116], in cross search movement strategy. One can conclude that this way of searching is less efficient for the MOSA scheme presented in this paper. Even with NB4S (none improvement), the obtained results are the same, when Q_2 figures are observed.

Going deeply into quality relations between these five manners of improvement, we have the $C(A,B)$ and $C(B,A)$ measures that make clear how many solutions provided by A are dominated by solutions from B, and vice versa. Tables 9, 10 and 11 show the comparison between each pair of techniques. On average, we can affirm that any improvement is better than none (NB4S shows the worst figures in Table 10). However the outperformance of JB4S over NB4S is negligible. IB4S, DB4S and CB4S are superior to JB4S (see Table 9). CB4S shows its prominence with respect to

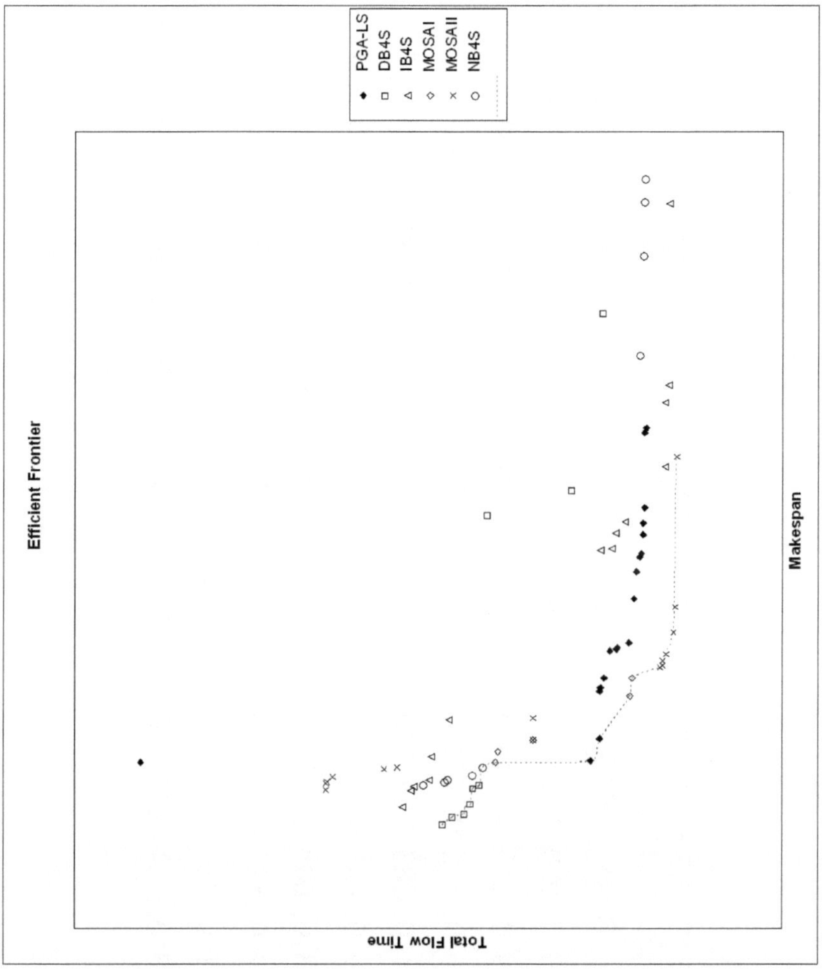

Fig. 1. Efficient Frontier obtained for Problem 10, size (50x20) of Taillard's benchmarks

Table 7. Average values of the cardinal measures Q_1 for IB4S, DB4S, CB4S, JB4S, IB2S, IB8S, DA4S, JA4S, NA4S, NB4S, JB2A and DB2A

n	*m*	IB4S	DB4S	CB4S	JB4S	IB2S	IB8S	DA4S	JA4S	NA4S	NB4S	JB2A	DB2A
20	5	0,2750	0,4050	0,3950	0,0425	0,0793	0,3739	0,0000	0,0425	0,0000	0,0793	0,1750	0,0091
20	10	0,2659	0,1210	0,4846	0,1595	0,2603	0,5097	0,0000	0,1595	0,0000	0,2603	0,1250	0,0733
20	20	0,3356	0,1530	0,2723	0,3333	0,3048	0,5739	0,0000	0,3333	0,0000	0,3048	0,0950	0,0500
50	5	0,1941	0,4283	0,2325	0,1409	0,1217	0,1703	0,0000	0,1409	0,0000	0,1217	0,0000	0,0200
50	10	0,2369	0,4641	0,1462	0,1544	0,2147	0,4578	0,0000	0,1544	0,0000	0,2147	0,0000	0,0167
50	20	0,0688	0,4609	0,2957	0,0419	0,0636	0,5096	0,0000	0,0419	0,0000	0,0636	0,0000	0,0200
100	5	0,1387	0,4525	0,5545	0,1475	0,1334	0,4700	0,0833	0,1475	0,0000	0,1334	0,0750	0,0000
100	10	0,1541	0,4385	0,5021	0,1105	0,1438	0,2958	0,0500	0,1105	0,0000	0,1438	0,0889	0,0889
100	20	0,2738	0,2404	0,3728	0,2016	0,2265	0,4173	0,0000	0,2016	0,0000	0,2265	0,0750	0,0750

Table 8. Average values of the cardinal measures Q_2 for IB4S, DB4S, CB4S, JB4S, IB2S, IB8S, DA4S, JA4S, NA4S, NB4S, JB2A and DB2A

n	m	IB4S	DB4S	CB4S	JB4S	IB2S	IB8S	DA4S	JA4S	NA4S	NB4S	JB2A	DB2A
20	5	0,1167	0,3428	0,1744	0,1167	0,1167	0,3761	0,0000	0,1167	0,0000	0,1167	0,0683	0,0250
20	10	0,2357	0,1311	0,3764	0,2357	0,2357	0,3351	0,0000	0,2357	0,0000	0,2357	0,0575	0,0767
20	20	0,2657	0,1156	0,1965	0,2657	0,2657	0,4221	0,0000	0,2657	0,0000	0,2657	0,0273	0,0067
50	5	0,1812	0,3686	0,2383	0,1812	0,1812	0,2287	0,0000	0,1812	0,0000	0,1812	0,0000	0,0100
50	10	0,1634	0,3719	0,0959	0,1634	0,1634	0,3625	0,0000	0,1634	0,0000	0,1634	0,0000	0,0063
50	20	0,0505	0,3494	0,1901	0,0505	0,0505	0,4033	0,0000	0,0505	0,0000	0,0505	0,0000	0,0033
100	5	0,1254	0,3753	0,2449	0,1254	0,1254	0,3275	0,0222	0,1254	0,0000	0,1254	0,0188	0,0000
100	10	0,0952	0,3350	0,2994	0,0952	0,0952	0,2231	0,0045	0,0952	0,0000	0,0952	0,0348	0,0348
100	20	0,1975	0,2030	0,2037	0,1975	0,1975	0,3759	0,0000	0,1975	0,0000	0,1975	0,0200	0,0200

Table 9. Average values of C metric for comparing algorithms with the proposed improvements with JB4S

N	*m*	C(JB4S,IB4S)	C(IB4S,JB4S)	C(JB4S,DB4S)	C(DB4S,JB4S)	C(JB4S,CB4S)	C(CB4S,JB4S)
20	5	0,14	0,55	0,13	0,80	0,02	0,85
20	10	0,29	0,51	0,40	0,28	0,27	0,61
20	20	0,30	0,49	0,58	0,39	0,51	0,35
50	5	0,18	0,60	0,09	0,87	0,26	0,67
50	10	0,22	0,53	0,19	0,59	0,40	0,44
50	20	0,42	0,36	0,44	0,48	0,29	0,55
100	5	0,27	0,44	0,21	0,74	0,26	0,57
100	10	0,54	0,33	0,46	0,53	0,39	0,54
100	20	0,33	0,55	0,33	0,53	0,46	0,48

Table 10. Average values of C metric for comparing algorithms with the proposed improvements with NB4S

N	*m*	C(IB4S,NB4S)	C(NB4S,IB4S)	C(DB4S,NB4S)	C(NB4S,DB4S)	C(CB4S,NB4S)	C(NB4S,CB4S)	C(JB4S,NB4S)	C(NB4S,JB4S)
20	5	0,58	0,20	0,81	0,12	0,71	0,04	0,45	0,37
20	10	0,40	0,29	0,41	0,45	0,61	0,31	0,43	0,37
20	20	0,49	0,15	0,41	0,35	0,45	0,44	0,43	0,41
50	5	0,46	0,32	0,76	0,11	0,64	0,19	0,37	0,51
50	10	0,51	0,41	0,72	0,26	0,43	0,39	0,46	0,44
50	20	0,40	0,36	0,55	0,25	0,57	0,27	0,49	0,41
100	5	0,44	0,35	0,57	0,33	0,55	0,39	0,39	0,49
100	10	0,32	0,50	0,60	0,34	0,54	0,28	0,50	0,41
100	20	0,58	0,31	0,48	0,30	0,49	0,33	0,44	0,37

Table 11. Average values of C metric for comparing algorithms with the proposed improvements between them

N	*m*	C(DB4S,IB4S)	C(IB4S,DB4S)	C(CB4S,DB4S)	C(DB4S,CB4S)	C(CB4S,IB4S)	C(IB4S,CB4S)
20	5	0,43	0,37	0,40	0,29	0,42	0,08
20	10	0,39	0,46	0,61	0,22	0,56	0,17
20	20	0,17	0,66	0,44	0,34	0,34	0,49
50	5	0,52	0,35	0,38	0,55	0,41	0,46
50	10	0,64	0,29	0,24	0,53	0,40	0,49
50	20	0,66	0,23	0,36	0,42	0,62	0,27
100	5	0,48	0,27	0,40	0,49	0,38	0,41
100	10	0,63	0,31	0,43	0,41	0,65	0,25
100	20	0,44	0,35	0,44	0,33	0,42	0,41

Table 12. Average values of C metric for comparing algorithms with different neighbouring generation techniques

n	*m*	C(DB4S,DA4S)	C(DA4S,DB4S)	C(JB4S,JA4S)	C(JA4S,JB4S)	C(NB4S,NA4S)	C(NA4S,NB4S)
20	5	0,87	0,00	0,83	0,00	0,98	0,00
20	10	0,92	0,00	0,92	0,00	0,94	0,00
20	20	0,97	0,00	0,86	0,01	0,98	0,00
50	5	0,68	0,00	0,93	0,00	0,95	0,00
50	10	0,85	0,00	0,92	0,00	1,00	0,00
50	20	0,90	0,00	0,92	0,00	0,94	0,00
100	5	0,82	0,00	0,86	0,00	0,96	0,00
100	10	0,85	0,00	0,81	0,00	0,89	0,00
100	20	0,95	0,01	0,97	0,00	1,00	0,00

IB4S and DB4S, and between these last two, DB4S performs better than IB4S. Observing the figures of Table 11, the superiority of DB4S and CB4S is evident, in the sense that they present the PE with more solutions that are efficient with respect to the reference set. Furthermore, DB4S outperforms CB4S for some instance sizes, while CB4S outperforms DB4S for others.

Referring to the perturbation techniques, scheme B is absolutely superior to scheme A. Table 12 shows how scheme A is incapable of obtaining solutions non-dominated by solutions obtained with the same algorithm, using scheme B.

With respect to the aggregated function, we can claim that in this MOSA scheme it does not work. Table 13 shows that the outputs are worse, in both reported cases and for all the tested instances, than the outputs yielded by simultaneous optimization. In the case of DB2S vs. DB2A the differences are more significant.

In order to evaluate the influence of the size of the initial solution set, we have obtained $C(A,B)$ for every pair of combinations between IB2S, IB4S and IB8S (Table 14). As could be expected, the larger the set, the better the results. Nevertheless, when compared with a reference set (see Table 8) the resulting efficient solutions are the same.

Table 13. Average values of C metric for evaluating algorithms with aggregation versus algorithms with simultaneous optimization

N	*m*	C(JB2A,JB4S)	C(JB2S,JB4A)	C(DB2A,DB4S)	C(DB2S,DB4A)
20	5	0.16	0.55	0.02	0.95
20	10	0.10	0.63	0.05	0.76
20	20	0.16	0.71	0.10	0.62
50	5	0.04	0.68	0.01	0.93
50	10	0.05	0.60	0.06	0.86
50	20	0.04	0.75	0.04	0.94
100	5	0.10	0.65	0.00	1.00
100	10	0.12	0.61	0.05	0.79
100	20	0.13	0.46	0.00	0.68

Table 14. Average values of C metric for comparing algorithms with different number of initial seeds

N	*m*	C(IB4S,IB2S)	C(IB2S,IB4S)	C(IB8S,IB4S)	C(IB4S,IB8S)	C(IB8S,IB2S)	C(IB2S,IB8S)
20	5	0.55	0.28	0.69	0.18	0.79	0.07
20	10	0.43	0.39	0.57	0.24	0.37	0.31
20	20	0.52	0.32	0.54	0.35	0.53	0.33
50	5	0.64	0.14	0.43	0.36	0.73	0.17
50	10	0.53	0.19	0.73	0.23	0.90	0.01
50	20	0.67	0.22	0.69	0.13	0.79	0.12
100	5	0.70	0.19	0.69	0.10	0.83	0.00
100	10	0.55	0.24	0.59	0.33	0.67	0.13
100	20	0.91	0.05	0.57	0.23	0.81	0.03

The benefit could be expected in a better distribution for the larger set (see Table 5). To complete this comparison requirement concepts must be considered.

A comparative study of the computational effort for the proposed algorithms has been made including CPU time consumption and the number of sequences generated during the entire search process for all the problem instances considered. Essential summaries are presented in Tables 15 and 16. The CPU time employed by IB2S is taken as the reference unity, because the algorithm with 2 seeds may correspond to the least time requiring for the presented battery of tests. Although eight-seed algorithm (IB8S) consumes almost four unities (predictable fact), the algorithms with four initial solutions always require less than twice as much. Even, IB4S nearly always consumes less or equal CPU time than IB2S. The DB2A and JB2A, besides giving non-efficient solutions, and with only two seeds, employed more time than IB2S.

Since the computational effort for a variant algorithm with eight seeds is considerably higher than the corresponding four-seed algorithm (see Table 14), the best trade off corresponds to the four-initial-solution version.

Table 15. CPU time required by the proposed algorithms, relative to the IB2S consumption

N	m	IB4S	DB4S	CB4S	JB4S	NB4S	IB2S	IB8S	JB2A	DB2A
20	5	0,55	1,59	0,97	1,59	1,81	1,00	4,21	1,40	0,84
20	10	0,71	0,75	1,32	1,25	1,44	1,00	3,73	1,77	1,06
20	20	0,98	0,81	1,24	1,24	1,23	1,00	4,09	1,96	1,18
50	5	1,17	2,05	2,02	2,34	2,43	1,00	4,17	2,02	1,21
50	10	1,87	2,41	8,84	2,43	2,64	1,00	4,12	1,80	1,08
50	20	1,00	1,42	1,91	1,72	1,18	1,00	3,53	1,00	0,60
100	5	1,00	1,54	1,95	1,68	1,47	1,00	2,59	1,62	1,04
100	10	0,98	1,89	2,89	2,61	2,29	1,00	3,69	1,51	0,92
100	20	1,60	1,21	1,34	0,65	0,62	1,00	2,20	1,62	0,81

Table 16. Average number of sequences generated for each proposed algorithm

N	m	IB4S	DB4S	CB4S	JB4S	NB4S	IB2S	IB8S	JB2A	DB2A	DA4S	JA4S	NA4S
20	5	4.409	3.144	4.065	4.270	3.561	1.673	42.174	45.429	49.173	188.273	188.253	188.274
20	10	4.563	5.366	4.453	4.838	5.350	1.855	142.977	114.096	128.215	189.860	189.854	189.850
20	20	4.500	3.405	3.912	3.459	3.884	1.619	140.623	102.856	139.505	191.339	191.342	191.326
50	5	9.709	9.934	10.922	10.922	10.922	4.520	115.425	324.751	106.955	194.083	194.083	194.083
50	10	19.173	15.338	15.311	13.179	14.637	7.074	736.424	418.993	146.924	194.486	194.449	194.478
50	20	23.187	21.071	15.822	30.200	17.940	12.685	272.971	960.152	355.239	194.707	194.741	194.714
100	5	12.979	10.863	12.409	11.654	11.796	5.313	318.605	273.743	266.539	194.758	194.748	194.749
100	10	14.083	13.090	13.087	13.500	15.322	8.946	587.628	535.067	502.796	195.185	195.189	195.185
100	20	15.320	27.223	26.205	29.173	24.006	12.148	285.386	660.537	154.851	195.435	195.436	195.427

It is possible to conclude that the influence of the improvement technique is crucial for the efficiency of the output, while a larger number of generated solutions (both by initial seeds or neighbouring generation) help to improve the diversification of the output with non-efficient solutions, increasing considerably the computational effort.

7 Conclusions

In this work we present new algorithms based on MOSA techniques for a hard multicriteria scheduling problem. Starting with initial permutations obtained by single criteria constructive algorithms, improvements are made by computing lower bounds on the partial scheduling of neighbors, reducing the objective search space. The selection is made according to a criterion that is the preferred at each iteration.

Due to the complexity of evaluating the quality of solutions, a set of different metrics have been computed, considering the different attributes of the methods. Furthermore, net set of non-dominated solutions for the benchmarks problems of Taillard [113] have been obtained. After an extensive computational analysis, including a comparison with other metaheuristic algorithms that have been published in the last few years, we can conclude that, though this kind of approach presents less percentage in the final net set (Q_2), it results in less percentage of non-efficient solutions in the potential efficient output set (Q_1).

Results of the computational experiment give support to the hypothesis which states that specially-developed algorithms, combining general metaheuristic techniques, for specified combinatorial problems, perform better than general methods. It is not realistic to hope for general meta-optimization methods that solve MOCO problems efficiently.

The main proposed algorithms (IB4S, DB4S, and CB4S) are appropriate to warrant a quick approximation output, which can serve as input for an interactive procedure. The search process should continue in the direction of the decision-maker preferences.

We are working now on developing similar approaches considering more than two criterion scheduling problems.

Acknowledgments

This research was in part supported by the Research Projects DPI2004-06366-C03-02 and ECO2008-05895-C02-02, Ministerio de Ciencia e Innovación, Spain.

The author is indebted to the referees for their helpful remarks and comments, and to Paul Alexander Ayres for his help in the correct use of English.

References

[1] Agrawal, S., Dashora, Y., Tiwari, M.K., et al.: Interactive Particle Swarm: A Pareto-Adaptive Metaheuristic to Multiobjective Optimization. IEEE T. Syst. Man Cy. A. 38(2), 258–277 (2008)
[2] Aickelin, U.: Genetic Algorithms for Multiple-Choice Problems. PhD Thesis. University of Wales, Swansea (1999)
[3] Akers, S.B.: A graphical approach to production scheduling problems. Oper. Res. 4, 244–245 (1956)
[4] Andrés, C.: Programación de la Producción en Talleres de Flujo Híbridos con Tiempos de Cambio de Partida Dependientes de la Secuencia: Modelos, Métodos y Algoritmos de Resolución: Aplicación a Empresas del Sector Cerámico. PhD Thesis. Universidad Politécnica de Valencia, Valencia (2001)
[5] Arroyo, J., Armentano, V.: Genetic local search for multi-objective flowshop scheduling problems. Eur. J. Oper. Res. 167, 717–738 (2005)
[6] Bagchi, T.P.: Multiobjective Scheduling by Genetic Algorithms. Kluwer Academic Publishers, Dordrecht (1999)
[7] Baker, K.R.: A comparative study of flow shop algorithms. Oper. Res. 23, 62–73 (1975)
[8] Blazewicz, J., Ecker, K., Pesch, E., et al.: Handbook on Scheduling. Springer, Berlin (2007)
[9] Brucker, P.: An efficient algorithm for the job-shop problem with two jobs. Computing 40, 353–359 (1988)
[10] Brucker, P.: Scheduling Algorithms. Springer, Berlin (2004)
[11] Bülbül, K., Kaminsky, P., Yano, C.: Flow shop scheduling with earliness, tardiness, and intermediate inventory holding costs. University of California, Berkeley (2003)

[12] Burke, E.K., Landa-Silva, J.D., Soubeiga, E.: Hyperheuristic Approaches for Multiobjective Optimization. In: Proceedings of the 5th Metaheuristics International Conference, Kyoto (2003)
[13] Campbell, H.G., Dudek, R.A., Smith, M.L.: A Heuristic Algorithm for the n-Job, m-Machine Sequencing Problem. Manag. Sci. 16(10), 630–637 (1970)
[14] Carlier, J., Rebaï, I.: Two branch and bound algorithms for the permutation flow shop problem. Eur. J. Oper. Res. 90, 238–251 (1996)
[15] Chang, P.C., Chen, S.H., Liu, C.H.: Sub-population genetic algorithm with mining gene structures for multiobjective flowshop scheduling problems. Expert. Syst. Appl. 33, 762–777 (2007)
[16] Chang, P.C., Hsieh, J.-C., Lin, S.G.: The development of gradual priority weighting approach for the multi-objective flowshop scheduling problem. Int. J. Prod. Econ. 79, 171–183 (2002)
[17] Chankong, V., Haimes, Y.Y.: Multiobjective Decision Making Theory and Methodology. Elsevier Science, New York (1983)
[18] Charnes, A., Cooper, W.: Management Models and Industrial Applications of Linear Programming. John Wiley and Sons, Chichester (1961)
[19] Coello, C., Mariano, C.: Algorithms and Multiple Objective. In: Ehrgott, M., Gandibleux, X. (eds.) Multiple Criteria Optimization. State of the Art Annotated Bibliographic Surveys. Kluwer Academic Publishers, Boston (2002)
[20] Czyzak, P., Jaszkiewicz, A.: Pareto Simulated Annealing – a metaheuristic technique for multiple objective combinatorial optimization. J. Multicriteria. Dec. Anal. 7, 34–47 (1998)
[21] Daniels, R.L., Chambers, R.J.: Multiobjective flow-shop scheduling. Nav. Res. Log. 37, 981–995 (1990)
[22] Dorn, J., Girsch, M., Skele, G., et al.: Comparison of iterative improvement techniques for schedule optimization. Eur. J. Oper. Res. 94, 349–361 (1996)
[23] Dudek, R.A., Panwalkar, S.S., Smith, M.L.: The lessons of flowshop scheduling research. Oper. Res. 40, 7–13 (1992)
[24] Eck, B.T., Pinedo, M.: On the minimization of the makespan subject to flowtime optimality. Oper. Res. 41, 797–801 (1993)
[25] Ehrgott, M.: Approximation algorithms for combinatorial multicriteria optimization problems. Int. T. Oper. Res. 7, 5–31 (2000)
[26] Ehrgott, M., Gandibleux, X.: Bounds and bound sets for biobjective Combinatorial Optimization problems. Lect. Notes Econ. Math., vol. 507, pp. 242–253 (2001)
[27] Ehrgott, M., Gandibleux, X.: Multiobjective Combinatorial Optimization: Theory, Methodology, and Applications. In: Ehrgott, M., Gandibleux, X. (eds.) Multiple Criteria Optimization: State of the Art Annotated Bibliographic Surveys. Kluwer Academic Publishers, Boston (2002)
[28] Ehrgott, M., Wiecek, M.: Multiobjective Programming. In: Figueira, J., Greco, S., Ehrgott, M. (eds.) Multiple Criteria Decision Analysis. Springer, New York (2005)
[29] Emelichev, V.A., Perepelista, V.A.: On cardinality of the set of alternatives in discrete many-criterion problems. Discrete. Math. Appl. 2(5), 461–471 (1992)
[30] Framinan, J.M., Leisten, R., Ruiz-Usano, R.: Efficient heuristics for flowshop sequencing with the objectives of makespan and flowtime minimisation. Eur. J. Oper. Res. 141, 559–569 (2002)
[31] French, S.: Sequencing and Scheduling: An Introduction to the Mathematics of the Job Shop. Ellis Horwood, Chichester (1982)
[32] Gandibleux, X., Mezdaoui, N., Fréville, A.: A tabu search procedure to solve multiobjective combinatorial optimization problems. Lect. Notes Econ. Math., vol. 455, pp. 291–300 (1997)

[33] Garey, M.R., Johnson, D.S.: Computers and Intractability: A Guide to the Theory of NP-Completeness. Freeman, San Francisco (1979)
[34] Geiger, M.: On operators and search space topology in multi-objective flow shop scheduling. Eur. J. Oper. Res. 181, 195–206 (2007)
[35] González, T., Johnson, D.B.: A new algorithm for preemptive scheduling of trees. J. Assoc. Comp. Mach. 27, 287–312 (1980)
[36] Gordon, V., Proth, J.M., Chu, C.: A survey of the state of the art of common due date assignment and scheduling research. Eur. J. Oper. Res. 139, 1–25 (2002)
[37] Grabowski, J., Wodecki, M.: Some local search algorithms for no-wait flow-shop problem with makespan criterion. Comp. Oper. Res. 32, 2197–2212 (2004)
[38] Graham, R.L., Lawler, E.L., Lenstra, J.K., et al.: Optimization and approximation in deterministic sequencing and scheduling: A survey. Ann. Discrete Math. 5, 287–326 (1979)
[39] Gupta, J.N.D.: Heuristic Algorithms for Multistage Flowshop Scheduling Problem. AIIE T. 4(1), 11–18 (1972)
[40] Gupta, J.N.D., Neppalli, V.R., Werner, F.: Minimizing total flow time in a two-machine flowshop problem with minimum makespan. Int. J. Prod. Econ. 69(3), 323–338 (2001)
[41] Hapke, M., Jaszkiewicz, A., Slowinski, R.: Interactive Analysis of multiple-criteria project scheduling problems. Eur. J. Oper. Res. 107(2), 315–324 (1998)
[42] Haupt, R.: A survey of priority rule-based scheduling. Oper. Res. Spektrum 11, 3–16 (1989)
[43] Ho, J.C., Chang, Y.-L.: A new heuristic for the n-job, m-machine flowshop problem. Eur. J. Oper. Res. 52, 194–202 (1991)
[44] Hoogeveen, H.: Multicriteria Scheduling. Eur. J. Oper. Res. 167, 592–623 (2005)
[45] Hoogeveen, J.A.: Single-Machine Bicriteria Scheduling. PhD Thesis. The Netherlands Technology, Amsterdam (1992)
[46] Horsky, D., Rao, M.R.: Estimation of attribute weights from preference comparison. Manag. Sci. 30(7), 801–822 (1984)
[47] Huang, G., Lim, A.: Fragmental Optimization on the 2-Machine Bicriteria Flowshop Scheduling Problem. In: Proceedings of 15th IEEE International Conference on Tools with Artificial Intelligence (2003)
[48] Ignall, E., Schrage, L.E.: Application of the branch-and-bound technique to some flowshop scheduling problems. Oper. Res. 13, 400–412 (1965)
[49] Isermann, H.: The enumeration of the set of all efficient solutions for a linear multiple objective program. Oper. Res. Quart. 28(3), 711–725 (1977)
[50] Ishibuchi, H., Misaki, S., Tanaka, H.: Modified simulated annealing algorithms for the flow shop sequencing problem. Eur. J. Oper. Res. 81, 388–398 (1995)
[51] Ishibuchi, H., Murata, T.: A multi-objective genetic local search algorithm and its application to flowshop scheduling. IEEE T. Syst. Man Cy. C. 28(3), 392–403 (1998)
[52] Jaszkiewicz, A.: A Comparative Study of Multiple-Objective Metaheuristics on the Bi-Objective Set Covering Problem and the Pareto Memetic Algorithm. Ann. Oper. Res. 131(1-4), 135–158 (2004)
[53] Jaszkiewicz, A., Ferhat, A.B.: Solving multiple criteria choice problems by interactive trichotomy segmentation. Eur. J. Oper. Res. 113(2), 271–280 (1999)
[54] Johnson, S.M.: Optimal two- and three-stage production schedules with setup times included. Nav. Res. Log. 1, 61–68 (1954)
[55] Jones, D.F., Mirrazavi, S.K., Tamiz, M.: Multi-objective meta-heuristics: An overview of the current state of the art. Eur. J. Oper. Res. 137, 1–9 (2002)
[56] Kirkpatrick, S., Gelatt, C.D., Vecchi, M.P.: Optimization by simulated annealing. Science 220, 671–680 (1983)

[57] Knowles, J., Corne, D.: On Metrics Comparing Nondominated Sets. In: Proceedings of the 2002 Congress on Evolutionary Computation Conference, pp. 711–716. IEEE Press, Los Alamitos (2002)
[58] Koulamas, C.: A new constructive heuristic for the flowshop scheduling problem. Eur. J. Oper. Res. 105, 66–71 (1998)
[59] van Laarhoven, P.J.M., Aarts, E.H.L.: Simulated Annealing: Theory and Practice. Kluwer Academic Publishers, Dordrecht (1987)
[60] Lageweg, B.J., Ixnstra, J.K., Rinnooy Kan, A.H.G.: A general bounding to minimize makespan/total flowtime of jobs. Eur. J. Oper. Res. 155, 426–438 (1978)
[61] Laha, D., Chakraborty, U.K.: An efficient heuristic approach to flowtime minimization in permutation flowshop scheduling. Int. J. Adv. Manuf. Technol. (2007) (DOI: 10.1007/s00170-007-1156-z)
[62] Laha, D., Chakraborty, U.K.: An efficient stochastic hybrid heuristic for flowshop scheduling. Engineering Applications of Artificial Intelligence 20, 851–856 (2007)
[63] Laha, D., Chakraborty, U.K.: A constructive heuristic for minimizing makespan in no-wait flowshop scheduling. Int. J. Adv. Manuf. Technol. (2008) (DOI: 10.1007/s00170-008-1454-0)
[64] Landa-Silva, J.D., Burke, E.K., Petrovic, S.: An Introduction to Multiobjective Metaheuristics for Scheduling and Timetabling. Lect. Notes Econ. Math., vol. 535, pp. 91–129 (2004)
[65] Lawler, E.L., Lenstra, J.K., Rinnooy Kan, A.H.G.: Sequencing and scheduling: Algorithms and complexity. In: Handbooks in Operations Research and Management Science, Logistics of Production and Inventory, vol. 4, pp. 445–524. North-Holland, Amsterdam (1993)
[66] Leung, J.Y.-T., Young, G.H.: Minimizing schedule length subject to minimum flow time. Siam. J.Comp. 18, 314–326 (1989)
[67] Liao, C.J., Yu, W.C., Joe, C.B.: Bicriterion scheduling in the two-machine flowshop. J. Oper. Res. Soc. 48, 929–935 (1997)
[68] Liu, J., Reeves, C.R.: Constructive and composite heuristic solutions to the $P//\sum C_i$ scheduling problem. Eur. J. Oper. Res. 132, 439–452 (2001)
[69] Lomnicki, A.: Branch-and-bound algorithm for the exact solution of the three-machine scheduling problem. Oper. Res. Quart. 16, 89–100 (1965)
[70] Loukil, T., Teghem, J., Tuyttens, D.: Solving multi-objective production scheduling problems using metaheuristics. Eur. J. Oper. Res. 161, 42–61 (2005)
[71] McMahon, G.B.: Optimal Production Schedules for Flow Shop. Can. Oper. Res. Soc. J. 7, 141–151 (1969)
[72] Monma, C.L., Rinnooy Kan, A.H.G.: A concise survey of efficiently solvable special cases of the permutation flow-shop problem. RAIRO-Rech. Oper. 17, 105–119 (1983)
[73] Murata, T., Ishibuchi, H., Tanaka, H.: Multi-Objective Genetic Algorithm and its Applications to Flowshop Scheduling. Comp. Ind. Eng. 30(4), 957–968 (1996)
[74] Nagar, H.J., Heragu, S.S.: Multiple and bicriteria scheduling: A literature survey. Eur. J. Oper. Res. 81, 88–104 (1995)
[75] Nawaz, M., Enscore Jr., E.E., Ham, I.: A heuristic algorithm for the m-machine, n-job flowshop sequencing problem. OMEGA-Int. J. Manage. S. 11, 91–95 (1983)
[76] Neppalli, V.R., Chen, C.L., Gupta, J.N.D.: Genetic algorithms for the two-stage bicriteria flowshop problem. Eur. J. Oper. Res. 95, 356–373 (1996)
[77] Nowicki, E., Zdrzałka, S.: A survey of results for sequencing problems with controllable processing times. Discrete Appl. Math. 26, 271–287 (1990)
[78] Ogbu, F.A., Smith, D.K.: The Application of the Simulated Annealing Algorithm to the Solution of the n/m/Cmax Flowshop Problem. Comp. Oper. Res. 17(3), 243–253 (1990)
[79] Onwubolu, G., Davendra, D.: Scheduling flow shops using differential evolution algorithm. Eur. J. Oper. Res. 171, 674–692 (2006)

[80] Osman, I.H., Potts, C.N.: Simulated Annealing for Permutation Flowshop Scheduling. OMEGA-Int. J. Manage. S. 17(6), 551–557 (1989)
[81] Panwalkar, S.S., Iskander, W.: A survey of scheduling rules. Oper. Res. 25, 45–61 (1977)
[82] Parker, R.G.: Deterministic Scheduling Theory. Chapman & Hall, New York (1995)
[83] Parthasarathy, S., Rajendran, C.: An experimental evaluation of heuristics for scheduling in a real-life flowshop with sequence-dependent setup times of jobs. Int. J. Prod. Econ. 49, 255–263 (1997)
[84] Pasupathy, T., Rajendran, C., Suresh, R.K.: A multi-objective genetic algorithm for scheduling in flow shops to minimize the makespan and total flow time of jobs. Int. J. Adv. Manuf. Technol. 27, 804–815 (2006)
[85] Pinedo, M.L.: Scheduling: Theory, Algorithms, and Systems. Prentice Hall, New Jersey (2002)
[86] Potts, C.N.: An adaptive branching rule for the permutation flow-shop problem. Eur. J. Oper. Res. 5, 19–25 (1980)
[87] Potts, C.N., Shmoys, D.B., Williamson, D.P.: Permutation vs. non-permutation flow shop schedules. Oper. Res. Lett. 10, 281–284 (1991)
[88] Rajendran, C.: Two-stage flowshop scheduling problem with bicriteria. J. Oper. Res. Soc. 43(9), 879–884 (1992)
[89] Rajendran, C.: Heuristic algorithm for scheduling in a flowshop to minimize total flow-time. Int. J. Prod. Econ. 29, 65–73 (1993)
[90] Rajendran, C.: Heuristics for scheduling in flowshop with multiple objectives. Eur. J. Oper. Res. 82, 540–555 (1995)
[91] Rajendran, C., Ziegler, H.: An efficient heuristic for scheduling in a flowshop to minimize total weighted flowtime of jobs. Eur. J. Oper. Res. 103, 129–138 (1997)
[92] Rajendran, C., Ziegler, H.: Ant-colony algorithms for permutation: flowshop scheduling. Eur. J. Oper. Res. 155, 426–438 (2004)
[93] Reeves, C.R.: Improving the Efficiency of Tabu Search for Machine Scheduling Problems. J. Oper. Res. Soc. 44(4), 375–382 (1993)
[94] Reeves, C.R.: A Genetic Algorithm for Flowshop Sequencing. Comp. Oper. Res. 22, 5–13 (1995)
[95] Rinnooy Kan, A.H.G.: Machine Scheduling problems: Classification, Complexity and Computations, Martinus Nijhoff, The Hague (1976)
[96] Ruiz, R: Técnicas Metaheurísticas para la Programación Flexible de la Producción. PhD Thesis. Universidad Politécnica de Valencia, Valencia (2003)
[97] Ruiz, R., Maroto, C.: A comprehensive review and evaluation of permutation flowshop heuristics. Eur. J. Oper. Res. 165, 479–494 (2005)
[98] Ruiz-Díaz, F.S.: A survey of multi-objective combinatorial scheduling. In: French, S., Hartley, R., Thomas, L.C., et al. (eds.) Multi-Objective Decision Making. Academic Press, New York (1983)
[99] Saaty, T.L.: The Analytic Hierarchy Process. McGrawHill, New York (1980)
[100] Sayin, S., Karabati, S.: A bicriteria approach to the two-machine flow shop scheduling problem. Eur. J. Oper. Res. 113, 435–449 (1999)
[101] Schulz, A.: Scheduling and Polytopes. PhD Thesis. Technical University of Berlin, Berlin (1996)
[102] Selen, W.J., Hott, D.D.: A mixed-integer goal-programming formulation of the standard flow-shop scheduling problem. J. Oper. Res. Soc. 12(37), 1121–1128 (1986)
[103] Serafini, P.: Simulated annealing for multiple objective optimization problems. In: Proceedings of the Tenth International Conference on Multiple Criteria Decision Making, Taipei (1992)
[104] Shmoys, D.B., Tardos, É.: An approximation algorithm for the generalized assignment problem. Math. Program. 62, 461–474 (1993)

[105] Sin, C.C.S.: Some topics of parallel-machine scheduling theory. Thesis. University of Manitoba (1989)

[106] Sivrikaya-Serifoglu, F.S., Ulusoy, G.: A bicriteria two machine permutation flowshop problem. Eur. J. Oper. Res. 107, 414–430 (1998)

[107] Srinivas, N., Deb, K.: Multiobjective function optimization using nondominated sorting genetic algorithms. Evol. Comp. 2(3), 221–248 (1995)

[108] T'kindt, V., Billaut, J.-C.: Multicriteria scheduling problems: a survey. RAIRO-Oper. Res. 35, 143–163 (2001)

[109] T'kindt, V., Billaut, J.-C.: Multicriteria scheduling: Theory, Models and Algorithms, 2nd edn. Springer, Berlin (2006)

[110] T'kindt, V., Gupta, J.N.D., Billaut, J.-C.: Two machine flowshop scheduling problem with a secondary criterion. Comp. Oper. Res. 30(4), 505–526 (2003)

[111] T'kindt, V., Monmarche, N., Tercinet, F., et al.: An ant colony optimization algorithm to solve a 2-machine bicriteria flowshop scheduling problem. Eur. J. Oper. Res. 142(2), 250–257 (2002)

[112] Taillard, E.: Some efficient heuristic methods for the flor shop sequencing problem. Eur. J. Oper. Res. 47, 67–74 (1990)

[113] Taillard, E.: Benchmark for basic scheduling problems. Eur. J. Oper. Res. 64, 278–285 (1993)

[114] Ulungu, E.L.: Optimisation Combinatoire MultiCritère: Détermination de l'ensemble des solutions efficaces et méthodes interactives. PhD Thesis. Université de Mons-Hainaut, Mons (1993)

[115] Ulungu, E.L., Teghem, J.: Multiobjective Combinatorial Optimization problems: A survey. J. Multicriteria Dec. Anal. 3, 83–104 (1994)

[116] Varadharajan, T.K., Rajendran, C.: A multi-objective simulated-annealing algorithm for scheduling in flowshops to minimize the makespan and total flowtime of jobs. Eur. J. Oper. Res. 167, 772–795 (2005)

[117] Wierzbicki, A.P.: A methodological guide to the multiobjective optimization. Lect. Notes Contr. Inf., vol. 1(23), pp. 99–123 (1980)

[118] Wilson, J.M.: Alternative formulation of a flow shop scheduling problem. J. Oper. Res. Soc. 40(4), 395–399 (1989)

[119] Wodecki, M., Bozejko, W.: Solving the Flow Shop Problem by Parallel Simulated Annealing. In: Wyrzykowski, R., Dongarra, J., Paprzycki, M., Waśniewski, J. (eds.) PPAM 2001. LNCS, vol. 2328, pp. 236–244. Springer, Heidelberg (2002)

[120] Yagmahan, B., Yenisey, M.M.: Ant. colony optimization for multi-objective flow shop scheduling problem. Comp. Ind. Eng. 54, 411–420 (2008)

[121] Zitzler, E.: Evolutionary Algorithms for Multiobjective Optimization: Methods and Applications. PhD Thesis. Swiss Federal Institute of Technology, Zurich (1999)

An Estimation of Distribution Algorithm for Minimizing the Makespan in Blocking Flowshop Scheduling Problems

Bassem Jarboui[1], Mansour Eddaly[1], Patrick Siarry[2], and Abdelwaheb Rebaï[1]

[1] FSEGS, route de l'aéroport km 4.5, B.P. No. 1088, Sfax 3018, Tunisie
bassem_jarboui@yahoo.fr, eddaly.mansour@gmail, abdelwaheb.rebai@fsegs.rnu.tn

[2] LiSSi, Université de Paris 12, 61 avenue du Général de Gaulle, 94010 Créteil, France
siarry@univ-paris12.fr

Summary. This chapter addresses to the blocking flowshop scheduling problem with the aim of minimizing the makespan. An Estimation of Distribution Algorithm, followed by a local search procedure, after the step of creating a new individual, was developed in order to solve this problem. Our comparisons were performed against representative approaches proposed in the literature related to the blocking flowshop scheduling problem. The obtained results have shown that the proposed algorithm is able to improve 109 out of 120 best known solutions of Taillard's instances. Moreover, our algorithm outperforms all competing approaches in terms of solution quality and computational time.

1 Introduction

In the nature, the evolution of species in a population, through the sexual reproduction, was formulated by Charles Darwin (T. Back, 1996). It can be modelled by means of three mechanisms: recombination (or crossover), mutation and selection. The process of recombination occurs during meiosis resulting from crossover between parental chromosomes. Through this process, the offspring inherit different combinations of genes from their parents. The mutation arises from errors of copying in genetic materials during cell division. It creates changes into offspring's chromosomes. Under selection, individuals with best traits tend to have more luck to survive and reproduce for further generations. Evolutionary algorithms (EAs) are a class of algorithms that use computers to simulate the natural evolution of species to solve hard optimization problems through evolving a population of candidate solutions. EAs have proved their performance against classical techniques of optimization (Fogel, 1995). Several algorithms are included in this class such as the Genetic Algorithm (GA), which is the most popular. Neighbouring nature-inspired approaches are Ant Colony Optimization, Particle Swarm Optimization, etc.

Recently, a new EA was introduced by Mühlenbein and Paaß in (Mühlenbein and Paaß, 1996), called Estimation of Distribution Algorithm (EDA). It constitutes a new tool of evolutionary algorithms (Larranaga P. and Lozano J.A., 2002), based on the

U.K. Chakraborty (Ed.): Comput. Intel. in Flow Shop and Job Shop Sched., SCI 230, pp. 151–167.
springerlink.com

probabilistic model learned from a population of individuals. Starting with a population of individuals (candidate solutions), generally randomly generated, this algorithm selects good individuals with respect to their fitness. Then a new distribution of probability is estimated from the selected candidates. Next, new offspring are generated from the estimated distribution. The process is repeated until the termination criterion is met. In the literature, diverse versions of EDAs were developed, depending on the chosen probabilistic model. The EDAs can be classified into three classes: EDAs with no dependencies between the variables, EDAs with two-order dependencies and EDAs with multiple dependencies between the variables.

EDAs have been employed for solving combinatorial optimization problems. So, several successful applications were proposed such as: quadratic assignment problem (Zhang et al., 2006), 0-1 knapsack problem (Hui Li et al., 2004), n-queen problem (Paul TK and Iba H, 2002), travelling salesman problem (Robles et al., 2006) and hybrid flowshop scheduling problem (Salhi et al., 2007). In recent works, the EDAs were devoted to solve multi-objective optimization problems (Zhang et al. 2008, Hui Li et al., 2004).

In this work, we propose to adopt this new technique for solving the blocking flowshop scheduling problem. In this variant of flowshop scheduling, there is a set of n jobs that must be processed on a set of m machines in the same order. While the storage is not allowed, when a job is completed on a machine, the latter is blocked until a free next machine becomes available. Blocking constraints takes place because of the automation of new production systems and the use of the robotic manufacturing. Typical areas are chemical and pharmaceutical industries, where a partially completed job cannot quit the machine on which it is processed, while downstream machines are busy (Grabowski and Pempera, 2007). Grabowski and Pempera (2000) have presented a real case of scheduling client orders in a building industry that produces concrete blocks. Also, Hall and Sriskandarajah (1996) have presented a review of applications of blocking scheduling models. They have indicated that blocking environment occurs from characteristics of the process technology itself or from the lack of the storage capacity between the machines. They have proved that this problem is strongly NP-complete for m=3, where the makespan $(C_{\max})$ is a measure of performance.

In the literature, various approaches were developed to solve the permutation flowshop scheduling problem under blocking constraints, including branch and bound algorithm (B&B) (Levner, 1969, Suhami and Mah, 1981, Ronconi, 2005, Company and Mateo, 2007), constructive heuristics (McCormick et al., 1989, Leisten, 1990, Abadi et al., 2000, Ronconi and Armentano, 2001, Ronconi, 2004), genetic algorithm (GA) (Caraffa et al., 2001) and tabu search (TS) (Grabowski and Pempera, 2007).

The remaining of this chapter is organized as follows: section 2 presents the Estimation of Distribution Algorithm and its variants; section 3 presents the existing works with EDA in combinatorial optimization. The blocking flowshop is described in section 4. Our proposed algorithm is presented in section 5. Section 6 presents the computational results and conclusion is given in section 7.

2 Estimation of Distribution Algorithm (EDA)

EDA is an evolutionary algorithm proposed by Mühlenbein and Paaß in 1996. Instead of recombination and mutation, EDA generates new individuals with respect to a probabilistic model, learned from the population of parents.

2.1 Basic EDA

The general framework of the basic EDA can be presented as follows (Mühlenbein and Paaß, 1996). Starting with a randomly generated initial population, one selects a subpopulation of M parent individuals through a selection method based on the fitness function. Next, one estimates the probability of distribution of the selected parents with a probabilistic model. Then, one generates new offspring, according to the estimated probability distribution. Finally, some individuals in the current population are replaced with new generated offspring. These steps are repeated until one stopping criterion is met. The pseudo-code of the canonical EDA is given in Figure 1.

Basic EDA

Generate an initial population of P individuals;

do

- Select a set of Q parents with a selection method;
- Build a probabilistic model for the set of selected parents;
- Create new P_1 offspring according to the estimated probability distribution;
- Replace some individuals in the current population with new individuals;

while a stopping criterion is not met

Fig. 1. Canonical version of EDA

Three classes of EDA were developed, according to the chosen probabilistic model. The first class consists of models which don't take into account the dependencies between variables of candidate solutions, i.e. all variables are independent. The second class assumes at most two-order dependencies between these variables and the last class assumes multiple dependencies between the variables.

2.2 EDAs with No Dependencies

Let X_i, $i = 1,2,.....,n$, be a random variable and x_i its possible realization and let $p(X_i = x_i) = p(x_i)$ the mass probability of X_i over the point x_i. By analogy, we denote by $\mathbf{X} = \{X_1, X_2, ..., X_n\}$ a set of n-dimensional random variables, $\mathbf{x} = \{x_1, x_2, ..., x_n\}$ its possible realizations and $p(\mathbf{X} = \mathbf{x}) = p(\mathbf{x})$ the joint mass probability of $\mathbf{X}$ over the point $\mathbf{x}$.

In this class of EDAs, it is assumed that the n-dimensional joint probability distribution is calculated through the product of the marginal probabilities of n variables, as follows:

$$p(x)=\prod_{i=1}^{n} p(x_i).$$

In other hand, the hypothesis of interaction between the variables is rejected.

Among the EDAs included in this class we can cite: Bit-Based Simulated Crossover (BBSC) of Syswerda (1993), Population-Based Incremental Learning (PBIL) of Baluja (1994), Compact Genetic Algorithm (CGA) of Harik et al. (1998) and Univariate Marginal Distribution Algorithm (UMDA) of Mühlenbein et al. (1998).

Although these approaches have provided better results for some problems, their assumption seems to be inexact for difficult optimization problems, where we cannot exclude the interdependencies between the variables completely (Paul TK and Iba H, 2002).

2.3 EDAs with Two-Order Dependencies

In this class, only paired interactions between the variables are taken into account. So, EDAs belonging to this group constitute an extension of the previous one. Therefore, the parametric learning of model, proposed in EDAs with no interaction, becomes structural.

In the literature, several approaches were developed in this class, such as: Mutual Information Maximization for Input Clustering (MIMIC) in De Bonet al. (1997), Combining Optimizers with Mutual Information Trees (COMIT) in Baluja and Davies (1997) and Bivariate Marginal Distribution Algorithm (BMDA) in Pelikan and Mühlenbein (1999).

2.4 EDAs with Multiple Dependencies

This last class of EDAs is the most general case, and the leaning process of models proposed here is more complex, because the estimation of joint probability is performed by taking into account an order of dependencies greater than two.

The following approaches of EDAs are included in this class: Factorized Distribution Algorithm (FDA) (Mühlenbein et al., 1999), Estimation of Bayesian Networks Algorithm (EBNA) (Etxeberria and Larranaga, 1999), Bayesian Optimization Algorithm (BOA) (Pelikan et al., 1999), Learning Factorized Distribution Algorithm (LFDA) (Mühlenbein and Mahning, 1999) and the Extended Compact Genetic Algorithm (ECGA) (Harik, 1999).

3 Some EDAs for Combinatorial Optimization Problems

Although, EDA was recently invented, the number of its applications in the field of combinatorial optimization increases rapidly. In this section, we will present some applications of EDA to combinatorial optimization problems and we will mainly focus on the constructed probabilistic model for each application.

The Jobshop Scheduling Problem (JSP) was addressed by J. Lozano et al. (in Larrañaga and Lozano, 2002). The authors have selected some variants of EDA and used both continuous and discrete versions. The selected algorithms are UMDA, BBSC, PBIL, MIMIC and EBNA.

The obtained results are comparable to those obtained using GA. In particular the continuous EDAs perform better than the discrete EDAs.

Paul TK and Iba H have proposed, in (Paul TK and Iba H, 2002), an UMDA to solve *n*-queen problem. The objective of this problem is to find a way of putting n_q queens ($n_q \geq 4$) on a $n_q \times n_q$ chessboard, such that none of them can capture any other, i.e. two queens cannot share the same row, column or diagonal. A problem's solution x was represented as follows: $\mathbf{x} = \{x_1, x_2, ..., x_{nq}\}$, where x_i, $1 \leq i \leq n_q$, denotes the column position in row i where the queen i can be put. The initial population was randomly generated while excluding cases where two queens are in the same column or row. The fitness of each individual is calculated as the number of queens that do not share the same diagonal. Next, the first 50% of individuals (best individuals) were selected according to their fitness. Then, the joint probability was selected using the marginal frequencies of each x_i and new individuals were generated according to it. Finally, the elitism was used for the replacement step and the algorithm was stopped when the fitness of the best individual was equal to n_q. The computational results show that this algorithm is able to reach a good solution in a reasonable amount of time.

Hui et *al.* (2004) have proposed a hybrid EDA for solving the multiobjective 0-1 knapsack problem. For modelling the joint probability distribution, an UMDA is used. At each generation *t*, an individual is selected, based on the following probability, depending on the set of selected individuals at generation *t*-1:

$$p(x,t) = p(x/\text{selected individuals}(t-1)) = \prod_{i=1}^{n_k} p(x_i,t)$$

where $x \in \{0,1\}^{n_k}$.

The results showed that the EDA performed better than the Genetic Algorithm, both in convergence and in diversity.

Salhi et *al.* (2007) have proposed an EDA for hybrid flowshop scheduling problem with respect to the makespan criterion. The joint probability $p_{ij}(t)$ denotes the probability that the job i is located on the position j at the generation t $(1 \leq i \leq n \text{ and } 1 \leq j \leq n)$.

This probability was initially set to $1/n^2$ and updated as follows:

$$p_{ij}(t) = (1-\beta)\frac{1}{N}\sum_{k=1}^{N} I_{ij}(\pi_k) + \beta p_{ij}(t-1)$$

where π_k is the kth solution of the population at the generation t $(1 \leq k \leq N)$, $I_{ij} = \begin{cases} 1 & if\, \pi(i) = j \\ 0 & otherwise \end{cases}$ and $(0 \leq \beta \leq 1)$.

The obtained results were compared with those provided by two heuristic algorithms, a Random Key Genetic Algorithm and a Genetic Algorithm. The results show that EDA outperforms these two algorithms for the considered instances.

4 Problem Description

In a blocking flowshop problem, there is a set of *n* jobs to be processed on a set of *m* machines in the same order, while having no intermediate buffers, i.e. a job $j \in \{1,2,....,n\}$ cannot pass from machine $k \in \{1,2,.....,m\}$ to machine k+1 while the latter is busy. Since the makespan is the criterion to be minimized in our case, this problem can be denoted by $F_m / blocking / C_{\max}$ (Graham et al., 1979).

Let $p_{[j]k}$ denote the processing time of the job in the jth position in the sequence on the machine *k* and $D_{[j]k}$ denote the departure time (starting time) of the job in the jth position in the sequence on the machine *k*.

The makespan $(C_{\max})$ can be found through the recursive expression of the departure time, as follows:

$$D_{[1]0} = 0;$$

$$D_{[1]k} = \sum_{i=1}^{k} p_{[1]i} \qquad k = 1,2,.....,m-1;$$

$$D_{[j]0} = D_{[j-1]1} \qquad j = 2,3,.....,n;$$

$$D_{[j]k} = \max\left\{D_{[j]k-1} + p_{[j]k}, D_{[j-1]k+1}\right\} \qquad j = 2,3,.....,n, \quad k = 1,2,.....,m-1;$$

$$D_{[j]m} = D_{[j]m-1} + p_{[j]m} \qquad j = 1,2,.....,n;$$

Thus,

$$C_{\max} = D_{[n]m-1} + p_{[n]m}$$

5 Hybrid EDA for BFSP

In this section we present in detail an EDA to solve the Blocking Flowshop Scheduling Problem (BFSP), which is aimed at makespan minimization.

5.1 Solution Representation

For encoding the solution, we use the well-known representation scheme for the PFSP, that is the permutation of *n* jobs, where the jth number in the permutation denotes the job located in position *j*.

5.2 Initial Population

For generating the initial population of P individuals, we propose to generate P-1 individuals randomly and we apply NEH algorithm, proposed by Nawaz et al. (1983), for the remaining element.

NEH can be described as follows:

Step1: The jobs are sorted with respect to the decreasing order of sums of their processing times.
Step2: Take the first two jobs and evaluate the two possible schedules containing them. The sequence with better objective function value is taken for further consideration.
Step 3: Take every remaining job in the permutation given in Step 1 and find the best schedule, by placing it at all possible positions in the sequence of jobs that are already scheduled.

5.3 Selection

In our algorithm, we adopted the same procedure of selection employed by Reeves (1995) for solving the flowshop scheduling problem. We describe this procedure as follows.

First, for each individual p, the fitness value $f\left(p\right)=\dfrac{1}{C_{\max}\left(p\right)}$ is calculated, second the individuals of the initial population are sorted in ascending order according to their fitness, i.e. the individual with a higher makespan value will be at the top of the list. Finally, a set of Q individuals are selected from the sorted list.

5.4 Construction of a Probabilistic Model and Creation of New Individuals

The probabilistic model constitutes the main issue for an EDA and the performance of the algorithm is closely related to it (Lozano J.A et al., 2006), the best choice of the model is crucial. This step consists in building an estimation of distribution for the subset of Q selected individuals.

In our algorithm, we select at random a sequence of jobs, denoted sr, from the set of 25% best solutions in the sorted list of sequences. Based on the priority rules of the order of the q first jobs in the s_r, we determine the estimation of distribution model while taking into account both the order of the jobs in the sequence and the similar blocks of jobs presented in the selected parents. In fact, the parameter q is an intensification parameter because, when it is possible, it leads to maintain the same structure of q first jobs and setting it to a constant value preserves the linearity of the algorithm.

Let:

– η_{jk} be the number of times of apparition of job j before or in the position k in the subset of the selected sequences augmented by a given constant δ_1. The value of η_{jk} refers to the importance of the order of the jobs in the sequence.

– $\mu_{j[k-1]}$ be the number of times of apparition of job j after the job in the position k-1 in the subset of the selected sequences augmented by a given δ_2 . $\mu_{j[k-1]}$ indicates the importance of the similar blocks of jobs in the sequences. In such way, we prefer to conserve the similar blocks as much as possible.

We note that δ_1 and δ_2 are two parameters used for the diversification of the solutions. Indeed, we employ these parameters in order to slow down the convergence of the algorithm.

– Let Ω_k be the set of q first jobs not already scheduled following their order in s_r until position k.

We define π_{jk} the probability of selection of the job j in the kth position by the following formula:

$$\pi_{jk} = \frac{\eta_{jk} \times \mu_{j[k-1]}}{\sum_{l \in \Omega_k} \eta_{lk} \times \mu_{l[k-1]}}$$

For each position k in the sequence of a new individual, we select a job j among the set of q first jobs not already scheduled, following their order in s_r by sampling from the probability distribution π_{jk} .

5.5 Replacement

Replacement is the last phase in the EDA, it consists in updating the population. Therefore, at each iteration, O offspring are generated from the subset of the selected parents. There are many techniques available to decide if the new individuals will be added to the population.

In our algorithm, we compare the new individual with the worst individual in the current population. If the offspring is best than this individual and the sequence of the offspring is unique, then the worst individual quits the population and is replaced with the new individual.

5.6 Stopping Criterion

The stopping condition indicates when the search will be terminated. Various stopping criteria may be listed, such as maximum number of generations, bound of time, maximum number of iterations without improvement, etc. In our algorithm, we set a maximum number of generations and a maximal computational time.

5.7 Local Search

To improve the performance of EDA, the successful way is to hybridize it with local search methods (Lozano J.A. et al., 2006). We propose to apply a local search algorithm as an improvement procedure, after the creation of a new individual.

We propose to restrict the application of the local search procedure to a part of individuals by employing a probability of improvement that depends on the quality of the subjected individual. We define this probability as follows:

Let $p^c = \exp\left(\frac{RD}{\alpha}\right)$ be the calculated probability for application of local search, where:

$$RD = \left(\frac{f\left(x_{current}\right) - f\left(x_{best}\right)}{f\left(x_{best}\right)}\right)$$

with $x_{current}$ denotes the created offspring and x_{best} denotes the best solution found by the algorithm. For each individual, we draw at random a number between 0 and 1. If this number is less than or equal to p^c, then we apply the local search procedure to the individual under consideration.

At each iteration of the local search procedure, we select one among two kinds of neighbourhoods randomly. The first one leads to choose two distinct positions (i, j) at random, following the uniform distribution in the range $[1,n]$, and the jobs on these positions are exchanged. The second one consists in selecting at random a job j from the sequence and inserting it on a random position i. This procedure will be repeated as far as reaching the maximal number of iterations $iter_{max}$.

6 Computational Results

In this section, we discuss the performance of our proposed algorithms: EDA (without hybridization) and H-EDA. All computations for blocking flowshop scheduling problem, with respect to the makespan criterion, were implemented using C++ program and carried out on an Intel Pentium IV 3.2 GHz, RAM 512 MB based computer, running under Windows XP. In order to evaluate the performances of the proposed algorithms, the Taillard's instances were used for the flowshop scheduling problem (Taillard E., 1993). These instances consist of a set of 120 problems with sizes m=5, 10 and 20 and n=20, 50, 100, 200 and 500. The performance measure employed in our numerical study was average relative percentage deviation in makespan $\Delta_{average}$:

$$\Delta_{average} = \frac{\sum_{i=1}^{R}\left(\frac{Heu_i - Best_{known}}{Best_{known}} \times 100\right)}{R}$$

where Heu_i is the solution given by any of the R replications of the considered algorithms and $Best_{known}$ is the best solution provided by a competing algorithm for the specified problem or by one of our algorithms.

The parameters of the algorithms were fixed after a set of preliminary experiments, as follows: P = 60, $\delta_1 = \delta_2 = 4/n$, the number of the selected parents Q = 3, q = 20, the number of generated offspring O = 3. Numerically, $p^c = 0.5$ leads to accepting a sequence with a makespan superior by 5% relatively to the best value of makespan found.

So, $\beta = \dfrac{RD}{\log(p^c)} = \dfrac{0.01}{\log(0.5)}$ thereafter we determined p^c according to this formula:

$$p^c = \exp\left(\frac{RD}{\beta}\right).$$

The maximum number of iterations of the local search procedure was set to $2n^2$.

6.1 Comparison with GA

For testing the efficiency of our proposed EDA (without local search) against another evolutionary algorithm, we have implemented the GA of Caraffa et al. (2001). For performing a meaningful comparison we have set the same stopping criterion of 1000 generations for both algorithms.

The obtained results for each class of instances, over R=10 replications, are given in Table 1. For the small instances, with $n < 200$, in average, EDA outperforms GA both in terms of $\Delta_{average}$ and Δ_{max}, so, EDA can find better results than GA in average and worst case. Regarding Δ_{min} the two algorithms provide almost the same results. Also, for these instances, the range of changes for EDA solutions, i.e. the difference between Δ_{min} and Δ_{max}, is smaller than that range for GA, in average, thus EDA is more robust than GA. For large instances, with n = 200 and 500, EDA confirms its superiority, in terms of $\Delta_{average}$ and Δ_{max}, and it is better than GA for finding the best results (Δ_{min}). Although EDA is better than GA in term of solution quality, the latter appears faster after 1000 generations (Table 6).

Table 1. Comparison between EDA and GA

instances	EDA			GA		
	Δ_{min}	Δ_{avg}	Δ_{max}	Δ_{min}	Δ_{avg}	Δ_{max}
20*05	0.01	0.02	0.03	0.01	0.03	0.05
20*10	0.01	0.02	0.03	0.01	0.03	0.05
20*20	0.01	0.01	0.02	0.01	0.02	0.03
50*05	0.02	0.03	0.04	0.03	0.04	0.06
50*10	0.02	0.03	0.04	0.02	0.04	0.06
50*20	0.03	0.03	0.04	0.01	0.03	0.05
100*05	0.04	0.05	0.06	0.05	0.06	0.07
100*10	0.03	0.04	0.04	0.03	0.04	0.06
100*20	0.02	0.03	0.03	0.02	0.03	0.04
200*10	0.04	0.04	0.05	0.06	0.07	0.08
200*20	0.03	0.03	0.03	0.04	0.04	0.05
500*20	0.02	0.02	0.02	0.05	0.05	0.05
average	0.02	0.03	0.03	0.03	0.04	0.05

Table 2. Results of H-EDA for 20 jobs instances

instances	Best known	RON	TS+M	H-EDA		
				Δ_{min}	Δ_{avg}	Δ_{max}
ta_20_5_01	**1374**	0.01	0.01	0.00	0.00	0.01
ta_20_5_02	1411	0.00	0.01	0.00	0.00	0.00
ta_20_5_03	**1280**	0.01	0.01	0.00	0.00	0.01
ta_20_5_04	1448	0.00	0.00	0.00	0.00	0.00
ta_20_5_05	**1342**	0.02	0.01	0.00	0.00	0.00
ta_20_5_06	1363	0.00	0.00	0.00	0.00	0.00
ta_20_5_07	1381	0.00	0.00	0.00	0.00	0.01
ta_20_5_08	**1379**	0.00	0.01	0.00	0.00	0.00
ta_20_5_09	**1373**	0.00	0.01	0.00	0.00	0.01
ta_20_5_10	1283	0.00	0.01	0.00	0.00	0.01
ta_20_10_01	1698	0.02	0.00	0.00	0.00	0.00
ta_20_10_02	**1833**	0.03	0.00	0.00	0.00	0.00
ta_20_10_03	**1659**	0.01	0.00	0.00	0.00	0.01
ta_20_10_04	**1535**	0.06	0.01	0.00	0.00	0.01
ta_20_10_05	**1617**	0.03	0.01	0.00	0.00	0.01
ta_20_10_06	**1592**	0.03	0.01	0.00	0.00	0.01
ta_20_10_07	**1622**	0.01	0.00	0.00	0.00	0.00
ta_20_10_08	**1731**	0.01	0.01	0.00	0.00	0.01
ta_20_10_09	**1747**	0.02	0.01	0.00	0.00	0.01
ta_20_10_10	1782	0.04	0.00	0.00	0.00	0.00
ta_20_20_01	**2436**	0.04	0.00	0.00	0.00	0.00
ta_20_20_02	**2234**	0.03	0.00	0.00	0.00	0.00
ta_20_20_03	**2480**	0.03	0.00	0.00	0.00	0.01
ta_20_20_04	2348	0.02	0.00	0.00	0.00	0.00
ta_20_20_05	**2435**	0.04	0.01	0.00	0.00	0.01
ta_20_20_06	**2389**	0.03	0.00	0.00	0.00	0.01
ta_20_20_07	2390	0.05	0.00	0.00	0.00	0.01
ta_20_20_08	**2328**	0.04	0.01	0.00	0.00	0.01
ta_20_20_09	**2363**	0.02	0.00	0.00	0.00	0.01
ta_20_20_10	**2323**	0.04	0.00	0.00	0.00	0.00
average		0.02	0.01	0.00	0.00	0.01

6.2 Performance of H-EDA

The performance of H-EDA is evaluated against the representative approaches developed for the same problem. The competing algorithms are the branch and bound algorithm of Ronconi (2005) and the Tabu Search of Grabowski and Pempera (2007), denoted by RON and TS+M respectively. We set the CPU time limit of each replication to $(n \times m) \times 20/3$ seconds.

Table 2 to Table 5 present the results found by our H-EDA. First, in total, our algorithm has improved 109 solutions out of 120 and, even for the 11 remaining instances,

Table 3. Results of H-EDA for 50 jobs instances

instances	Best known	RON	TS+M	H-EDA		
				Δ_{min}	Δ_{avg}	Δ_{max}
ta_50_5_01	**3055**	0.03	0.04	0.00	0.01	0.01
ta_50_5_02	**3249**	0.05	0.03	0.00	0.01	0.01
ta_50_5_03	**3056**	0.04	0.04	0.00	0.01	0.01
ta_50_5_04	**3170**	0.05	0.04	0.00	0.01	0.01
ta_50_5_05	**3200**	0.03	0.05	0.00	0.01	0.01
ta_50_5_06	**3224**	0.06	0.04	0.00	0.00	0.01
ta_50_5_07	**3079**	0.05	0.03	0.00	0.00	0.01
ta_50_5_08	**3097**	0.06	0.05	0.00	0.01	0.01
ta_50_5_09	**2963**	0.06	0.04	0.00	0.00	0.01
ta_50_5_10	**3160**	0.04	0.04	0.00	0.01	0.01
ta_50_10_01	**3737**	0.06	0.02	0.00	0.00	0.01
ta_50_10_02	**3562**	0.06	0.02	0.00	0.01	0.02
ta_50_10_03	**3554**	0.05	0.01	0.00	0.00	0.01
ta_50_10_04	**3754**	0.04	0.02	0.00	0.00	0.01
ta_50_10_05	**3698**	0.06	0.01	0.00	0.01	0.02
ta_50_10_06	**3678**	0.05	0.03	0.00	0.01	0.01
ta_50_10_07	**3765**	0.06	0.01	0.00	0.01	0.01
ta_50_10_08	**3632**	0.04	0.02	0.00	0.01	0.01
ta_50_10_09	**3604**	0.05	0.02	0.00	0.01	0.01
ta_50_10_10	**3691**	0.06	0.01	0.00	0.01	0.01
ta_50_20_01	**4591**	0.07	0.01	0.00	0.00	0.01
ta_50_20_02	**4373**	0.07	0.01	0.00	0.01	0.01
ta_50_20_03	**4354**	0.07	0.01	0.00	0.01	0.02
ta_50_20_04	**4448**	0.05	0.01	0.00	0.01	0.01
ta_50_20_05	**4353**	0.03	0.01	0.00	0.00	0.01
ta_50_20_06	**4368**	0.04	0.00	0.00	0.01	0.01
ta_50_20_07	**4386**	0.04	0.00	0.00	0.00	0.01
ta_50_20_08	**4415**	0.07	0.01	0.00	0.01	0.01
ta_50_20_09	**4400**	0.03	0.00	0.00	0.00	0.01
ta_50_20_10	**4502**	0.08	0.03	0.00	0.01	0.01
average		0.05	0.02	0.00	0.01	0.01

it can reach 9 upper bounds found by TS+M. Additionally, the most important improvement occurs for the instances with the size larger than 20. Especially when n = 50, 100 and 200, H-EDA has improved all upper bounds provided by previous approaches. In other hand, concerning the CPU time, in average, when we take into account the difference between the computer characteristics, H-EDA is faster than the TS+M approach (Table 6).

Table 4. Results of H-EDA for 100 jobs instances

instances	Best known	RON	TS+M	H-EDA		
				Δ_{min}	Δ_{avg}	Δ_{max}
ta_100_5_01	**6256**	0.01	0.06	0.00	0.00	0.01
ta_100_5_02	**6075**	0.00	0.06	0.00	0.01	0.01
ta_100_5_03	**6018**	0.01	0.05	0.00	0.00	0.01
ta_100_5_04	**5832**	0.00	0.05	0.00	0.01	0.02
ta_100_5_05	**6055**	0.02	0.06	0.00	0.00	0.01
ta_100_5_06	**5914**	0.00	0.05	0.00	0.01	0.02
ta_100_5_07	**6073**	0.00	0.05	0.00	0.00	0.01
ta_100_5_08	**5981**	0.00	0.06	0.00	0.00	0.01
ta_100_5_09	**6210**	0.00	0.06	0.00	0.01	0.01
ta_100_5_10	**6226**	0.00	0.05	0.00	0.01	0.01
ta_100_10_01	**7190**	0.02	0.02	0.00	0.00	0.01
ta_100_10_02	**6890**	0.03	0.04	0.00	0.01	0.01
ta_100_10_03	**7073**	0.01	0.03	0.00	0.00	0.01
ta_100_10_04	**7282**	0.06	0.02	0.00	0.01	0.02
ta_100_10_05	**6956**	0.03	0.03	0.00	0.01	0.02
ta_100_10_06	**6811**	0.03	0.03	0.00	0.00	0.01
ta_100_10_07	**6933**	0.01	0.03	0.00	0.01	0.01
ta_100_10_08	**6934**	0.01	0.02	0.00	0.02	0.02
ta_100_10_09	**7223**	0.02	0.02	0.00	0.00	0.01
ta_100_10_10	**7054**	0.04	0.03	0.00	0.01	0.02
ta_100_20_01	**8000**	0.04	0.01	0.00	0.01	0.01
ta_100_20_02	**8021**	0.03	0.02	0.00	0.00	0.01
ta_100_20_03	**8014**	0.03	0.01	0.00	0.01	0.01
ta_100_20_04	**8023**	0.02	0.01	0.00	0.01	0.01
ta_100_20_05	**8004**	0.04	0.01	0.00	0.01	0.01
ta_100_20_06	**8079**	0.03	0.02	0.00	0.00	0.01
ta_100_20_07	**8152**	0.05	0.02	0.00	0.01	0.01
ta_100_20_08	**8209**	0.04	0.02	0.00	0.00	0.01
ta_100_20_09	**8116**	0.02	0.01	0.00	0.00	0.01
ta_100_20_10	**8160**	0.04	0.01	0.00	0.01	0.01
average		0.02	0.03	0.00	0.01	0.01

Table 5. Results of H-EDA for 200 and 500 jobs instances

instances	Best known	RON	TS+M	H-EDA		
				Δ_{min}	Δ_{avg}	Δ_{max}
ta_200_10_01	**13718**	0.03	0.04	0.00	0.01	0.02
ta_200_10_02	**13618**	0.04	0.04	0.00	0.01	0.02
ta_200_10_03	**13779**	0.06	0.04	0.00	0.00	0.01
ta_200_10_04	**13718**	0.06	0.04	0.00	0.01	0.01
ta_200_10_05	**13763**	0.04	0.05	0.00	0.00	0.01
ta_200_10_06	**13472**	0.04	0.05	0.00	0.01	0.01
ta_200_10_07	**13869**	0.06	0.03	0.00	0.01	0.01
ta_200_10_08	**13848**	0.04	0.04	0.00	0.01	0.01
ta_200_10_09	**13580**	0.04	0.04	0.00	0.01	0.02
ta_200_10_10	**13712**	0.05	0.02	0.00	0.01	0.01
ta_200_20_01	**15122**	0.03	0.01	0.00	0.01	0.02
ta_200_20_02	**15379**	0.03	0.03	0.00	0.01	0.01
ta_200_20_03	**15528**	0.04	0.03	0.00	0.00	0.01
ta_200_20_04	**15331**	0.05	0.02	0.00	0.01	0.02
ta_200_20_05	**15295**	0.05	0.01	0.00	0.00	0.01
ta_200_20_06	**15387**	0.04	0.01	0.00	0.01	0.01
ta_200_20_07	**15370**	0.04	0.02	0.00	0.01	0.01
ta_200_20_08	**15386**	0.05	0.01	0.00	0.01	0.01
ta_200_20_09	**15279**	0.04	0.02	0.00	0.01	0.02
ta_200_20_10	**15375**	0.05	0.04	0.00	0.01	0.01
average		0.04	0.03	0.00	0.01	0.01
ta_500_20_01	**37530**	0.03	0.02	0.00	0.01	0.01
ta_500_20_02	**37942**	0.03	0.01	0.00	0.00	0.01
ta_500_20_03	**37637**	0.03	0.01	0.00	0.00	0.00
ta_500_20_04	**37888**	0.03	0.02	0.00	0.00	0.01
ta_500_20_05	**37622**	0.04	0.02	0.00	0.00	0.01
ta_500_20_06	**37950**	0.02	0.01	0.00	0.00	0.01
ta_500_20_07	37561	0.03	0.01	0.00	0.01	0.01
ta_500_20_08	37750	0.03	0.01	0.00	0.00	0.01
ta_500_20_09	**37521**	0.03	0.01	0.00	0.00	0.01
ta_500_20_10	**37869**	0.03	0.02	0.00	0.00	0.00
average		0.03	0.02	0.00	0.00	0.01

Table 6. Computational times

instances	EDA			GA			TS+M	H-EDA		
	min	average	max	min	average	max		min	average	max
20*05	0.24	0.87	1.43	0.00	0.02	0.08	2.70	0.03	0.28	0.60
20*10	0.21	0.93	1.57	0.00	0.02	0.08	4.60	0.10	0.67	1.24
20*20	0.32	0.95	1.74	0.01	0.05	0.17	7.60	0.13	1.37	2.43
50*05	2.54	4.14	4.99	0.08	0.16	0.27	6.20	0.19	0.85	1.55
50*10	2.34	4.35	5.30	0.11	0.26	0.46	10.80	0.45	1.74	2.95
50*20	2.89	4.61	5.76	0.22	0.51	0.92	19.30	0.53	3.14	5.98
100*05	6.51	9.50	10.94	0.48	0.57	0.61	12.40	0.77	1.97	3.28
100*10	8.11	10.22	11.59	0.81	1.03	1.12	22.10	1.49	3.90	6.44
100*20	6.34	10.26	12.41	1.47	1.88	2.04	39.40	3.00	8.20	12.88
200*10	17.76	22.26	24.38	1.99	2.15	2.22	44.30	9.54	12.40	13.34
200*20	18.39	24.18	26.77	3.74	4.05	4.23	79.40	18.10	24.22	26.67
500*20	48.20	70.65	82.86	10.30	10.71	11.78	209.00	66.67	66.67	66.67
average	9.49	13.58	15.81	1.60	1.78	2.00	38.15	8.42	10.45	12.00

7 Conclusion

In this chapter, we have proposed a hybrid EDA algorithm to minimize the makespan in the blocking flowshop scheduling problem. The probabilistic model built for our EDA depends on both the order of the jobs in the sequence and the similar blocks of jobs presented in the set of selected parents. A local search procedure is added to the EDA as an improvement phase, after creating a new individual. The computational results show that our proposed EDA, without hybridization, performs better than a GA previously developed for the same problem in terms of solution quality. However the GA outperforms our algorithm when 1000 generations are set as stopping criterion for both algorithms.

Also, by comparing the hybrid algorithm against competing approaches available in the literature, it's seen that our algorithm is better than these approaches, both in terms of solution's quality and computational times and it seems able to improve best known solutions.

References

Abadi, I.N.K., Hall, N.G., Sriskandarajah, C.: Minimizing cycle time in a blocking flowshop. Operations Research 48, 177–180 (2000)

Back, T.: Evolutionary Algorithms in Theory and Practice. Oxford University Press, Oxford (1996)

Baluja, S.: Population-based incremental learning: a method for integrating genetic search based function optimization and competitive learning, Technical Report, Carnegie Mellon Report, CMU-CS: 94-163 (1994)

Baluja, S., Davies, S.: Using optimal dependency trees for combinatorial optimization: Learning the structure of search space. Technical Report No. CMU-CS-97-107, Carnegie Mellon University, Pittsburgh, Pennsylvania (1997)

Caraffa, V., Ianes, S., Bagchi, T.P., Sriskandarajah, C.: Minimizing makespan in a blocking flowshop using genetic algorithms. International Journal of Production Economics 70, 101–115 (2001)

Companys, R., Mateo, M.: Different behaviour of a double branch-and-bound algorithm on Fm|prmu|Cmax and Fm|block|Cmax problems. Computers and Operations Research 34, 938–953 (2007)

DeBonet, J.S., Isbell, C.L., Viola, P.: MIMIC: Finding optima by estimating probability densities. In: Mozer, M., Jordan, M., Petsche, T. (eds.) Advances in Neural Information Processing Systems, vol. 9 (1997)

Fogel, D.B.: Evolutionary Computation. In: Toward a New Philosophy of Machine Intelligence. IEEE Press, Piscataway (1995)

Grabowski, J., Pempera, J.: Sequencing of jobs in some production system. European Journal of Operational Research 125, 535–550 (2000)

Grabowski, J., Pempera, J.: The permutation flow shop problem with blocking. A tabu search approach. OMEGA The International Journal of Management Science 35, 302–311 (2007)

Graham, R.L., Lawler, E.L., Lenstra, J.K., Rinnooy Kan, A.H.G.: Optimization and approximation in deterministic sequencing and scheduling: a survey. Annals of Discrete Mathematics 5, 287–326 (1979)

Hall, N.G., Sriskandarajah, C.: A survey of machine scheduling problems with blocking and no-wait in process. Operations Research 44, 510–525 (1996)

Harik, G., Lobo, F.G., Golberg, D.E.: The compact genetic algorithm. In: Proceedings of the IEEE Conference on Evolutionary Computation, pp. 523–528 (1998)

Li, H., Zhang, Q., Tsang, E., Ford, J.A.: Hybrid Estimation of Distribution Algorithm for Multi-objective Knapsack Problem. In: The 4th European Conference on Evolutionary Computation in Combinatorial Optimization, Coimbra, Portugal, 5-7 April (2004)

Larrañaga, P., Etxeberria, R., Lozano, J.A., Pena, J.M.: Combinatorial Optimization by learning and simulation of Bayesian networks. In: Proceedings of the Sixteenth Conference on Uncertainty in Artificial Intelligence, Stanford, pp. 343–352 (2000)

Larrañaga, P., Lozano, J.A.: Estimation of Distribution Algorithms. In: A New Tool for Evolutionary Computation. Kluwer Academic Publishers, Dordrecht (2002)

Leistein, R.: Flowshop sequencing with limited buffer storage. International Journal of Production Research 28, 2085–2100 (1990)

Levner, E.M.: Optimal Planning of Parts Machining on a Number of Machines. Automation and Remote Control 12, 1972–1978 (1969)

Lozano, J., Larraanaga, P., Inza, I., Bengoetxea, E.: Towards a New Evolutionary Computation: Advances in the Estimation of Distribution Algorithms. Springer, Heidelberg (2006)

Lozano, J.A., Mendiburu, A.: EDAs applied to the job shop scheduling problem. In: Lozano, J.A., Larraanaga, P., Inza, I., Bengoetxea, E. (eds.) Towards a New Evolutionary Computation: Advances in the Estimation of Distribution Algorithms, pp. 231–240. Springer, Heidelberg (2002)

McCormick, S.T., Pinedo, M.L., Shenker, S., Wolf, B.: Sequencing in an assembly line with blocking to minimize cycle time. Operations Research 37, 925–935 (1989)

Mühlenbein, H.: The equation for response to selection and its use for prediction. Evolut. Comput. 5, 303–346 (1998)

Mühlenbein, H., Mahnig, T.: The Factorized Distribution Algorithm for additively decomposed functions. In: Proceedings of the 1999 Congress on Evolutionary Computation, pp. 752–759. IEEE press, Los Alamitos (1999)
Mühlenbein, H., Paaß, G.: From Recombination of Genes to the Estimation of Distributions I. Binary Parameters. PPSN, 178–187 (1996)
Nawaz, M., Enscore Jr., E.E., Ham, I.: A heuristic algorithm for the m-machine, n-job flowshop sequencing problem. OMEGA The International Journal of Management Science 11, 91–95 (1983)
Paul, T.K., Iba, H.: Linear and Combinatorial Optimizations by estimation of Distribution Algorithms. In: 9th MPS Symposium on Evolutionary Computation, IPSJ, Japan (2002)
Pelikan, M., Mühlenbein, H.: The bivariate marginal distribution algorithm. In: Roy, R., Furuhashi, T., Chandhory, P.K. (eds.) Advances in Soft Computing-Engineering Design and Manufacturing, pp. 521–535. Springer, Heidelberg (1999)
Pelikan, M., Goldberg, D.E., Cantpaz, E.: Linkage Problem, Distribution Estimation and Bayesian Networks. Evolutionary Computation 8(3), 311–340 (2000)
Reeves, C.R.: A genetic algorithm for flowshop sequencing. Computers and Operations Research 22, 5–13 (1995)
Ronconi, D.P.: A note on constructive heuristics for the flowshop problem with blocking. International Journal of Production Economics 87, 39–48 (2004)
Ronconi, D.P.: A branch-and-bound algorithm to minimize the makespan in a flowshop problem with blocking. Annals of Operations Research 138, 53–65 (2005)
Ronconi, D.P., Armentano, V.A.: Lower bounding schemes for flowshops with blocking in-process. Journal of the Operational Research Society 52, 1289–1297 (2001)
Salhi, A., Rodriguez, J.A.V., Zhang, Q.: An Estimation of Distribution Algorithm with Guided Mutation for a Complex Flow Shop Scheduling Problem GECCO 2007, London, England, United Kingdom, July 7–11 (2007)
Suhami, I., Mah, R.S.H.: An Implicit Enumeration Scheme for the Flowshop Problem with No Intermediate Storage. Computers and Chemical Engineering 5, 83–91 (1981)
Syswerda, G.: Simulated crossover in genetic algorithms. In: Foundations of Genetic Algorithms, vol. 2, pp. 239–255. Morgan Kaufmann, San Francisco (1993)
Taillard, E.: Benchmarks for basic scheduling problems. European Journal of Operational Research 64, 278–285 (1993)
Zhang, Q., Sun, J., Tsang, E.P.K., Ford, J.: Estimation of Distribution Algorithm with 2-opt Local Search for the Quadratic Assignment Problem. to be appeared in a book on Estimation of Distribution Algorithm. In: Lozano, J., Larraanaga, P., Inza, I., Bengoetxea, E. (eds.) Towards a New Evolutionary Computation: Advances in the Estimation of Distribution Algorithms, pp. 281–291. Springer, Heidelberg (2006)
Zhang, Q., Zhou, A., Jin, Y.: RM-MEDA: A Regularity Model Based Multiobjective Estimation of Distribution Algorithm. IEEE Trans. Evolutionary Computation 12, 41–63 (2008)

A Scatter Search Method for Multiobjective Fuzzy Permutation Flow Shop Scheduling Problem: A Real World Application

Orhan Engin[1], Cengiz Kahraman[2], Mustafa Kerim Yilmaz[3]

[1] Department of Industrial Engineering, Selçuk University, Konya, Turkey
orhanengin@yahoo.com
[2] Department of Industrial Engineering, İstanbul Technical University, İstanbul, Turkey
kahramanc@itu.edu.tr
[3] Department of Industrial Engineering, Selçuk University, Konya, Turkey
m_kerim@hotmail.com

Summary. In this chapter, a scatter search (SS) method is proposed to solve the multiobjective permutation fuzzy flow shop scheduling problem. The objectives are minimizing the average tardiness and the number of tardy jobs. The developed scatter search method is tested on real-world data collected at an engine piston manufacturing company. Using the proposed SS algorithm, the best set of parameters is used to obtain the optimal or near optimal solutions of multiobjective fuzzy flow shop scheduling problem in the shortest time. These parameters are determined by full factorial design of experiments (DOE). The feasibility and effectiveness of the proposed scatter search method is demonstrated by comparing it with the hybrid genetic algorithm (HGA).

1 Introduction

Metaheuristic search techniques have been applied with success to several optimization problems like scheduling problems. In the last few decades, several effective metaheuristic search techniques have been proposed for solving these hard combinatorial optimization problems. Typical examples of such metaheuristic search techniques are Genetic Algorithms (Goldberg 1989), Simulated Annealing (Kirkpatrick et al. 1983), Tabu Search (Glover and Laguna 1997), Ant Colony Optimization (Dorigo and Gambardella 1997), Artificial Immune System (Forrest et al. 1994; Dasgupta and Forrest 1996; De Castro and Von Zuben 1999), and Scatter Search (Glover 1977).

In the recent years, the SS has been successfully applied to several scheduling problems in the literature. Sevaux and Thomin (2002) proposed a SS algorithm for solving one machine scheduling problem. The proposed approach was compared with Genetic Algorithm on several sets of instances in OR-LIB. Dell'Amico et al. (2004) introduced greedy heuristic, local search and a SS approach for the $P//C_{max}$ parallel processors scheduling problem with makespan criteria. Maenhout and Vanhoucke (2006) presented a SS algorithm for the nurse scheduling problem with the total preference cost of the nurses and the total penalty cost from violations of the soft constraints. Nowicki and Smutnicki (2006) proposed a new algorithm for flow shop

U.K. Chakraborty (Ed.): Comput. Intel. in Flow Shop and Job Shop Sched., SCI 230, pp. 169–189.
springerlink.com

scheduling problems that uses some elements of the SS, the path relinking technique and some properties on neighborhoods. Later, Noorul et al. (2007) proposed a new SS algorithm for general flow shop scheduling problem. The algorithm was compared with the Tabu search approach on the benchmark problems in the literature. Rahimi-Vahed et al. (2008) designed a multi-objective SS method for bi-criteria no-wait flow shop scheduling problem. The propose algorithm was compared with a multi-objective Genetic Algorithm.

Scatter search is an evolutionary method and it may be called a population-based algorithm. The recent researches have shown that SS has a great potential for solving hard combinatorial optimization problems such as scheduling problems. In this study, a scatter search method is generated for the multiobjective fuzzy permutation flow shop scheduling problem.

To the best of our knowledge, there is no scatter search method applied to multiobjective fuzzy permutation flow shop scheduling problem in the literature. This is the first attempt for a real world application for multiobjective fuzzy permutation flow shop scheduling. Also this is the first attempt to use those two approaches: the possibility measure introduced by Dubois and Prade (1988) and the area of intersection introduced by Sakawa and Kubota (2000) for multiobjective fuzzy flow shop scheduling problem.

The rest of the chapter is organized as follows. Section 2 presents the formulation of the multiobjective fuzzy permutation flow shop scheduling problem. Sections 3 and 4 are devoted to the scatter search and hybrid genetic algorithm methods. Section 5 describes the performance of the SS on real-world data and Section 6 presents the main conclusion and suggestions for future research.

2 The Multiobjective Fuzzy Permutation Flow Shop Scheduling Problem

In a static permutation flow shop, the processing time for each job and due dates are usually assumed to be known exactly, but in many real world applications, processing times and due dates vary dynamically due to human factors or operating faults. In the literature, fuzzy sets are used to model the uncertain processing times and due dates for the flow shop scheduling problems. The recent research in terms of fuzzy permutation flow shop scheduling problem are given as follows.

Yao and Lin (2002) constructed a fuzzy flow shop sequencing model based on statistical data, which uses level ($1-\alpha$, $1-\beta$) interval-valued fuzzy numbers to present the unknown job processing time. Temiz and Erol (2004) modified the branch and bound algorithm of Ignall and Schrage (1965) and rewrote for three-machine flow shop problem with fuzzy processing time. Niu and Gu (2006) proposed a genetic-based particle swarm optimization for no idle permutation flow shops with fuzzy processing time. Zhu et al. (2006) studied fuzzy flow shop scheduling problem with distinct due window. Fuzzy time is denoted using triangular fuzzy numbers. Petrovic and Song (2006) generated a new optimization algorithm based on Johnson's (1954) algorithm to two machine flow shop problem with uncertain processing times. Nezhad and Assadi (2008) developed a method to approximate maximum operator in the form of a triangular fuzzy number, applied in flow shop scheduling and they modified Campbell Dudek and Smith's (1970) algorithm by using this maximum operator number. To the best of our knowledge, there isn't any study in the literature about solving the multiobjective fuzzy permutation flow shop scheduling problem by metaheuristics methods. This will be the first attempt for the solution of the multiobjective fuzzy permutation flow shop scheduling problem.

The multiobjective fuzzy permutation flow shop scheduling problem can be formulated as follows:

First of all, some assumptions are made for multiobjective fuzzy permutation flow shop scheduling problem. These assumptions are; (1) The number of jobs and machines are known and fixed during the schedule; (2) All processing times and due dates are fuzzy positive integers numbers; (3) Each machine can carry out at most one job at the same time; (4) The jobs must be carried out in a non preemptive way; (5) The processing times contain the set up times for every job at every operation and (6) For carrying out these jobs all machines are continuously available.

Where *n* and *m* are represent the number of jobs to be scheduled and the number of machines, respectively; $\tilde{t}_{ij}$ and $\tilde{d}_j$ represent the fuzzy processing times of job *i* on machine *j*, and the fuzzy due date of job *j*, respectively; and $\tilde{T}_j$ and $\tilde{C}_j$ represent the fuzzy tardiness of job *j* and the fuzzy completion time of job *j*, respectively.

Fuzzy processing times $\tilde{t}_{ij}$ are modeled by triangular membership functions and represented by a triplet $(t_{ij}^1, t_{ij}^2, t_{ij}^3)$, where t_{ij}^1 and t_{ij}^3 are lower and upper bounds of the processing time and t_{ij}^2 is the most possible processing time. The membership function of a triangular fuzzy processing time is shown in Fig 1. The due date $\tilde{d}_j$ of each job is modeled by a trapezoidal fuzzy set and represented by a doublet $\left(d_j^1, d_j^2\right)$, where its fuzzy membership function is shown in Fig 2.

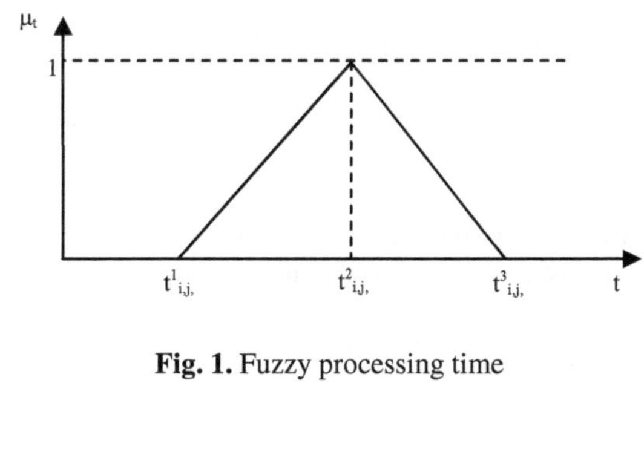

Fig. 1. Fuzzy processing time

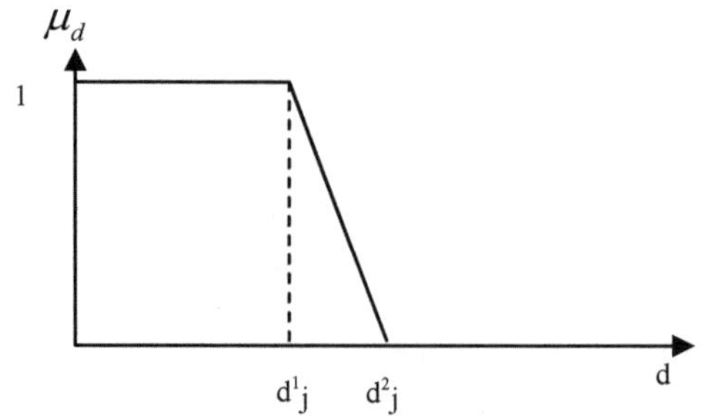

Fig. 2. Fuzzy due date

The following two objectives (Fayad and Petrovic 2003) are considered to minimize in this study.

(1) to minimize the average tardiness C_{AT}:

$$C_{AT} = \frac{1}{n}\sum_{j=1}^{n} T_j; \qquad T_j = \max\{0, \tilde{C}_j - \tilde{d}_j\} \qquad j = 1,\ldots\ldots, n \tag{1}$$

(2) to minimize the number of tardy jobs C_{NT}:

$$C_{NT} = \sum_{j=1}^{n} u_j \qquad u_j = 1 \;\; \text{if} \;\; T_j > 0, \;\; \text{otherwise} \;\; u_j = 0 \tag{2}$$

3 Scatter Search Method

Scatter search method was first introduced in Glover (1977) as a heuristic for integer programming (Marti et al. 2006). SS is an evolutionary method that has been successfully applied to combinatorial optimization problems. SS uses a reference set to combine its solutions and construct others. SS generates a reference set from a population of solutions. Then the solutions in this reference set are combined to get starting solutions to run an improvement procedure, whose result may indicate an updating of the reference set and even an updating of the population of solutions (Herrera et al. 2006). The schematic of the proposed SS method is presented Fig 3.

In the proposed SS method, the initial population is generated based on a memetic algorithm (Bajestani et al. 2009). The steps of the used memetic algorithm are given as follows:

Step 1. Generate the initial population randomly.

Step 2. Apply two-point crossover procedure to couple of chromosomes in the initial population.

Step 3. Apply the neighborhood based mutation procedure to all chromosomes in the population.

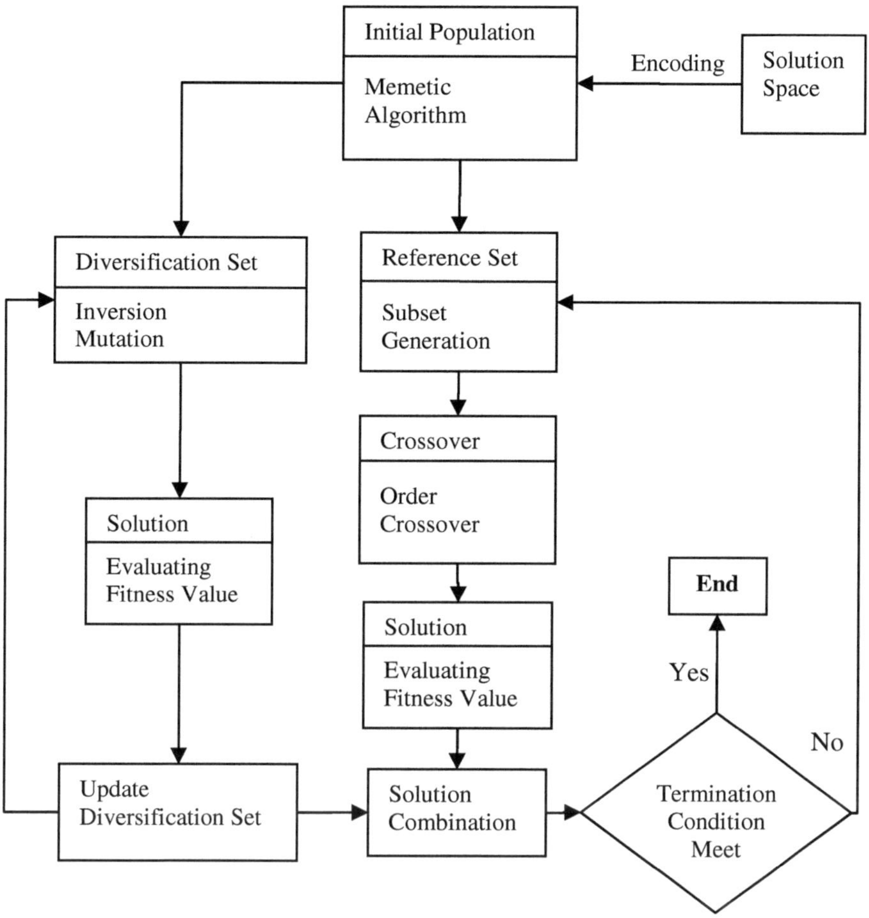

Fig. 3. Flow Chart of the proposed SS method

Step 4. Sort the chromosomes in ascending order depending on the fitness function value.

Step 5. Select chromosomes as many as initial population sizes.

The proposed SS method consists of five methods (Silva et al. 2006). These are

- Diversification-generation method,
- Improvement method,
- Reference set update method,
- Subset generation method and
- Solution combination method.

Diversification- generation method generates a collection of diverse trial solutions, using an arbitrary trial solution (or seed solution) as an input (Russel and Chiang 2006).

Improvement method transforms a trial solution into one or more enhanced trial solutions (Neither the input nor the output solutions are required to be feasible, though the output solutions will usually be expected to be so. If no improvement occurs in the input trial solution, the "enhanced" solution is considered to be the same as the input solution) (Marti et al. 2006).

Reference set update method builds and maintains a reference set consisting of the *b* "best" solutions (where the value of *b* is typically small, $b< 20$), organized to provide efficient accessing by the other parts of the method. Solutions gain membership degrees to the reference set according to their quality or their diversity (Marti et al. 2006).

Subset generation method operates on the reference set to produce a subset of its solutions as a basis for creating combined solutions (Hung and Song 2001).

Solution combination method transforms a given subset of solutions produced by the subset generation method into one or more combined solution vectors (Hung et al. 2002).

4 Hybrid Genetic Algorithms

Genetic Algorithm (GA) was invented by John Holland (Goldberg 1989). GA is one of the best known metaheuristic methods for solving a flow shop scheduling problem (Reeves 1995). GA uses a collection of solutions called population. Each individual in the population is called a chromosome (a string of symbols) and a chromosome represents a solution to the problem (Kahraman et al. 2008).

The role of local search in the context of genetic algorithms has been receiving serious consideration and many successful applications are strongly in favor of such a hybrid approach (Cheng et al. 1999). The hybridization can be done in a variety of ways including (Cheng et al. 1999):

1. Incorporate heuristics into initialization to generate well-adapted initial population. In this way, a hybrid genetic algorithm with elitism can guarantee to do no worse than the conventional heuristic does.
2. Incorporate heuristics into evaluation function to decode chromosomes to schedules.
3. Incorporate local search heuristic as an add-on extra to the basic loop of genetic algorithm, working together with mutation and crossover operators, to perform quick and localized optimization in order to improve offspring before returning it to be evaluated.

The feasibility and effectiveness of the proposed scatter search method is demonstrated by comparing it with HGA. The structure of the used HGA is given in Fig 4.

There are many studies on genetic algorithms for solving the multiobjective flow shop scheduling problems in the literature. Some of them are given as follows:

Pasupathy et al. (2006) proposed a Pareto genetic algorithm for the problem of permutation flow shop scheduling with the objectives of minimizing the makespan and total flow time of jobs. Chang et al. (2007) developed a sub-population genetic algorithm with mining gene structures for multiobjective flow shop scheduling problems.

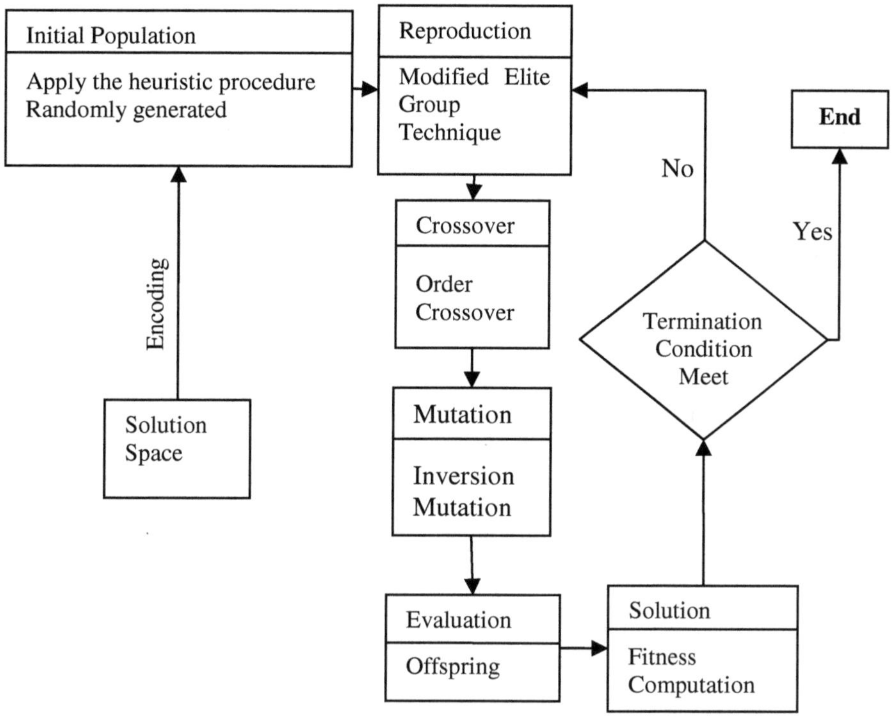

Fig. 4. The flow diagram of the used hybrid genetic algorithm (HGA)

The used HGA is based on a permutation representation of the n jobs. The details of our implementation for the HGA are given as follows.

This study adopts the job based encoding method. In this coding, a chromosome represents a job schedule.

The proposed algorithm utilizes a modified elite group technique (Chung et al. 2009). An elitism technique preserves the best chromosome from the current generation to the next to improve the local search (Chung et al. 2009). The elite group tries to maintain both diversity and quality of solutions. It works as follows. A parent pool, a pool of chromosomes generated from the parent pool by the crossover operation and a pool of mutations generated by the mutation operators are merged to form a combined pool (Choi at al. 2003). Then the chromosomes in the combined pool are sorted according to the fitness values and grouped in three clusters. For instance, top 50%, next 40% and the last 10% of the chromosomes in the combined pool form three groups (Choi at al. 2003). In the HGA, the Chung et al. (2009)'s modified elite group technique is used. In the modified elite group, the best two chromosomes are preserved in the next generation without changes in its genes. The population size is kept constant through the generations. A heuristic procedure has been used to obtain initial population. This procedure is divided into simple steps:

1. Calculate the total fuzzy processing times of all jobs
2. Jobs are sorted in descending order of the total processing times.

The remaining chromosomes are randomly generated.

We assessed the performance of HGA by comparing it with the Engin (2001)'s simple genetic algorithm. The used HGA found a better solution from the simple genetic algorithm. HGA parameters are determined by full factorial design of experiments as in Table 1.

Table 1. HGA parameters

GA Parameter			Value
Initial population			50
Selection operator			Modified elite group tecnique
	Superior	%50	25
Group Proportion %	Middle	%15	6
	Inferior	%35	14
Crossover operator			Order Crossover
Mutation operator			Inversion mutation
Probability of crossover			0.20
Probability of mutation			0.50
Termination condition			50

The selection is made by fitness values in the modified elite group technique.

Order Crossover
Select a substring from one string at random,

Produce a new string by copying the substring into the position corresponding to those in the string,

Delete all of the symbols from the second string. The resulting sequence contains the symbols the new string needs,

Place the symbols into unfixed positions in the new string from left to right according to the order of the sequence to produce an offspring.

Inversion mutation
The inversion mutation can be seen from Fig 5.

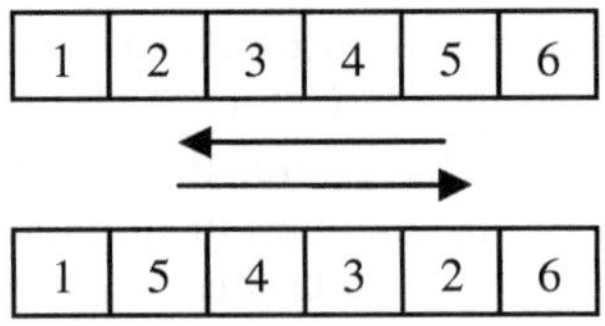

Inversion

Fig. 5. The illustration of the Inversion mutation operators

4.1 Scatter Search Method vs. Genetic Algorithms

Both Scatter search and Genetic algorithm are evolutionary method and the main features are population-based. In contrast to genetic algorithms, scatter search is founded on the premise that systematic designs and methods for creating new solutions afford significant benefits beyond those derived from resource to randomization. It uses strategies for search diversification and intensification that have proved effective in a variety of optimization problems (Marti et al. 2006).

In Genetic algorithms, parents are chosen following a random sampling schema. By contrast, in SS, the selection of parents in made using a deterministic method called subset generation method (Herrera et al. 2006).

In GA, two solutions are randomly sampled from a fairly large population and combined to generate a new offspring (Chakraborty et al. 1996), SS selects two or more elements from a smaller population set in a systematic way to be combined new solution generation (Glover et al. 2003).

SS also allows one to incorporate special forms of adaptive memory programming usually associated with the Tabu search metaheuristic along with mechanisms for exploring that memory. This makes SS very attractive for the design of a heuristic search method (Djan- Sampson and Sahin 2004).

The SS and GA can also be compared according to intensification and diversification as in Table 2 (Sevaux and Thomin 2002).

Table 2. Comparison of SS with GA according to intensification and diversification

Metahuristic methods	Intensification	Diversification
	Inner Loop	
SS	Crossover	Diverse Replacement
	Local Search	
	Selection	
GA	Crossover	Mutation
	Replacement	

5 Performance of the SS on Real-World Data

5.1 An Engine Piston Manufacturing Process

The developed SS method is tested on the real-world data collected at an engine piston manufacturing firm in Konya industry area in Turkey. The engine pistons are shown in Fig 6.

Piston is one of the most important moving components in the engine. It provides the necessary vacuuming (sucking stroke) process required for filing the fuel-air mixture into the motor rotation and compression (compression stroke) process to form the necessary pressure to combust the mixture instantly by utilizing the inert power of the crankshaft (Kaya and Engin 2007). The engine pistons are processed on the machines

Fig. 6. Picture of piston

which are equipped with the computer controlled machinery using the latest technology. They are processed by 6 different operations. These operations are explained roughly as shown in Fig 7.

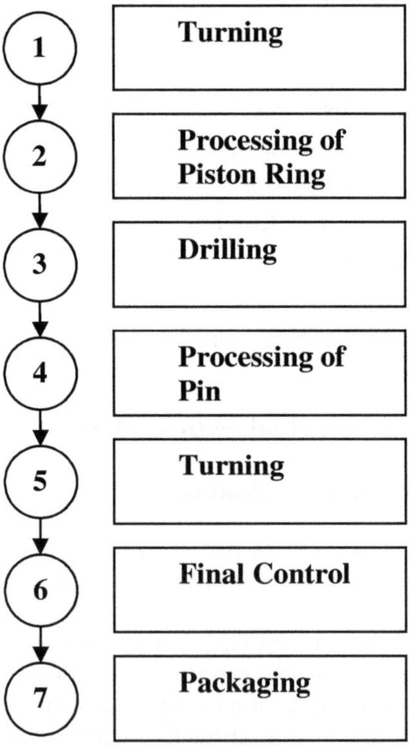

Fig. 7. Operations of engine piston

5.2 The Multiobjective Value

The multiobjective value aggregates the Satisfaction Index (SI) of two objectives. The satisfaction indexes are calculated taking into consideration the completion times of jobs. The question arises how to compare a fuzzy completion time of a job with its fuzzy due date, i.e. how to calculate the likelihood that a job is tardy. In this study, two approaches are used;

The approach based on the possibility measure (PM) introduced by Dubois et al. (1988) and the approach based on the area of intersection measure (AIM) introduced by Sakawa and Kubota (2000).

1. The possibility measure

The possibility measure approach was used by Itoh and Ishii (1999). The possibility measure $\pi_{\tilde{C}_j}(\tilde{d}_j)$ of a fuzzy event, $\tilde{C}_j$ on a fuzzy set $\tilde{d}_j$ is defined as follows (Itoh and Ishii 1999).

$$\pi_{\tilde{C}_j}(\tilde{d}_j) = \sup \min\{\mu_{\tilde{C}_j}(t), \quad \mu_{\tilde{d}_j}(t)\} \qquad j = 1,\ldots,n \tag{3}$$

It is used to measure the satisfaction grade of a fuzzy completion time $SG_T(\tilde{C}_j)$ of job j :

$$SG_T(\tilde{C}_j) = \pi_{\tilde{C}_j}(\tilde{d}_j) \tag{4}$$

Where $\mu_{\tilde{C}_j}(t)$ and $\mu_{\tilde{d}_j}(t)$ are the membership functions of fuzzy sets $\tilde{C}_j$ and $\tilde{d}_j$ respectively (Fayad and Petrovic 2003). The possibility measure of the fuzzy due date $\tilde{d}_j$ is illustrated in Fig 8.

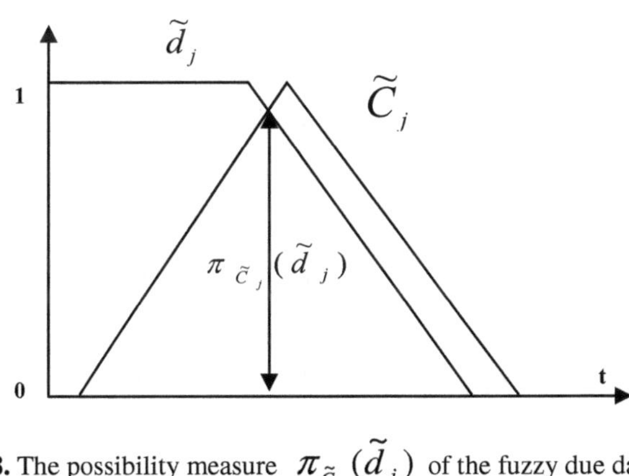

Fig. 8. The possibility measure $\pi_{\tilde{C}_j}(\tilde{d}_j)$ of the fuzzy due date $\tilde{d}_j$

2. Area of intersection measures approach

For the fuzzy completion for each job expressed as a triangular membership functions, $\tilde{C}_j$, as an index showing the portion of $\tilde{C}_j$ that meets the fuzzy due date $\tilde{d}_j$

(Sakawa and Kubota 2000). The area of intersection measures the portion of $\tilde{C}_j$, that is completed by the due date $\tilde{d}_j$. It is shown in Fig 9. The satisfaction grade of a fuzzy completion time of job *j* is defined as follows (Fayad and Petrovic 2003).

$$SG_T(\tilde{C}_j) = {(area\tilde{C}_j \cap \tilde{d}_j)} \Big/ {(area\tilde{C}_j)} \tag{5}$$

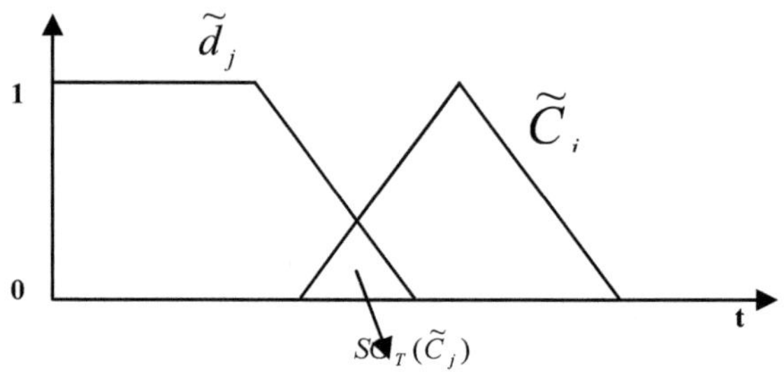

Fig. 9. Satisfaction grade of completion time using area of intersection

The objectives given in (3) and (4) are transformed into the objectives to maximize their corresponding satisfaction grades as follow (Fayad and Petrovic 2003):

1. Satisfaction grade of Average Tardiness

$$S_{AT} = \frac{1}{n}\sum_{j=1}^{n} SG_T(\tilde{C}_j) \tag{6}$$

2. Satisfaction grade of number of tardy jobs: A parameter λ is introduced such that a job j , j=1,...,n is considered to be tardy if $SG_T(\tilde{C}_j) \le \lambda, \quad 0 \le \lambda \le 1$. After calculating the number of tardy jobs n tardy, the satisfaction grade S_{NT} is given as (Fayad and Petrovic 2003):

$$S_{NT} = \begin{cases} 1 & if \quad nTardy = 0 \\ (n'' - nTardy)/n'' & if \quad 0 < nTardy < n'' \\ 0 & if \quad nTardy > n'' \end{cases} \tag{7}$$

n'' =15 % of the total number of jobs.

In the study, three different aggregation operators are investigated. These are given (Fayad and Petrovic 2003):

1. Average of the satisfaction grades:

$$F_1 = \frac{(S_{AT} + S_{NT})}{2} \tag{8}$$

2. Minimum of the satisfaction grades:

$$F_2 = \min(S_{AT}, S_{NT}) \tag{9}$$

3. Average weighted sum of the satisfaction grades:

$$F_3 = 1/2\left(w_1 S_{AT} \quad + \quad w_2 S_{NT}\right) \tag{10}$$

Where $w_k \in [0,1]$, $k = 1,2$, are normalized weights randomly chosen used in the GA and changed in every iteration in order to explore different areas of the search space (Fayad and Petrovic 2003; Murata et al. 1996).

5.3 Experiments

The proposed SS procedure is given below;

```
Set initial values:
   Number of job;
   Number of operation;
   Fuzzy processing time;
   Fuzzy due date;
   Order quantity;
Set the SS value:
   Initial population size (PopSize);
   Reference Set size(Ref Set Size);
   Sub set size (Sub Set Size);
   Stopping Criterion 1;
   Stopping Criterion 2;
   Stopping Criterion 3;
begin
   repeat
       Create initial population based on a memetic algorithm (PopSize)
       Repeat
           Generate Reference Set (Ref Set Size);
           Repeat
               Select Subset (Sub Set Size);
               Combine Solutions;
               Improve Solutions;
           Until (Stopping Criterion 1);
           Update Reference Set;
       Until (Stopping Criterion 2);
   Until (Stopping Criterion 3);
   End.
```

Fig. 10. The proposed SS procedure

The proposed SS algorithms are tested on real-world data collected at an engine piston manufacturing firm over monthly periods along six months. The load of each month is given in Table 3.

Table 3. The load of each month

Month	Number of jobs
1	15
2	13
3	25
4	19
5	17
6	21

The fuzzy processing time of each operation is estimated according to the types of machines in use. While some machines are semi-automated and can be operated at different speeds, others are staff-operated and therefore the processing times are staff dependent (Fayad and Petrovic 2003).

Using the proposed SS algorithm, the best set of parameters is used to obtain the optimal or near optimal solutions of multiobjective fuzzy flow shop scheduling problem in the shortest time. These parameters are determined as follows:

In this study, full factorial DOE has been used. The application involves five parameters (factors) with different possible values each. These parameters are given in Table 4.

Table 4. The parameters proposed by SS for multiobjective fuzzy permutation flow shop scheduling problem

Parameter	**Range**
Initial population size	10, 20, 30, 40, 50
Reference Set size(Ref Set Size)	0.1, 0.2, 0.3, 0.4, 0.5, 0.6, 0.7, 0.8, 0.9
Sub set size	2, 4, 6, 8, 10
Stopping Criterion 1	25, 75, 125, 175, 225, 250
Stopping Criterion 2	25, 75, 125, 175, 225, 250

The best parameter set for the proposed SS is given in Table 5.

Table 5. The best parameter set for the proposed SS algorithm

Parameter	**Value**
Initial population size	40
Reference Set size(Ref Set Size)	0.5
Sub set size	2
Stopping Criterion 1	75
Stopping Criterion 2	125

In the study, the stopping criterion 3 is selected to be 25 (iteration number) as a constant.

Also in the study, two values are tested for λ : $\lambda = 0.4$ and 0.7 (Fayad and Petrovic 2003). The experimental result shows, the lower value of λ ($\lambda = 0.4$) find the better solution for the three aggregation operators. In the study we used the lower value of λ ($\lambda = 0.4$).

The algorithm was implemented in Borland Delphi and the computational experiments were performed on a Pentium 4 with 3 GHz processor and 512 MB memory.

Multiobjective fuzzy permutation flow shop scheduling problems are formulated by two objectives. These are to minimize the average tardiness and to minimize the number of tardy jobs. In the study the fitness value of the proposed SS aggregates the satisfaction index of these two objectives. To compare the fuzzy completion time of a job with its fuzzy due date, two approaches are used. These are PM introduced by Dubuois and Prade (1998) and AIM introduced by Sakawa and Kubota (2000). In this research, three different aggregation operators are investigated. These are the average of the satisfaction grades F_1, the minimum of the satisfaction grades F_2 and the average weighted sum of the satisfaction grades F_3.

The multiobjective fuzzy permutataion flow shop scheduling problems are solved by the proposed SS and HGA. These three aggregation operators' averages, standard deviations, and maximum values are presented in Tables 6.- 11. The improvement rate is calculated as

$$\textit{improvement rate} = \frac{SS - HGA}{HGA} \tag{11}$$

Table 6. First month's problem: the average and best values of satisfaction grades for SS and HGA

Fitness value		HGA		Proposed SS algorithm		Improvement Rate	
		AIM	PM	AIM	PM	AIM	PM
F_1 - Average of the satisfaction grades	Average	0.3893	1.0000	0.3818	1.0000	HGA Better	0.00
	Std. Dev.	0.0159	0.0000	0.0407	0.0000	1.56	-
	Max	0.4000	1.0000	0.4889	1.0000	0.22	0.00
F_2-Minimum of the satisfaction grades	Average	1.0000	1.0000	1.0000	1.0000	0.00	0.00
	Std. Dev.	0.0000	0.0000	0.0000	0.0000	-	-
	Max	1.0000	1.0000	1.0000	1.0000	0.00	0.00
F_3- Average weighted sum of the satisfaction grades	Average	0.3256	0.4926	0.4974	0.4615	0.53	HGA Better
	Std. Dev.	0.1877	0.1886	0.1974	0.2156	0.05	0.14
	Max	0.7320	0.7641	0.7809	0.8503	0.07	0.11

Table 7. Second month's problem: the average and best values of satisfaction grades for SS and HGA

Fitness value		HGA		Proposed SS algorithm		Improvement Rate	
		AIM	PM	AIM	PM	AIM	PM
F_1 - Average of the satisfaction grades	Average	0.3846	1.0000	0.3246	1.0000	0.18	0.00
	Std. Dev.	0.0000	0.0000	0.0370	0.0000	-	-
	Max	0.3846	1.0000	0.3846	1.0000	0.40	0.00
F_2-Minimum of the satisfaction grades	Average	1.0000	1.0000	1.0000	1.0000	0.00	0.00
	Std. Dev.	0.0000	0.0000	0.0000	0.0000	-	-
	Max	1.0000	1.0000	1.0000	1.0000	0.00	0.00
F_3- Average weighted sum of the satisfaction grades	Average	0.3645	0.5331	0.3511	0.5309	HGA Better	0.00
	Std. Dev.	0.1338	0.1916	0.2068	0.2401	0.55	0.25
	Max	0.5403	0.9487	0.8014	0.8765	0.48	HGA Better

Table 8. Third month's problem: the average and best values of satisfaction grades for SS and HGA

Fitness value		HGA		Proposed SS algorithm		Improvement Rate	
		AIM	PM	AIM	PM	AIM	PM
F_1 - Average of the satisfaction grades	Average	0.4344	1.0000	0.4552	1.0000	0.05	0.00
	Std. Dev.	0.0398	0.0000	0.0601	0.0000	0.51	-
	Max	0.5400	1.0000	0.5400	1.0000	0.00	0.00
F_2-Minimum of the satisfaction grades	Average	1.0000	1.0000	1.0000	1.0000	0.00	0.00
	Std. Dev.	0.0000	0.0000	0.0000	0.0000	-	-
	Max	1.0000	1.0000	1.0000	1.0000	0.00	0.00
F_3- Average weighted sum of the satisfaction grades	Average	0.4568	0.4705	0.4717	0.4848	0.03	0.03
	Std. Dev.	0.1812	0.2710	0.2090	0.1660	0.15	HGA Better
	Max	0.8003	0.9461	0.8634	0.8348	0.08	HGA Better

The improvement rates of the proposed SS with respect to HGA for each aggregation operator are presented in Tables 6- 11.

For the average of the satisfaction grades F_1; the proposed SS method found a better PM average value for all the six months problems and found a better AIM average value for the five months problems. HGA found a better AIM average value for only the one month problems.

Table 9. Fourth month's problem: the average and best values of satisfaction grades for SS and HGA

Fitness value		HGA		Proposed SS algorithm		Improvement Rate	
		AIM	PM	AIM	PM	AIM	PM
F_1 - Average of the satisfaction grades	Average	0.4126	1.0000	0.4193	1.0000	0.02	0.00
	Std. Dev.	0.0125	0.0000	0.0547	0.0000	3.37	-
	Max	0.4211	1.0000	0.5965	1.0000	0.42	0.00
F_2-Minimum of the satisfaction grades	Average	1.0000	1.0000	1.0000	1.0000	0.00	0.00
	Std. Dev.	0.0000	0.0000	0.0000	0.0000	-	-
	Max	1.0000	1.0000	1.0000	1.0000	0.00	0.00
F_3- Average weighted sum of the satisfaction grades	Average	0.3065	0.5501	0.3943	0.5102	0.29	HGA Better
	Std. Dev.	0.1517	0.2060	0.1449	0.2344	HGA Better	0.14
	Max	0.5310	0.8443	0.6930	0.9008	0.31	0.07

Table 10. Fifth month's problem: the average and best values of satisfaction grades for SS and HGA

Fitness value		HGA		Proposed SS algorithm		Improvement Rate	
		AIM	PM	AIM	PM	AIM	PM
F_1 - Average of the satisfaction grades	Average	0.4118	1.0000	0.4267	1.0000	0.04	0.00
	Std. Dev.	0.0000	0.0000	0.0559	0.0000	HGA Better	-
	Max	0.4118	1.0000	0.5490	1.0000	0.33	0.00
F_2-Minimum of the satisfaction grades	Average	1.0000	1.0000	1.0000	1.0000	0.00	0.00
	Std. Dev.	0.0000	0.0000	0.0000	0.0000	-	-
	Max	1.0000	1.0000	1.0000	1.0000	0.00	0.00
F_3- Average weighted sum of the satisfaction grades	Average	0.4531	0.5411	0.4139	0.5554	HGA Better	0.03
	Std. Dev.	0.2634	0.1662	0.1975	0.2161	HGA Better	0.30
	Max	0.9234	0.7943	0.8179	0.9434	HGA Better	0.19

For the minimum of the satisfaction grades F_2; the proposed SS method and the used HGA found the same average value of PM and AIM.

For the average weighted sum of the satisfaction grades F_3; the proposed SS method found a better PM average value for the three months problems and found a

Table 11. Sixth month's problem: the average and best values of satisfaction grades for SS and HGA

Fitness value		HGA		Proposed SS algorithm		Improvement Rate	
		AIM	PM	AIM	PM	AIM	PM
F_1 - Average of the satisfaction grades	Average	0.4238	1.0000	0.4616	1.0000	0.09	0.00
	Std. Dev.	0.0238	0.0000	0.0817	0.0000	2.43	-
	Max	0.4524	1.0000	0.6349	1.0000	0.40	0.00
F_2-Minimum of the satisfaction grades	Average	1.0000	1.0000	1.0000	1.0000	0.00	0.00
	Std. Dev.	0.0000	0.0000	0.0000	0.0000	-	-
	Max	1.0000	1.0000	1.0000	1.0000	0.00	0.00
F_3- Average weighted sum of the satisfaction grades	Average	0.3925	0.5459	0.4519	0.5056	0.15	HGA Better
	Std. Dev.	0.1908	0.2428	0.2044	0.2137	0.07	HGA Better
	Max	0.7478	0.9187	0.8104	0.8820	0.08	HGA Better

better AIM average value for the four months problems. HGA found a better PM average value for the three months problems and found a better AIM average value for only the two months problems.

As it is seen in Tables 6- 11, the proposed SS found the better solutions for the two aggregation operators. These are the average of the satisfaction grades and the average weighted sum of the satisfaction grades.

6 Conclusion and Directions for Future Research

The SS methodology is very flexible since each of its elements can be implemented in a variety of ways and degrees of sophistication (Marti et al. 2006). In this study, we applied scatter search method to multiobjective permutation fuzzy flow shop scheduling problem. The considered problem is a well known NP-hard problem. Two objectives which are average tardiness and the number of tardy jobs are minimized. The multiobjective approach aggregates the satisfaction index of two objectives. For calculating the satisfaction index, two approaches, which are possibility measure and area of intersection measure, are used. The proposed SS method is tested on real-world data collected at an engine piston manufacturing company. The result of the proposed SS method is compared with the HGA solutions. The proposed SS method outperformed HGA. The results show that the proposed SS method is a good problem solving technique for fuzzy multiobjective flow shop scheduling problem. For further research, the proposed SS method can be applied to other multiobjective scheduling problems.

List of Abbreviations

SS	Scatter Search
DOE	Design of Experiments
CDS	Campbell Dudek and Smith
$\tilde{t}_{ij}$	Fuzzy Processing Times
$\tilde{d}_j$	Fuzzy Due Date
$\tilde{T}_j$	Fuzzy Tardiness
$\tilde{C}_j$	Fuzzy Completion Time
C_{NT}	Number of Tardy Jobs
C_{AT}	Average Tardiness
b	Best Solution
GA	Genetic Algorithm
HGA	Hybrid Genetic Algorithm
SI	Satisfaction Index
PM	Possibility Measure
AIM	Area of Intersection Measure
SG_T	Satisfaction Grade of Fuzzy Completion Time
S_{AT}	Satisfaction Grade of Average Tardiness
S_{NT}	Satisfaction Grade
n''	15 % of the Total Number of Jobs
F_1	Average of the Satisfaction Grades
F_2	Minimum of the Satisfaction Grades
F_3	Average Weighted Sum of the Satisfaction Grades
PopSize	Population Size
Ref Set Size	Reference Set Size

References

Bajestani, M.A., Rabbani, M., Rahimi-Vahed, A.R.: A multi-objective scatter search for a dynamic cell formation problem. Computers & operations research 36, 777–794 (2009)

Campbell, H.G., Dudek, R.A., Smith, M.L.: An heuristic algorithm for the n-job m-machine sequencing problem. Management science 16(B), 630–637 (1970)

Chakraborty, U.K., Deb, K., Chakraborty, M.: Analysis of selection algorithms: A Markov chain approach. Evolutionary Computation 4, 133–167 (1996)

Chang, P.C., Chen, S.H., Liu, C.H.: Sub-population genetic algorithm with mining gene structures for multiobjective flowshop scheduling problems. Expert systems with applications 33, 762–771 (2007)

Cheng, R., Gen, M., Tsujimura, Y.: A tutorial survey of job shop scheduling problems using genetic algorithms Part II hybrid genetic search strategies. Computers & Industrial engineering 36, 343–364 (1999)

Choi, I.C., Kim, S.I., Kim, H.S.: A genetic algorithm with a mixed region search for the asymmetric traveling salesman problem. Computers &operations research 30, 773–786 (2003)

Chung, J.W., Oh, S.M., Choi, I.C.: A hybrid genetic algorithm for train sequencing in the Korean railway. Omega the international journal of management science 37, 555–565 (2009)

Dasgupta, D., Forrest, S.: Novelty detection in time series data using ideas from immunology. In: Proceedings of the ISCA 1996, Reno. Nevada. June 19-21 (1996)

De Castro, L.N., Von Zuben, F.J.: Artificial immune systems. Part 1. Basic theory and applications. Technical Report. TR-DCA 01/99 (1999)

Dell'Amico, M., Iori, M., Martello, S.: Heuristic algorithms and scatter search for cardinality constrained P//Cmax problem. Journal of Heuristic 10(2), 169–204 (2004)

Djan-Sampson, P.O., Sahin, F.: Structural learning of Bayesian networks from complete data using the scatter search documents. In: IEEE International Conference on Systems Man and Cybernetics, pp. 3619–3624 (2004)

Dorigo, M., Gambardella, L.M.: Ant Colony System: A cooperative learning approach to the traveling salesman problem. IEEE Transactions on Evolutionary Computation 1(1), 53–66 (1997)

Dubois, D., Prade, H.: Possibility theory: an approach to computerized processing of uncertainty, New York (1988)

Engin, O.: To increase the performance of flow shop scheduling problems solving with genetic algorithms: a parameters optimization. PhD. Thesis. Istanbul Technical University. Institute of Science and Technology. Istanbul. Turkey (2001)

Fayad, C., Petrovic, S.: A fuzzy genetic algorithm for real- World job shop scheduling. University of Nottingham (2003), http://www.cs.nott.ac.uk/~cxf~sxp

Forrest, S., Perelson, A., Allen, L., Cherukuri, R.: Self-nonself discrimination in a computer. In: Proceedings of the IEEE Symposium on Research in Security and Privacy, pp. 200–212. IEEE Computer Society Press, Los Alamitos (1994)

Glover, F., Laguna, M., Marti, R.: Scatter search. In: Advances in Evolutionary Computation: Theory and Applications, pp. 519–537. Springer, New York (2003)

Glover, F.: Heuristics for integer programming using surrogate constraints. Decision Sciences 8, 156–166 (1977)

Glover, F., Laguna, M.: Tabu Search. Kluwer Academic Publishers, Boston (1997)

Goldberg, D.E.: Genetic Algorithms in Search. In: Optimization and Machine Learning. Addision Wesley, London (1989)

Herrera, F., Lozano, M., Molina, D.: Continuous Scatter Search: An analysis of the integration of some combination methods and improvement strategies. European Journal of Operational Research 169, 450–476 (2006)

Hung, W.N.N., Song, X.: BDD Variable ordering by scatter search. IEEE, 368–373 (2001)

Hung, W.N.N., Song, X., Aboulhamid, E.M., Driscoll, M.: BDD Minimization by scatter search. IEEE Transactions on Computer-Aided Design of Integrated Circuits and Systems 21(8), 974–979 (2002)

Ignall, E., Schrage, L.: Applicant of the branch and bound technique to some flow shop scheduling problems. Operations research 13, 401–412 (1965)

Itoh, T., Ishii, H.: Fuzzy due date scheduling problem with fuzzy processing time. International transaction in operations research 6, 639–647 (1999)

Johnson, S.M.: Optimal two and three stage production schedules with setup times included. Naval Research Logistics Quarterly 1, 61–68 (1954)

Kahraman, C., Engin, O., Kaya, İ., Yılmaz, M.K.: An application of effective genetic algorithms for solving hybrid flow shop scheduling problems. International Journal of Computational Intelligence Systems 1(2), 134–147 (2008)

Kaya, I., Engin, O.: A New Approach to Define Sample Size at Atributes Control Chart in Multistage Processes: an Application in Engine Piston Manufacturing Process. Journal of Materials Processing Technology 183, 38–48 (2007)

Kirkpatrick, S., Gelatt, C.D., Vecchi, M.P.: Optimization by Simulated Annealing. Science 220(4598), 671–680 (1983)

Maenhout, B., Vanhoucke, M.: New computational results for the nurse scheduling problem: A scatter search algorithm. In: Gottlieb, J., Raidl, G.R. (eds.) EvoCOP 2006. LNCS, vol. 3906, pp. 159–170. Springer, Heidelberg (2006)

Marti, R., Laguna, M., Glover, F.G.: Principles of scatter search. European Journal of Operational Research 169, 359–372 (2006)

Murata, T., Ishibuchi, H., Tanaka, H.: Multiobjective genetic algorithm and its applications to flowshop scheduling. Computers Industrial engineering 30(4), 957–968 (1996)

Nezhad, S.S., Assadi, R.G.: Preference ratio based maximum operator approximation and its application in fuzzy flow shop scheduling. Applied soft computing 8, 759–766 (2008)

Niu, O., Gu, X.S.: An improved genetic-based particle swarm optimization for no idle permutation flow shops with fuzzy processing time. In: Yang, Q., Webb, G. (eds.) PRICAI 2006. LNCS (LNAI), vol. 4099, pp. 757–766. Springer, Heidelberg (2006)

Noorul, H.A., Saravanan, M., Vivekrajc, A.R.: A scatter search approach for general flow shop scheduling problem. Int. J. Adv. Manuf. Technol. 31, 731–736 (2007)

Nowicki, E., Smutnicki, C.: Some Aspects of Scatter Search in the Flow-Shop Problem. European Journal of Operational Research 169, 654–666 (2006)

Pasupathy, T., Rajendran, C., Suresh, R.K.: A multi- objective genetic algorithm for scheduling in flow shops to minimize the makespan and total flow time of jobs. Int. J. Adv. Manuf. Technol. 27, 804–815 (2006)

Petrovic, S., Song, X.: A new approach to two machine flow shop problem with uncertain processing times. Optimization and Engineering 7(3), 329–342 (2006)

Rahimi-Vahed, A.R., Javadi, B., Rabbani, M., Tavakkoli-Moghaddam, R.: A multi-objective scatter search for a bi-criteria no-wait flow shop scheduling problem. Engineering Optimization 40(4), 331–346 (2008)

Reeves, C.R.: Genetic algorithm for flow shop sequencing. Computers & operations research 15, 5–23 (1995)

Russell, R.A., Chiang, W.C.: Scatter Search for the Vehicle Routing Problem with Time Windows. European Journal of Operational Research 169, 606–622 (2006)

Sakawa, M., Kubota, R.: Fuzzy programming for multiobjective job shop scheduling with fuzzy processing time and fuzzy duedate through genetic algorithms. European journal of operational research 120, 393–407 (2000)

Sevaux, M., Thomin, P.: Scatter Search and Genetic Algorithm: a one machine scheduling problem comparison. In: The sixteenth triennial conference of international federation of operational research societies. IFORS, Edinburgh. UK. juillet (2002)

Silva, C.G.D., Climaco, J., Figueira, J.: A Scatter Search Method for Bi-criteria {0.1}-Knapsack Problems. European Journal of Operational Research 169, 373–391 (2006)

Temiz, I., Erol, S.: Fuzzy branch and bound algorithm for flow shop scheduling. Journal of intelligent manufacturing 15, 449–454 (2004)

Yao, J.S., Lin, F.T.: Constructing a fuzzy flow shop sequencing model based on statistical data. International journal of approximate reasoning 29, 215–234 (2002)

Zhu, J., Du, G., Wang, L.: Artificial immune algorithm for fuzzy flow shop scheduling problem. Dynamics of continuous discrete and impulse systems-series B- applications & algorithms 13, 383–386 (2006)

[illegible]
[illegible] accelerated [illegible] (2006)
[illegible]
[illegible] based [illegible] for [illegible] (2007)
[illegible]
[illegible] [illegible] of Operations Research 40, [illegible] (2006)
[illegible]
[illegible] International [illegible] Conference on [illegible] (2009)
[illegible] programming [illegible] (1998)
[illegible] Journal of Operational Research [illegible] (2006)
Zhu, L. [illegible] hybrid [illegible] algorithm for [illegible] problems [illegible] and [illegible] [illegible] algorithms LNCS 3[illegible], [illegible] (2006)

Genetic Algorithm for Job Shop Scheduling under Uncertainty

Deming Lei

School of Automation, Wuhan University of Technology, Wuhan City,
Hubei Province, P.R. China
deminglei11@163.com

Summary. This chapter first presents job shop scheduling problems (JSSP) with fuzzy processing time and fuzzy trapezoid or doublet due-date. An efficient random key genetic algorithm (RKGA) is suggested, in which a random key representation and a new decoding strategy are proposed and two-point or discrete crossover are used. Performance analyses on random key representation are done and RKGA is compared with other algorithm. Computations results validate the effectiveness of random key representation and the promising advantage of RKGA on fuzzy scheduling.

This chapter then presents flexible job shop scheduling problem (fJSSP) with fuzzy processing time. An efficient decomposition-integration genetic algorithm (DIGA) is developed, which uses two-string representation, an effective decoding method and a main population. In each generation, the main population is decomposed into two sub-populations for sub-problems of fJSSP, sub-populations evolve independently and a new main population is obtained by storing the best half of the population formed with two evolved sub-populations and their copies. DIGA is tested and compared with another algorithm. Computational results show good performance of DIGA.

Job shop scheduling problem with stochastic processing time is finally considered. The Giffler-Thompson (GT) procedure is extended in the stochastic context and some operations on the stochastic processing time are defined. A genetic algorithm (GA) is presented to minimize the maximum completion time, in which a permutation-based representation method and a modified crossover are used. The proposed algorithm is tested on a set of benchmark problems and compared with a hybrid method. Computational results demonstrate the effectiveness of the proposed algorithm.

1 Introduction

This chapter is made up of six sections. The introduction is done in Section 1. The second section summarizes the literature on JSSP under uncertainty. The third section is about random key scheduling algorithm for fuzzy job shop scheduling, in which a random key representation method and a direct decoding procedure are proposed. The minimum agreement index and the maximum fuzzy completion time are regarded

U.K. Chakraborty (Ed.): Comput. Intel. in Flow Shop and Job Shop Sched., SCI 230, pp. 191–228.
springerlink.com

respectively as an objective. The forth section presents flexible job shop scheduling problem with fuzzy processing time and an efficient decomposition-integration genetic algorithm is proposed, in which the main population is decomposed into two sub-populations that evolve independently and are combined for a new main population. The objective is to minimize the maximum fuzzy completion time. JSSP with stochastic processing time is considered in the fifth section. The extended GT procedure and some operations on stochastic processing time are first suggested to build a complete schedule. An efficient GA is then proposed, in which a permutation-based representation is utilized. The objective is the makespan itself and not the expected makespan. In the final section, some conclusions are drawn and new trends of job shop scheduling with uncertainty are discussed.

2 Literature Review

Manufacturing systems are often subject to some uncertain events which may disturb their working process [1]: machine failure, operator unavailability, out-of-stock condition, changes in availability date and the latest completion time. It is realistic to consider a system in an uncertain context; however, the research on job shop scheduling under uncertainty is still in infancy.

2.1 Single Objective Scheduling: Fuzzy Case

In general, the various factors of job shop scheduling are treated as crisp value; however, this assumption is not realistic in many cases. In order to reflect the real-world situations, it may be more appropriate to consider fuzzy processing time due to man-made factors and fuzzy due-date tolerating a certain amount of delay in due-date.

In the past decade, some results have been obtained for fuzzy job shop scheduling problem (FJSSP). Kuroda and Wang [2] discussed the static JSSP and dynamic JSSP with fuzzy information. A branch-and-bound algorithm is used to solve the static JSSP and the methods for dynamic JSSP are also considered. Sakawa and Mori [3] presented an efficient GA by incorporating the concept of similarity among individuals and matrix representation method. Song et al. [4] presented a combined strategy of GA and ant colony optimization. They also designed a new neighborhood search method and an improved tabu search to improve the local search ability of the hybrid algorithm. Niu et al. [5] redefined a particle swarm optimization (PSO), combined PSO with genetic operators and applied the combined PSO to job shop scheduling with fuzzy processing time.

2.2 Single Objective Scheduling: Stochastic Case

Stochastic job shop scheduling problem (SJSSP) is an important aspect of manufacturing systems in stochastic context. It is the extended version of JSSP and presents some difficulties for its nature. (1) The objective evaluation is very time-consuming, especially, when multiple objectives are optimized simultaneously, the sorting of objective vectors is very expensive in time. (2) Many approaches used in the deterministic case cannot be directly extended to the stochastic context. The optimal solution obtained without taking into account random events may present no interest in a stochastic context. Some concepts and methods are required to be defined or designed again. For instance, some coding and decoding methods of JSSP cannot be applied to represent the solution of SJSSP again.

There are many stochastic scheduling results which establish the rules to determine the sequence of parts to minimize an expected objective function on single machine [6,7]. Not many results have been obtained for the stochastic scheduling of more than two machines [8], as the problems are considerably harder. Few results have been obtained for SJSSP. Luh [9] presented an effective approach for JSSP considering uncertain arrival times, processing time, due-date and part priority. A solution methodology based on a combined Lagrangian relaxation and the stochastic dynamic programming is developed to obtain the dual solutions. Ginzburg [10] considered three sets of costs in JSSP with stochastic processing time in normal, exponential and uniform distributions and treated the problem as the identification of the earliest start time in order to minimize the average cost of storage and tardiness from the delivery time. Tavakkoli-Moghaddam [11] proposed a hybrid method based on neural network and simulated annealing (SA) for SJSSP. The method uses a neural network approach to generate an initial feasible solution and then a SA to improve the quality of the initial solution. Lei et al. [12] provided a stochastic order-based approach to compute the stochastic objective and suggested an efficient GA for SJSSP.

2.3 Multi-Objective Scheduling: Uncertain Case

Not many results have been obtained for uncertain scheduling problems involving multiple objectives. Sakawa and Kubota [13] considered FJSSP and presented a GA incorporating the concept of similarity among individuals by using Gantt charts. The objective is to maximize the minimum agreement index and the average agreement index and to minimize the maximum fuzzy completion time. Li et al.[14] proposed a GA for FJSSP with alternative machines by adopting two-chromosome presentation and the extended version of GT Procedure (Giffler and Thompson [15]).

Lei [16] addressed the fuzzy problem with objectives of the minimum agreement index, the maximum fuzzy completion time and the mean fuzzy completion time. He presented an efficient Pareto archive particle swarm optimization, in which the global best position selection is combined with the crowding measure-based archive maintenance. Xing et al.[17] presented a multi-objective genetic algorithm for fuzzy scheduling problem with objectives of the minimum agreement index and the average agreement index. Ghrayeb [18] presented a bi-criteria genetic algorithm to minimize the integral value and the uncertainty of the fuzzy makespan.

Javadi et al. [19] developed a fuzzy multi-objective linear programming model for multi-objective no-wait flow shop scheduling in a fuzzy environment. The proposed model attempts to simultaneously minimize the weighted mean completion time and the weighted mean earliness. Lei and Xiong [20] addressed the problem of stochastic job shop scheduling, in which the processing time is modeled by a random variable. They first presented a permutation-based representation method and then designed an efficient multi-objective evolutionary algorithm to minimize the expected makespan and the expected total tardiness.

3 Fuzzy Job Shop Scheduling

3.1 Problem Description

$n \times m$ FJSSP can be described as follows: given n jobs $J_i\,(i=1,2,\cdots,n)$, each composed of several operations o_{ij} that must be processed on machines $M_j\,(j=1,2,\cdots,m)$. The processing time of operation o_{ij} is represented as triangular fuzzy number (TFN) $\tilde{p}_{ij}=\left(a_{ij}^{1},a_{ij}^{2},a_{ij}^{3}\right)$. For job J_i, doublet due-date $\left(d_i^1,d_i^2\right)$ and trapezoid due-date $d_i=\left(e_i^1,e_i^2,d_i^1,d_i^2\right)$ are respectively considered. Other constraints of JSSP are still suitable to FJSSP. For instance, it is assumed that only one operation can be processed on each machine at a time and each operation cannot be commenced if the precedent operation is still being processed.

In the deterministic context, tardiness or earliness are used to describe the grad of the satisfaction of the customer for delivery. The agreement index can be regarded as the extended version of the above objective in the fuzzy case. The agreement index AI_i of job J_i is defined as follows.

$$AI_i = area\left(C_i \cap d_i\right)/area\left(C_i\right) \tag{1}$$

Where the fuzzy completion time of job J_i is expressed as TFN C_i.

For trapezoid due-date $d_i=\left(e_i^1,e_i^2,d_i^1,d_i^2\right)$, if the completion time of job J_i belongs to the interval $\left[e_i^2,d_i^1\right]$, the grad of satisfaction is equal to 1. In other cases, the grad of the satisfaction diminishes with the increase of the tardiness or earliness. Fig. 1 describes the fuzzy processing time and fuzzy due-date.

Two objectives are considered respectively.

$$AI_{min} = \min_{i=1,2\cdots n} AI_i \tag{2}$$

$$C_{max} = \max_{i=1,2,\cdots n} C_i\,. \tag{3}$$

where C_{max} is the maximum fuzzy completion time and AI_{min} is the minimum agreement index.

3.2 Operations on Fuzzy Processing Time

In fuzzy context, some operations of fuzzy number are required to be redefined to build a schedule. These operations involve addition operation and max operation of two fuzzy numbers as well as the ranking methods of fuzzy numbers. Addition operation is used to calculate the fuzzy completion time of operation. Max operation is to determine the fuzzy beginning time of operation and the ranking method is to compare the maximum fuzzy completion time.

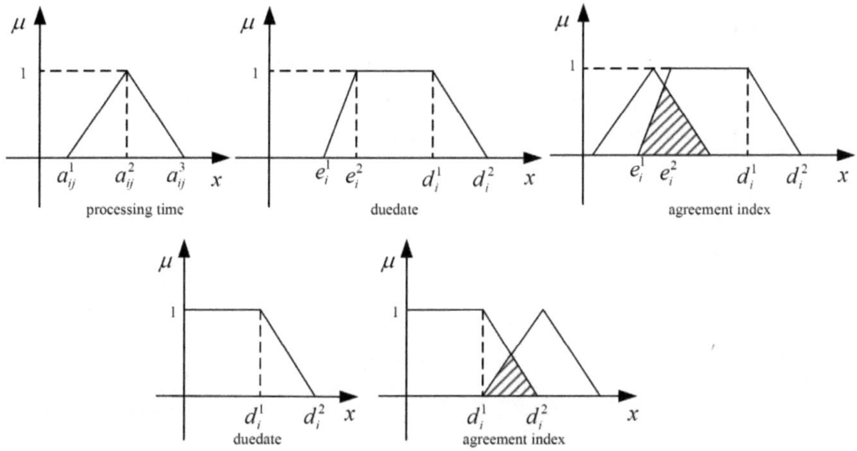

Fig. 1. Fuzzy processing time, fuzzy due-date and agreement index

For two TFNs $\tilde{s}=(s_1,s_2,s_3)$ and $\tilde{t}=(t_1,t_2,t_3)$, the addition of them is shown by the following formula:

$$\tilde{s}+\tilde{t}=(s_1+t_1,s_2+t_2,s_3+t_3) \tag{4}$$

The following criteria are adopted to rank $\tilde{s}=(s_1,s_2,s_3)$ and $\tilde{t}=(t_1,t_2,t_3)$.

Criterion 1: If $c_1(\tilde{s})=\dfrac{s_1+2s_2+s_3}{4}>(<)c_1(\tilde{t})=\dfrac{t_1+2t_2+t_3}{4}$, then $\tilde{s}>(<)\tilde{t}$;

Criterion 2: If $c_1(\tilde{s})=c_1(\tilde{t})$, then $c_2(\tilde{s})=s_2$ is compared with $c_2(\tilde{t})=t_2$ to rank them;

Criterion 3: If they have the identical c_1 and c_2, the difference of spread $c_3(\tilde{s})=s_3-s_1$ is chosen as a third criterion.

For $\tilde{s}=(s_1,s_2,s_3)$ and $\tilde{t}=(t_1,t_2,t_3)$, membership function $\mu_{\tilde{s}\vee\tilde{t}}(z)$ of $\tilde{s}\vee\tilde{t}$ is defined as follows.

$$\mu_{\tilde{s}\vee\tilde{t}}(z)=\sup_{z=x\vee y}\min(\mu_{\tilde{s}}(x),\mu_{\tilde{t}}(y)) \tag{5}$$

In this chapter, the max of two TFNs $\tilde{s}$ and $\tilde{t}$ is approximated with the following criterion:

$$\text{if } \tilde{s}>\tilde{t}\text{, then } \tilde{s}\vee\tilde{t}=\tilde{s}\text{ ; else } \tilde{s}\vee\tilde{t}=\tilde{t}$$

The criterion $\tilde{s}\vee\tilde{t}\approx(s_1\vee t_1,s_2\vee t_2,s_3\vee t_3)$ is first used by Sakawa and Mori [3] and named Sakawa criterion for simplicity. Sakawa criterion has been extensively applied to build a complete scheduling of the fuzzy problem. Fig. 2 shows the real max of two fuzzy numbers and two criteria for the approximate max. Compared with sakawa criterion, the new criterion has the following features:

(1) For $\tilde{s}$ and $\tilde{t}$, the approximate max of them is either $\tilde{s}$ or $\tilde{t}$;

(2) Only three pairs of special points (s_i, t_i) are compared in Sakawa criterion and three criteria to rank them are used in the new criterion. The approximate max obtained by the new criterion approaches the real max better than that of Sakawa criterion.

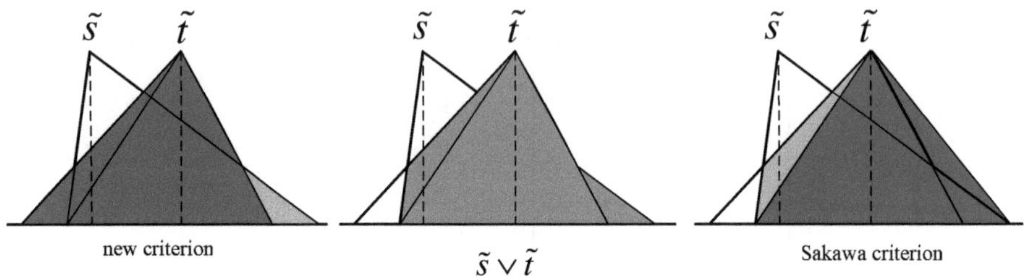

Fig. 2. Comparison between real max and approximate max

3.3 Random Key Genetic Algorithm

Based on random key representation, a new decoding procedure, elite strategy, binary tournament selection, two-point crossover (TPX) or discrete crossover (DX) and swap mutation, RKGA is designed. Compared with the GA with the operation-based representation, RKGA has the following features: the chromosome is a real string; however, RKGA can obtain an operation-based integer string finally. The implementation of RKGA is very simple. It is easy to apply TPX or DX and the illegal individual never occurs in the search process.

The framework of RKGA is described as follows.

(1) Randomly generate an initial population *P* with *N* individuals.
(2) Perform binary tournament selection on *P* .
(3) Perform TPX or DX and swap mutation on population *P* .
(4) If the termination condition is met, stop the search; otherwise, go to step (2).

Table 1. Example of 4 jobs 2 machines FJSSP

Job	Operations					
	Processing time		Processing sequence		Actual processing time	
1	1, 2, 3	3, 4, 5	M_1	M_2	2	4
2	2, 3, 4	2, 4, 6	M_2	M_1	3	4
3	2, 4, 5	3, 5, 7	M_1	M_2	4	5
4	1, 3, 4	2, 3, 5	M_2	M_1	4	3

3.3.1 Random Key Representation

The choice of representation or encoding affects the performance of GA; see, for example, Rothlauf[21], Chakraborty and Janikow[22]. Random key representation is first proposed by Bean [23], which encodes a solution of JSSP with random numbers. For $n \times m$ JSSP, each gene consists of two parts: an integer in set $\{1, 2, \cdots, m\}$ and a fraction generated randomly from $(0,1)$. The integer part of the random key is interpreted as the machine assignment for job and the decimal part is used to construct the operation sequence on each machine.

The above representation method is seldom considered for job sequences violating the precedence constraints and the requirement of the special genetic operators.

In this chapter, we present a new random key representation, which encodes a schedule of $n \times m$ FJSSP as a real string $(p_1, p_2, \cdots, p_n, \cdots, p_{mn})$ with $n \times m$ random numbers in the same interval $[a,b]$.

To obtain a feasible schedule, the following decoding procedure is adopted.

(1) Divide the interval $[a,b]$ into a group of sub-intervals $[a_1, a_2), \cdots, [a_i, a_{i+1}), \cdots, [a_l, a_{l+1}]$, Classify all genes of the chromosome into l groups and make the genes of each group in the same subinterval; where $a = a_1 < a_2 < \cdots a_l < a_{l+1} = b$;

(2) Let $t = 1, h = 0$, start with the first group, choose the gene from small to big and assign the chosen gene a new value of t and let $h = h + 1$, if $h = m$, then $t = t + 1$ and $h = 0$; repeat the above procedure until each gene is assigned a new value and a integer string is obtained;

(3) Translate the integer sting into a list of ordered operations;

(4) The first operation of the list is arranged first, and then the second operation and so on; each operation is allocated in the best available processing time for the required machine of the operation. The procedure is repeated until a schedule is obtained. The procedure is identical to the one proposed by Cheng et al. [24] except the processing time is fuzzy.

In the deterministic context, two strategies can be used to translate individual to schedule of JSSP. The first strategy is to obtain a schedule in terms of the ordered operation list or job permutation et al. The second is to build a schedule with GT procedure; however, only the second strategy using the extended GT procedure is adopted in the fuzzy case. The first decoding strategy is applied in this chapter.

Suppose a chromosome of the 4×2 example in Table 1 is (0.1, 1.3, 2.5, 2.7, 3.9, 1.1, 4.5, 0.8) and p_i is in $[0,5]$. The interval is first divided into five subintervals and then genes are separated into five groups; the integer string (1 2 3 3 4 2 4 1) is obtained after step 2 and the chromosome is finally converted into a ordered operation list $(o_{11}, o_{21}, o_{31}, o_{32}, o_{41}, o_{22}, o_{42}, o_{12})$ in step 3. Fig.3 (a) describes the operation sequence on each machine. The chart can be regarded as the modified version of Gantt chart in fuzzy context and called fuzzy Gantt chart. The TFN above the line is the fuzzy completion time of operation and the TFN under the line is fuzzy beginning time of operation. When the actual processing time of operations are decided and all sequences of operations keep invariant, the actual schedule is obtained. Fig.3 (b) shows the Gantt chart based on the actual processing time in Table 1. The schedule produced in the above procedure is always feasible.

Compared with the Bean's representation, the gene of the new representation also consists of the integer part and the decimal part, however, the random key of the new representation has different meaning, the chromosome of the new representation can be converted into the list of the ordered operation, the list never violates the precedence constraints and no special genetic operators are necessary for the new representation-based GA.

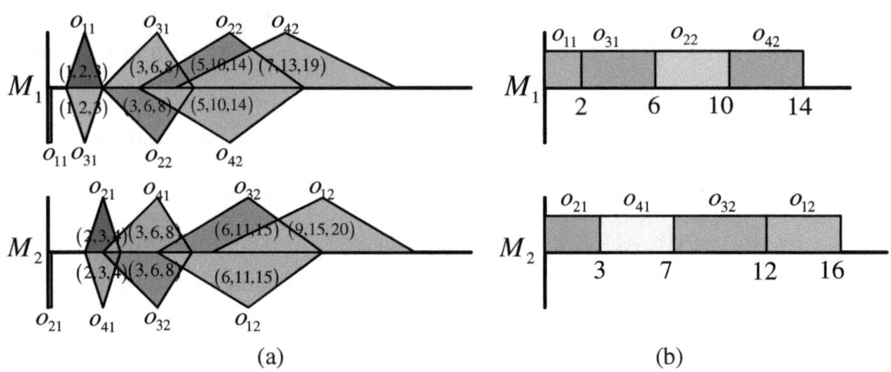

Fig. 3. Fuzzy Gantt chart and Gantt chart

3.3.2 Fitness, Elitism and Genetic Operators

In this chapter, we make fitness function of an individual be equal to its objective function. The classical elite strategy is used, in which the optimal solution produced by RKGA is stored as an elite individual, moreover, the elite individual is always added into population before reproduction.

Roulette wheel reproduction and breeding pool reproduction cannot be applied for the maximum fuzzy completion time, so tournament selection is used. Tournament selection, introduced by Brindle [25] and analyzed by Chakraborty et al. [26], is performed in the following way: first two individuals are randomly selected from the population, and then an individual is chosen if the individual has smaller fitness than the other individual. Finally, the selected individuals go back to the population and can be chosen again.

TPX and DX are frequently used in the real-coded GA. TPX is shown below: first randomly select two cut-off points and then exchange genes between the chosen points. DX is done in the following way: produce the random number s following the uniform distribution on $[0,1]$, if $s < 0.5$, select the gene of one parent; otherwise, select the gene of another parent; repeat the above step until an offspring is obtained.

Mutation is just used to produce small perturbations on chromosomes in order to maintain the diversity of population. The swap mutation is adopted and described as follows: randomly choose two genes and then exchange them.

3.4 Computational Experiments

In this section, performance analyses on random key representation are first done and then RKGA is compared with the GA proposed by Sakawa and Mori [3]. We call the latter SMGA. Ten benchmark problems are used. Problem 1, 2, 3, 5, 6 and 7 are designed by Sakawa and Mori [3] and problem 4 and 8 by Sakawa and Kubota [13]. Problem 1,2,3 and 4 are 6×6 FJSSP and problem 5,6,7,8 are 10×10 FJSSP. Two 15×10 problems 9 and 10 are designed.

3.4.1 Performance Analyses on Random Key Representation

By comparing RKGA with the GA with the operation-based representation (OPGA), the performance analyses on random key are done. OPGA has the same parameter settings and the flow with RKGA. Binary tournament selection and swap mutation like RKGA are also adopted. Generalized order crossover (GOX) and precedence preservative crossover (PPX) are respectively considered in OPGA.

GOX is proposed by Bierwirth [27]. First randomly select a substring ℓ of the first parent, determine the position of the first element of ℓ on the second parent and remove the substring ℓ from the second parent. By inserting ℓ into the position of its first element, the offspring is obtained.

PPX is suggested by Bierwirth et al.[28]. A string of equal length as the chromosome is filled at random with the elements of set $\{1, 2\}$. This string defines the order in which the genes are successively drawn from parent 1 and 2. The offspring is initially empty. A gene which occurs leftmost in two parents is selected. The chosen gene is appended to the offspring and deleted from two parents. This step is repeated until a complete offspring is obtained.

Two variants of RKGA are produced, which RKGA1 denotes RKGA with TPX and RKGA2 is RKGA with DX. OPGA also has two variants. OPGA1 represents OPGA with GOX and OPGA2 is OPGA with PPX. The parameters of four algorithms are as follows: crossover probability of 0.8, mutation probability of 0.1, population scale of 100, the maximum generation of 200 is chosen for 6×6 problems and 300 for 10×10 FJSSP and 500 for other instances.

Four algorithms are implemented by using Microsoft Visual C++ 7.0 and run on Pentium 2.0G PC. All algorithms randomly run 20 times with respect to each instance and the computational results are recoded in Table 2, in which *avg.* indicates the average value of the best solutions found in all runs and *opt.* denotes the best solutions. In each data grid, there are three kinds of data related to the first objective using doublet and trapezoid and the second respectively. The computational times of each algorithm are shown in Table 3. Problem 5, 6, 7 and 8 have the same number of job and machine, the search process of each algorithm for these problems nearly spend the same duration. So Table 3 only describes the average value of the computational times. The values in the parentheses are the average computational times about the first objective using the trapezoid due-date.

When the doublet due-date is considered, two variants of RKGA respectively obtain the maximum value of the first objective of 2 and 4 problems. OPGA1 and OPGA2 cannot approximate the best solutions of any instances. With respect to the

Table 2. Computational results of four algorithms

Problem	RKGA1		RKGA2	
	avg.	*opt.*	*avg.*	*opt.*
5	0.5320	0.6943	0.6875	0.8024
	0.4910	0.5602	0.5515	0.5909
	95.8,131.5,163.1	96,129,160	95.1,130.9,162.2	96,129,160
6	0.7873	0.9000	0.8539	0.900
	0.6382	0.7394	0.6733	0.7397
	94,128.4,164.2	95,125,164	93,126.2,163.6	89,123,158
7	0.2628	0.6032	0.4393	0.6061
	0.3580	0.4801	0.4542	0.5854
	85,117.3,147.9	85,116,143	84.6,115.9,148.6	85,116,143
8	0.9010	0.9687	0.9113	0.9688
	0.8640	0.9675	0.9315	0.9675
	28.7,47.9,65.6	27,47,64	28.4,48,64.1	28,47,62
Problem	OPGA1		OPGA2	
	avg.	*opt.*	*avg.*	*opt.*
5	0.6146	0.6943	0.2897	0.5050
	0.5141	0.5784	0.3272	0.4762
	94.9,130.8,162	96,129,160	96.9,135,164.4	95,133,161
6	0.8320	0.8690	0.6255	0.7951
	0.6610	0.7258	0.5764	0.6926
	94.1,125.9,164.9	95,125,164	96,128.8,165.9	95,125,164
7	0.3772	0.5055	0.1887	0.2561
	0.3609	0.4915	0.2216	0.3164
	85.3,115.4,147.5	85,116,143	86.1,118,147.8	85,116,143
8	0.9065	0.9512	0.8880	0.9675
	0.8760	0.9394	0.8943	0.9686
	28.4,47.8,64.3	28,47,62	29.1,48.2,64.3	26,47,64

Table 3. Computational times of four algorithms

RKGA1 t/s		RKGA2 t/s		OPGA1 t/s		OPGA2 t/s	
AI_{min}	C_{max}	AI_{min}	C_{max}	AI_{min}	C_{max}	AI_{min}	C_{max}
7.53(6.84)	7.45	7.40(6.98)	7.60	7.01(7.15)	6.99	8.01(7.68)	7.60

Table 4. Computational results of two algorithms on the first objective

Problem	RKGA		SMGA	
	avg.	*opt.*	*avg.*	*opt.*
1	0.867418	0.868072	0.826914	0.868072
	0.557809	0.609420	0.531258	0.609420
2	0.972718	0.984321	0.951325	0.984321
	0.747367	0.770032	0.732537	0.770032
3	0.923943	0.933824	0.923943	0.933824
	0.674154	0.700000	0.600433	0.700000
4	0.692308	0.692308	0.692308	0.692308
	0.692308	0.692308	0.692308	0.692308
5	0.687534	0.802372	0.290757	0.495251
	0.551560	0.590909	0.330201	0.476190
6	0.853892	0.90000	0.615423	0.795152
	0.673270	0.739734	0.563191	0.692553
7	0.439251	0.606061	0.176851	0.256061
	0.454249	0.585366	0.231725	0.326532
8	0.911335	0.968750	0.889673	0.941176
	0.931485	0.96748	0.864826	0.96748
9	0.883247	0.953747	0.637628	0.761536
	0.693155	0.780091	0.496435	0.601665
10	0.737256	0.842843	0.512367	0.584548
	0.752662	0.792570	0.505022	0.634521

average value of the first objective, it can be concluded that RKGA2 obtains better results than two variants of OPGA for four instances. The corresponding results of RKGA1 are also better than those of OPGA2.

When the trapezoid due-date is considered, RKGA2 produces the best results of 3 instances, OPGA2 approximate the best solution of one problem and both RKGA1 and OPGA1 cannot obtain the maximum objective value for any instances; however, for problem 5, 6 and 7, the maximum value of the first objective generated by OPGA2 is less than that of RKGA1, RKGA2 and OPGA1. With respect to the average results, RKGA2 performs better than other algorithms for four instances and OPGA2 is inferior to any other algorithms.

With respect to the second objective, RKGA2 and OPGA1 obtain the similar average maximum completion time for four 10×10 instances and these average results are smaller than the corresponding results of RKGA1and OPGA2. On the other hand, RKGA2 finds the best solution of 4 instances, especially for problem 6; the best solution is only generated by this algorithm. Both RKGA1 and OPGA1 converge to the

Table 5. Computational results of two algorithms on the second objective

prob-lem	RKGA		SMGA	
	avg.	*opt.*	*avg.*	*opt.*
1	56,80,103	56,80,103	56,80,103	56,80,103
2	52.2,71,87.6	51,70,86	52.6,71.5,88.5	51,70,86
3	50,65,84	50,65,84	50,65,84	50,65,84
4	28.9,36,43.1	29,36,43	28.2,36.1,44.4	29,36,43
5	95.1,130.9,162.2	96,129,160	96.8,134.9,164.7	95,133,161
6	93,126.2,163.6	89,123,158	96.5,129.7,168.3	93,129,168
7	84.6,115.9,148.6	85,116,143	86.1,118,147.8	88,115,146
8	28.4,48,64.1	28,47,62	29.1,48.3,64.5	28,47,66
9	144.7,211.2,274.7	142,207,271	149.1,216.1,279.6	146,212,272
10	122.7,176.2,227.3	118,170,223	125.9,180.2,231.7	121,176,231

best solutions of 2 problems and OPGA2 only approximates the best solution of one instance. Table 3 shows that the computational times of RKGA1 and RKGA2 are close to or smaller than those of OPGA1 and OPGA2. Thus, it can be concluded that two RKGAs have better performance than or similar performance with the GAs with the operation-based representation when spending the nearly equal times. This conclusion proves that the new representation is effective.

3.4.2 Results and Discussions

RKGA is tested on ten instances and compared with SMGA. We adopt the parameter settings proposed by Sakawa and Mori [3] except the number of objective function evaluation. The parameters and DX described in section 4.1 are used. The newly defined max operation is used in two algorithms. Table 4 shows the computational results of RKGA and SMGA on the first objective. Table 5 depicts the comparison between two algorithms on the second objective.

From Table 4 and 5, it can be concluded that RKGA performs better than SMGA for all instances. For four simple problems, two algorithms have similar performance. For other instances, the results generated by SMGA are notably worse than those of RKGA.

Two different representations and two decoding strategies are respectively used in two algorithms. The new solutions produced by RKGA are always feasible, the combination of random key representation and DX makes RKGA excel in fuzzy scheduling. On the other hand, SMGA cannot guarantee the feasibility of new solutions and its low performance mainly results from its restricted optimization ability caused by the shortcoming of matrix representation.

4 Flexible Job Shop Scheduling with Fuzzy Processing Time

4.1 Problem Description

fJSSP is composed of n jobs $J_i\ (i=1,2,\cdots,n)$ and m machines $M_k\ (k=1,2,\cdots,m)$. Each job consists of several operations. Each operation can be processed on more than one machine. There are several constraints on jobs and machines, such as:

Each machine can process at most one operation at a time,
No jobs may be processed on more than one machine at a time,
Operation cannot be interrupted,
Setup times and remove times are included in the processing times.

In this chapter, fJSSP with fuzzy processing time is considered. The processing time of the *jth* operation of J_i on machine M_k is represented as TFN $\tilde{p}_{ijk} = \left(a_{ijk}{}^1, a_{ijk}{}^2, a_{ijk}{}^3\right)$.

The problem is to assign each operation to an appropriate machine (machine assignment problem), and to sequence the operations on the machines (operation sequence problem) in order to optimize the maximum fuzzy completion time.

$$C_{max} = \max_{i=1,2,\cdots n} C_i \tag{6}$$

where C_{max} is the maximum fuzzy completion time and C_i is the fuzzy completion time of job J_i.

Table 6. Example of 4 jobs 2 machines fJSSP with fuzzy processing time

Job	M_1	M_2	Job	M_1	M_2
J_1 o_{11}	1, 2, 3	3, 4, 5	J_3 o_{31}	3, 4, 6	2, 4, 5
o_{12}	2, 3, 4	2, 4, 6	o_{32}	1, 3, 4	3, 5, 8
J_2 o_{21}	2, 4, 5	3, 5, 7	J_4 o_{41}	1, 2, 4	4, 5, 7
o_{22}	1, 3, 4	2, 3, 5	o_{42}	2, 3, 5	4, 6, 9

Table 6 shows an example, in which rows correspond to operations and columns correspond to machines. The entries of the input table are the processing times. In this example, we have total flexibility. In a partial flexibility scenario, an empty entry in the table means that a machine cannot execute the corresponding operation, i.e., it does not belong to the subset of compatible machines for that operation.

4.2 Decomposition-Integration Genetic Algorithm

Two methods are often used to solve fJSSP. The first is the separation method, in which two sub-problems of fJSSP are considered in turn. The second is the integration method, which integrate operation sequence problem and machine assignment problem together. In this section, we present a different approach with population decomposition and integration.

The overall structure of DIGA can be described as follows.

(1) Randomly generate initial main population AB and evaluate its individual,
(2) Decompose population AB into sub-populations A and B, make the ith individual of two sub-populations have the same fitness as that of AB, $i = 1, 2, \cdots, N$;
(3) Generate populations A_1 and B_1 as the copy of A and B respectively;
(4) Perform binary tournament selection, crossover and mutation on population A;
(5) Perform binary tournament selection, crossover and mutation on population B;
(6) Construct a new population AB based on two evolved sub-populations and their copies, calculate the fitness of its individuals and delete the worst half of the population.
(7) If the termination condition is met, stop the search; otherwise, go to step (2).

In step 2, the main population is decomposed into A and B in the following way: for individual i of AB, its sequencing string becomes the ith individual of A, its assigning string is the ith individual of B. $i = 1, 2, \cdots, N$, N is population size.

In step 6, AB with $2N$ individuals is obtained: for $i = 1, 2, \cdots, N$, the ith individual of AB is made up of that of A and B_1; for $i = N+1, \cdots, 2N$, the ith individual of AB is the integration of that of A_1 and B.

DIGA is unique in two respects.

(1) It calculates the fitness of individuals in the main population with $2N$ individuals;
(2) The main population itself doesn't evolve; however, it is updated twice in each generation. The update is first done by the independent evolution of two sub-populations and then executed by storing the best half of the population based on two sub-populations and their copies.

In next three subsections, the different steps of DIGA are described in detail.

4.2.1 Two-String Representation

A two-string representation is used to decode a schedule of fJSSP with two integer strings $\left(p_1, p_2, \cdots, p_{h_1}, \cdots, p_h\right)$ and $\left(q_{11}, q_{12}, \cdots, q_{1h_1}, \cdots, q_{nh_n}\right)$, $h = \sum_{i=1}^{n} h_i$. The first string is used for job sequencing, in which 1 occurs h_1 times, 2 occurs h_2 times and so on. The second is for machine assigning. Each gene $q_{ij} \in \left[1, u_{ij}\right]$ corresponds to the jth operation o_{ij} of job J_i and u_{ij} indicates the maximum number of machines on which the operation o_{ij} can be processed. If q_{ij} is equal to l, the operation o_{ij} is processed on the lth

machine of the total u_{ij} machines. Fig.4 describes a chromosome of the example shown in Table 6.

To obtain a feasible schedule, the following decoding procedure is adopted:

(1) Translate the first string into a list of the ordered operations and assign a machine for each operation according to the second string;

(2) The first operation of the operation list is arranged first, and the second operation and so on. Each operation is allocated in the best available processing time for the required machine of the operation. The procedure is repeated until a schedule is obtained. The procedure is identical to the one proposed by Cheng et al.[25] except the processing time is fuzzy.

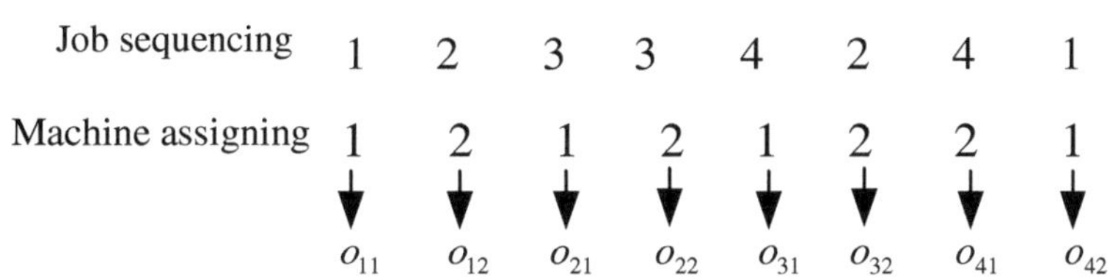

Fig. 4. An illustration of the two-string representation

The max operation and the decoding strategy for job sequencing string shown in section 3 are adopted in this section. For the chromosome in Fig. 4, the sequencing string (1 2 3 3 4 2 4 1) is converted into a list of the ordered operations $(o_{11}, o_{21}, o_{31}, o_{32}, o_{41}, o_{22}, o_{42}, o_{12})$, the second string is (1 2 1 2 1 2 2 1). Fig. 5 shows fuzzy Gantt chart of the final schedule.

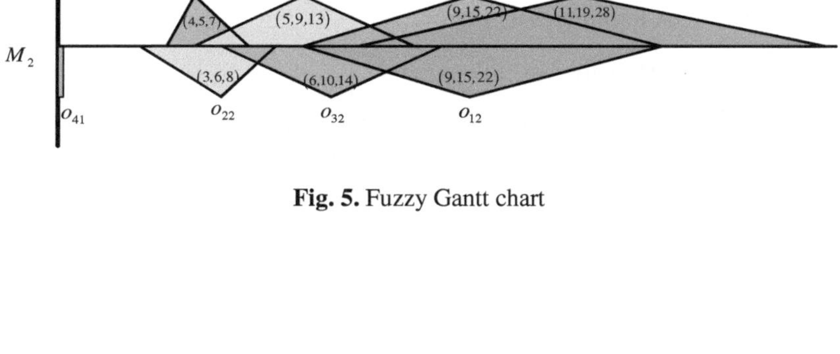

Fig. 5. Fuzzy Gantt chart

Parent 1	2	4	3	1	2	3	1	3	1	2	4	4
Parent 2	3	2	**1**	4	**2**	**1**	1	**3**	4	2	3	4
Offspring 1	3	2	**1**	**2**	**3**	**1**	4	1	4	2	3	4

GOX

Parent 1	2	4	3	1	2	3	1	3	1	2	4	4
Parent 2	3	2	**1**	4	**2**	**1**	1	**3**	4	2	3	4
Offspring 1	3	2	4	**1**	**2**	**3**	**1**	1	4	2	3	4

GPX

Parent 1	2	4	3	1	2	3	1	3	1	2	4	4
Parent 2	3	2	1	4	2	1	1	3	4	2	3	4
String	1	2	2	1	1	2	1	2	1	1	2	2
Offspring 1	2	3	1	4	2	1	3	1	4	2	3	4

GPPX

Fig. 6. Example of three crossovers of DIGA

4.2.2 Crossover, Mutation and Termination Condition

We consider TPX for population B and three crossovers for population A respectively. The first crossover is generalized order crossover (GOX) and has been shown in 3.4.1.

The second is generalized position crossover (GPX) (Mattfeld [29]). GPX is similar to GOX. The main difference between them is that the insertion of a sub-string in the second parent is done according to its position in the first parent for GPX.

The third is a generalization of precedence preservative crossover (GPPX). A string is filled at random with h elements of set $\{1, 2\}$. This string defines the order in which the genes are successively drawn from parent 1 and 2. The offspring is initially empty. When a gene θ is selected, it is appended to the offspring. If the gene θ comes from parent 1(2), then the same gene in parent 2(1) is deleted. This step is repeated until the chromosome of two parents are empty and an offspring is obtained. Fig. 6 describes an illustration of the crossovers of DIGA.

The swap operator acts as the mutation. When the predetermined number of generations is met, DIGA terminates its search.

4.3 Computational Results

fJSSP with fuzzy processing time is seldom considered and the numerical examples are hard to found. In this section, we first provide three numerical examples, which are 10 jobs 10 machines fJSSP. The total number of operations of instances 1 and 2 is 40 and the corresponding number of instances 3 is 50. We then test the impact of three crossovers GOX, GPX and GPPX on the performance of DIGA for the high complexity of job sequencing problem. Finally, we compare our results with those obtained by other algorithms.

4.3.1 Results of DIGA

Three variants of DIGA are considered. DIGA1 is defined as DIGA with GOX, DIGA2 denotes DIGA using GPPX and DIGA3 represents DIGA with GPX. These algorithms have the same flow and parameters except crossover operators. Two-point crossover is applied to the population B . We set the same parameters for the evolution of two sub-populations: crossover probability of 0.8, mutation probability of 0.1, population size of 100 and the maximum generation of 500. All algorithms randomly run 20 times with respect to each instance and the computational results are shown in Table 7. Figure 6, 7 and 8 shows the results in form of fuzzy Gantt chart.

Table 7. Computational results of DIGA

Inst-ance	DIGA 1		DIGA 2		DIGA 3	
	avg.	*opt.*	*avg.*	*opt.*	*avg.*	*opt.*
1	26.30,37, 47.79	21,33, 43	23.90,35.56, 46.64	23,32, 45	23.89,35.41, 47.61	22,33, 44
2	38.00,51.43, 65.22	34,48, 63	36.71,51.34, 66.11	32,47, 57	36.54,51.13, 66.04	37,47, 58
3	40.19,56.43, 73.14	35,53, 68	40.14,55.53, 72.82	36,51, 65	38.90,55.49, 72.00	36,50, 69

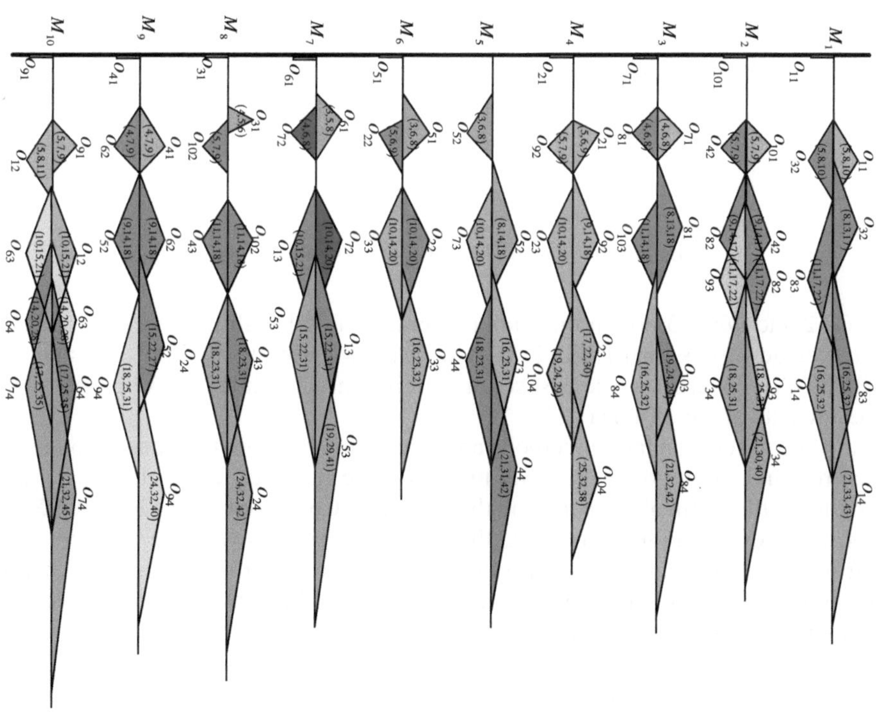

Fig. 7. Fuzzy Gantt chart of a solution of instance 1

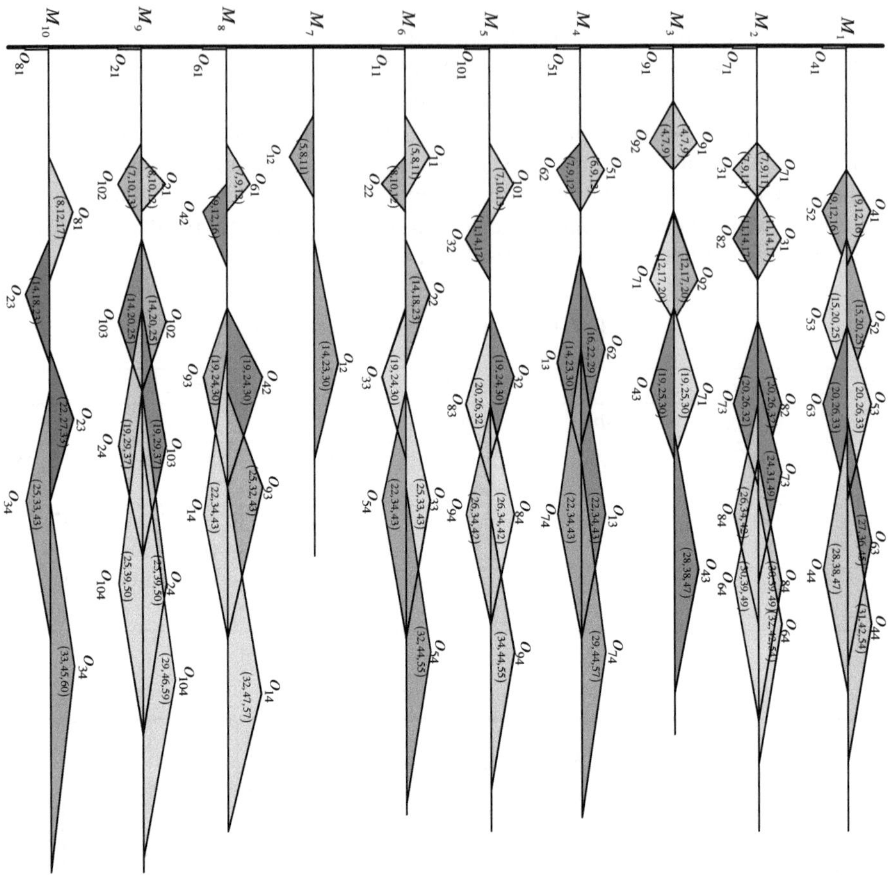

Fig. 8. Fuzzy Gantt chart of a solution of instance 2

DIGA2 and DIGA3 obtain the similar average value for all instances and produce better average results than DIGA1. DIGA2 reaches the best solutions of instances 2 and 3, DIGA1converges to the best solution of instance 1, while DIGA3 cannot obtain the best results of any instances.

The most of representation methods are redundant and more than one chromosome can correspond to one objective. GA must produce many different chromosomes to approximate the optimal schedule of the problem. Compared with GOX, GPX and GPPX has stronger ability to produce the new chromosomes with the different structure from the old ones, meanwhile, some genes of the old individual is still remained in the new one. DIGA2 and DIGA3 may keep the good balance between exploration and exploitation; as a result, they perform better than DIGA1.

4.3.2 Comparative Results

We compare DIGA with the algorithm proposed by Pezzella et al. [30], which are labeled as PEGA. We adopt all parameters shown in [30] except population size of 100 and maximum generation of 1000. Binary tournament for selection and three dispatching rules for initial population are still used in PEGA. We make use of the parameters in section 3.1 and GPPX for DIGA. Two algorithms randomly run 20 times on each instance. The corresponding computational results are listed in Table 8.

Table 8. Computational results of DIGA and PEGA

Algorithm	Instance 1		Instance 2		Instance 3	
	avg.	*opt.*	*avg.*	*opt.*	*avg.*	*opt.*
DIGA	23.90,35.56, 46.64	23,32, 45	36.71,51.34, 66.11	32,47, 57	40.14,55.53, 72.82	36,51, 65
PEGA	25.00,35.67, 47.78	23,34, 44	37.51,51.75, 66.75	34,45, 60	40.62,56.43, 73.29	38,51, 66

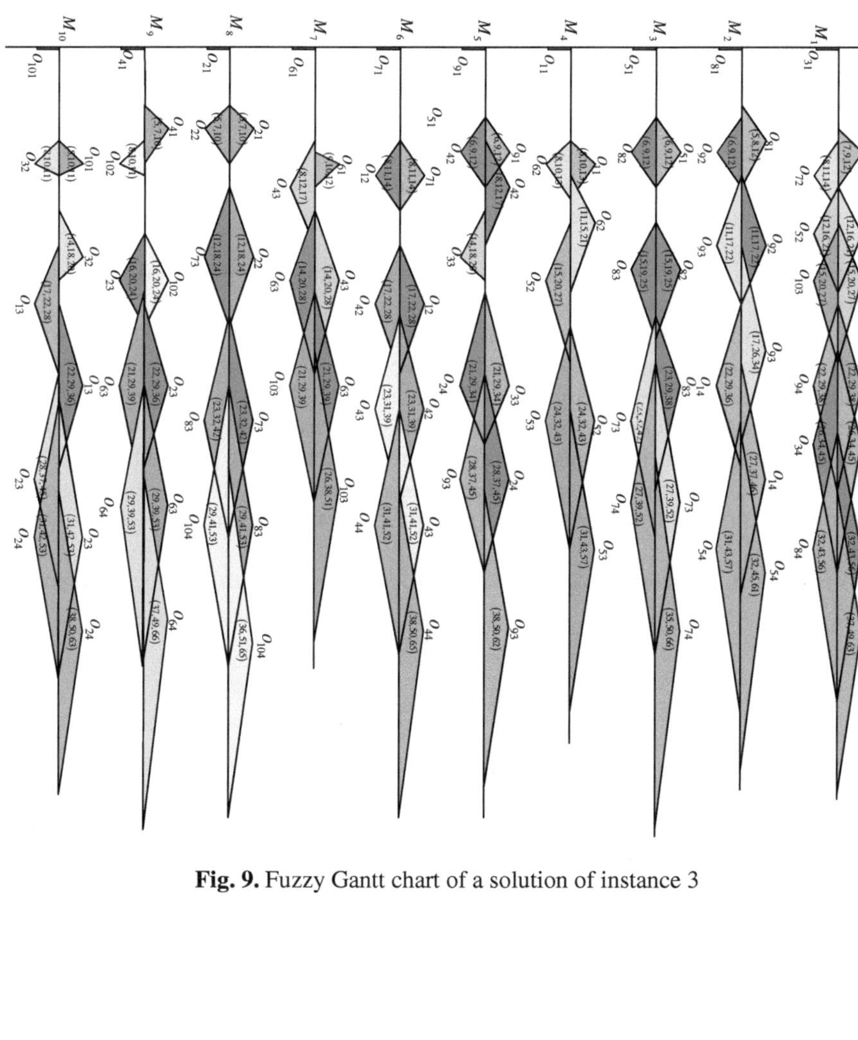

Fig. 9. Fuzzy Gantt chart of a solution of instance 3

It can be concluded that DIGA performs better than PEGA. The best solutions of DIGA are always better than those of PEGA. Like DIGA, PEGA generates new population by using different crossover and mutation on two parts of each individual; however, two parts of some individuals may be changed simultaneously and population is only renewed one time in a generation of PEGA. Compared with PEGA, DIGA updates only one string of individuals in the main population and renews the main population twice. DIGA maintains good balance between exploration and exploitation; as a result, DIGA has better performance than PEGA.

5 Job Shop Scheduling with Stochastic Processing Time

5.1 Problem Formulation

JSSP is composed of n jobs $J_i\,(i=1,2,\cdots,n)$ and m machines $M_j\,(j=1,2,\cdots,m)$. Each job consists of several operations and each operation o_{ij} is processed during a fixed duration. There are several constraints on jobs and machines, such as:

Each machine can process at most one operation at a time,
No jobs may be processed on more than one machine at a time,
Operation cannot be interrupted,
Operations of a given job have to be processed in a given order,
Setup times and remove times are included in the processing times.

In this section, JSSP with stochastic processing time is considered, in which the processing time of each operation is modeled by an independent random variable with a given probability distribution. In general, processing time is indicated by using the normal, exponential or uniform distribution. We suppose that processing time follows normal distribution. Other constraints of JSSP are still valid in the stochastic context.

In the deterministic context, makespan is the most frequently considered objective and many efficient heuristics and meta-heuristics such as GA [31], tabu search [32] and particle swarm optimization [33] have been proposed to minimize makespan. In the stochastic context, the expected makespan is often regarded as the objective [6,7,8].

In this section, makespan itself is still regarded as the objective of the problem.

$$\tilde{C}_{max} = \max_{i=1,2,\cdots n} \tilde{C}_i \tag{7}$$

Where $\tilde{C}_i$ is the stochastic completion time of job J_i.

5.2 Active Schedule Generating Algorithm

For JSSP and SJSSP, we can expect that an optimal schedule is within the set of active schedules since inclusion of idle time is not preferable. The GT procedure can be extended from the deterministic context to the stochastic case. In this section, some operations on processing time are first defined and then the extended GT procedure is proposed.

5.2.1 Operations on Stochastic Processing Time

In the stochastic context, the max operation and ranking operation of random variables are required to be defined again to decide the earliest beginning time and the completion time of jobs on each machine.

The ranking of random variables is based on stochastic dominance theory. Many kinds of stochastic dominance including expectation dominance and almost sure dominance et al. have been defined; some of them such as expectation dominance and almost sure dominance can be easily implemented on computer. In this chapter, almost sure dominance is considered.

Definition 1. X_1 and X_2 are random variables. X_1 almost surely dominates X_2 if $P(X_1 \geqslant X_2)=1$.

For random variable $X_1 \sim N(\mu_1,\sigma_1^2)$ and $X_2 \sim N(\mu_2,\sigma_2^2)$, $\sigma_2 > \sigma_1$

The following procedure is used to rank X_1 and X_2:

(1) Set $a_1 = \mu_1 - 3\sigma_1$, $a_2 = \mu_1 + 3\sigma_1$, $b_1 = \mu_2 - 3\sigma_2$, $b_2 = \mu_2 + 3\sigma_2$;

(2) $X_1 > X_2$ if $a_1 \geqslant \mu_2$ or ($\mu_1 > \mu_2$, $a_2 \geqslant b_2$) $\mu_2 - \mu_1 \geqslant max\{\sigma_2 - \sigma_1, \alpha\}$

$X_2 > X_1$ if ($\mu_2 > \mu_1$, $b_1 \geqslant a_1$) or $a_2 \leqslant \mu_2$; if X cannot be determined, the following criterion is used;

(3) $s = 0$,first produce L pairs of random numbers s_1, s_2 and for each pair of s_1, s_2, if $s_1 > s_2$, $s = s+1$; then $s \leftarrow s/L$,

finally $X_2 < X_1$ if $s \geqslant 0.9$, or $X_1 < X_2$ if $s \leqslant 0.1$.

The max of these variables is calculated as follows:

(1) If X_1 and X_2 meet the special conditions shown the second step of the max procedure, then directly determine $X = X_1$ or $X = X_2$; else go to (2);

(2) $s = 0$, first produce L pairs of random numbers s_1, s_2 and for each pair of s_1, s_2 ,if $s_1 > s_2$, $s = s+1$; then $s \leftarrow s/L$,

finally if $s \notin (0.1, 0.9)$, $X = X_1$ if $(s \geqslant 0.9)$ or $X = X_2$ if $(s \leqslant 0.1)$;

else if $X \sim N(\mu_1 + 0.5\sigma_2, \sigma_1^2)$ $(s > 0.6)$ or $X \sim N(\mu_2 + 0.5\sigma_2, \sigma_2^2)$ $(s < 0.4)$; else $X \sim N(\mu_2 + \sigma_2, \sigma_2^2)$

where $L(\geqslant 30)$ and $\alpha(\geqslant 9)$ are a constant. s_1 and s_2 respectively follow X_1 and X_2.

With respect to the ranking procedure, we conduct the following explanations: (1) the consideration of those special cases is to save computation time; (2) If X_1 is smaller than X_2, it can be proved that the actual value of X_1 is smaller than that of X_2 at a high probability, which is close to 0.9.

Random variable X obtained in the max procedure is very close to the real max of X_1, X_2 ($max(X_1, X_2)$) for the following equation:

$$P(max(X_1, X_2) \leqslant z) = P(X_1 \leqslant z)P(X_2 \leqslant z) \approx P(X \leqslant z) \tag{8}$$

All random variables are always independent in SJSSP, as a result, the completion time of jobs are or nearly are normal distribution variables even if the processing time of operation is not normal distribution variable in terms of central limit theorem. Thus, the above two procedures can also be applied to JSSP with processing time following other probability distribution.

5.2.2 The Extended GT Procedure

The GT procedure is well-known as the algorithm for generating active schedules. In the stochastic context, the GT procedure is defined in the following way:

(1) Let $t = 1$, $PS_t = \Phi$, determine S_t;

(2) Calculate $EC_{min} = min\{EC(o_{ij}) | o_{ij} \in S_t\}$, record the corresponding machine M_{j^*} and part J_{i^*};

(3) Define $c_t = \{o_{ij^*} \in S_t | EB(o_{ij^*}) < EC_{min}\} \cup \{o_{i^*j^*}\}$, choose $o_{uj^*} \in c_t$, $PS_{t+1} = PS_t \cup \{o_{uj^*}\}$, delete o_{uj^*} from S_t and add the next operation of job J_u into S_t and form S_{t+1};

(4) $t = t + 1$, go to (2) until a complete scheduling plan is obtained.

where Φ is an empty set, S_t is a set of operations which can be scheduled in the *t-th* iteration and PS_t is a set of operations which have been scheduled in the *t-th* iteration. $EC(o_{ij})$ and $EB(o_{ij})$ respectively indicate the earliest completion time and the earliest beginning time of operation o_{ij}. c_t is the conflict set which consist of all operations competing for the same machine.

$c_t \setminus \{o_{i^*j^*}\}$ may be empty for the random feature of $EC(o_{ij})$ and $EB(o_{ij})$. To avoid this case, the conflict set must include operation $o_{i^*j^*}$.

5.3 Genetic Algorithm for SJSSP

Representation is the key to solve scheduling problem using GA. There are a number of representation methods such as job-based representation and priority rule based representation in the deterministic context. However, some of them cannot be or are hard to be used in the stochastic context. The main obstacle is the decoding process. In this section, based on permutation-based coding and decoding, a GA is suggested.

The framework of the GA is described as follows.

(1) Randomly generate an initial population P.
(2) Perform binary tournament selection on P.
(3) Perform crossover and mutation on population P.
(4) If the termination condition is met, stop the search; otherwise, go to step (2).

5.3.1 Representation Method

A permutation-based representation method is considered. For $n\times m$ JSSP with the stochastic processing time, a chromosome $(p_{11}, p_{21}, \cdots, p_{n1}, \cdots, p_{mn})$ is composed of m permutations, each for one machine. Each gene p_{ij} corresponds to the operation o_{ij} of part J_i processed on machine M_j. Take 4×2 JSSP as an instance, a chromosome may be (2, 3, 1, 4, 4, 2, 3, 1), in which the genes of the first permutation (2, 3, 1, 4) corresponds to o_{11}, o_{21}, o_{31}, o_{41} and the genes of the second permutation to $o_{12}, o_{22}, o_{32}, o_{42}$.

The chromosome $(p_{11}, p_{21}, \cdots, p_{n1}, \cdots, p_{mn})$ is decoded using the extended GT procedure. When several genes compete for a machine, the one with the minimum value $p_{uj^*} = min\left\{p_{ij^*} \middle| o_{ij^*} \in c_t\right\}$ is preferably chosen from the conflict set.

The above representation can be regarded as the modified version of preference-list representation [34] in the stochastic context. The main difference between them lies in the decoding procedure: to build a schedule, the gene occurring leftmost in a preference list is always preferably chosen, while the gene with minimum value in a permutation is given the highest priority.

5.3.2 Fitness, Elitism and Genetic Operators

Like section 3 and 4, fitness function of an individual is equal to its objective function. The classical elite strategy and tournament selection shown in section 3 are also adopted.

Because each individual consists of several permutations, a two-phase crossover is developed to adapt the special structure of chromosome. β permutations are first randomly chosen in the first phase, and then crossover operator is performed on each chosen permutation. By modifying precedence preservative crossover (PPX) [28], we obtain a new crossover operator called MPPX.

PPX is performed in the following way: for each permutation, a string is filled at random with n elements of set {1, 2} and the remaining procedure is identical with the one shown in 3.4.1.

When the leftmost gene θ of parent 2 is appended to the offspring, the same gene θ of parent 1 is deleted and the leftmost gene θ_1 of parent 1 is inserted to the position of the gene θ. This is the main difference between PPX and MPPX. Fig.1 shows the difference of these crossovers.

Two-phase mutation is also used. A permutation is randomly chosen and then an operator is performed on the chosen permutation. The swap operator and insertion operator are considered. When the predetermined number of generations is met, the search is terminated.

5.4 Computational Results

In this section, the proposed GA is first compared with other GAs using different crossover and mutation and then it is compared with the hybrid method developed by [13]. 24 benchmark problems are used. These problems are the extension of ORB1-10, ABZ5-6, FT10, FT20, LA11-20 et al. The processing times of these deterministic problems are the mean value of the stochastic processing time. The corresponding standard variances are taken randomly from the uniform distribution in [2, 11]. To simplify, these extended problems are still called ORB1-10, ABZ5-6, FT10, TF20, LA11-20.

5.4.1 Performance Analyses on the GA

Four crossover operators of PPX, MPPX, partially mapping crossover (PMX) [35] and order crossover (OX) [36] and two mutations of insertion and swap are considered to test the effectiveness of the modified crossover.

We construct seven GAs having the same flow and the same parameters as the proposed GA except crossover and mutation. To simplify, we label these algorithms with their crossover and mutation. For eight GAs, population scale of 100, crossover probability of 0.9 and mutation probability of 0.1 are used. For each problem, all algorithms randomly run 20 times and the search terminates when the number of objective function evaluation reaches to 30000. Table 9 shows the best solution of each problem obtained by all algorithms. A combination of crossover and mutation represents a GA. Each group of data consist of two parts: the first part is the mean value and the second part is the standard variance.

From table 9, it can be concluded that the GA with MPPX performs better than the GA with any other crossovers. The GA with MPPX generates the minimum makespan of 11 problems, while the GA with PMX only converges to the best results of 6 problems. The combination of MPPX and swap is also better than any other combinations. The proposed GA approximates to the best solution of 8 problems. Thus, it is reasonable and effective to select MPPX and swap in the proposed GA.

5.4.2 Results and Discussion

Tavakkoli-Moghaddam et al. [11] proposed a hybrid method, in which an initial feasible solution is generated by a neural network approach and then the initial solution is improved by SA. The hybrid method adopts the following parameter settings: the initial temperature $T_0 = 1.0$, cooling rate is 0.05. In each temperature, the number of the movements is 150. The proposed GA uses the parameters shown in sub-section 5.1. Both algorithms have the same stopping criterion shown in section 5.1 and randomly run 20 times on each instance. Table 10 shows the computational results obtained by two algorithms.

Table 9. Computational results of eight variants

Problem	Mutation	PPX	PMX	OX	MPPX
ORB6	insertion	1159.31, 33.898	1128.80, 31.730	1209.06, 34.332	1132.13, 30.084
	swap	1157.93, 34.146	1115.07, 31.084	1168.14, 33.681	1106.40, 30.749
ORB5	insertion	976.76, 31.674	969.56, 28.371	1010.67, 31.450	971.00, 29.904
	swap	986.99, 30.047	974.10, 29.803	1013.13, 29.007	944.52, 28.303
ORB4	insertion	1104.00, 30.862	1088.00, 30.815	1155.63, 33.337	1092.00, 33.066
	swap	1113.79, 30.627	1088.15, 28.605	1161.92, 30.353	1087.00, 28.936
ORB3	insertion	1191.66, 31.597	1119.22, 34.726	1204.00, 33.039	1137.35, 33.605
	swap	1172.98, 31.496	1099.47, 34.732	1176.91, 36.971	1080.51, 27.162
ORB2	insertion	1007.20, 30.155	968.00, 29.751	1001.87, 26.898	964.94, 21.522
	swap	979.20, 25.762	952.31, 21.716	986.48, 30.546	947.18, 30.031
ORB1	insertion	1200.09, 33.856	1160.06, 31.961	1220.00, 33.598	1172.00, 30.679
	swap	1202.51, 33.723	1143.99, 32.776	1236.73, 33.721	1129.17, 33.173
FT20	insertion	1343.00, 30.580	1244.00, 36.595	1346.64, 36.504	1256.08, 33.709
	swap	1313.31, 38.796	1244.00, 33.880	1326.00, 34.187	1242.46, 33.386
LA19	insertion	923.00, 25.802	905.00, 26.360	954.78, 25.818	912.00, 26.119
	swap	938.14, 28.535	929.22, 30.975	963.02, 30.283	897.14, 27.469
LA18	insertion	902.00, 30.468	926.78, 30.396	957.69, 28.586	915.28, 27.998
	swap	916.09, 26.979	902.99, 25.037	950.50, 29.060	923.28, 29.329

Table 9. (*continued*)

LA17	insertion	877.55, 21.083	812.00, 25.341	855.43, 26.167	813.14, 26.532
	swap	853.175, 25.654	816.75, 26.656	849.00, 27.252	834.39, 24.346
LA16	insertion	1002.51, 27.495	1008.00, 27.450	1039.86, 29.434	1008.00, 27.495
	swap	1029.00, 24.963	1012.00, 27.495	1021.60, 30.212	1022.00, 28.799
LA11	insertion	1260.19, 33.875	1238.00, 30.191	1287.00, 34.718	1222.00, 30.120
	swap	1230.00, 31.694	1222.00, 30.120	1268.13, 30.138	1229.87, 32.503
LA12	insertion	1047.00, 29.769	1039.00, 29.041	1080.30, 32.130	1039.00, 29.041
	swap	1059.08, 26.815	1039.82, 31.554	1106.33, 28.892	1040.48, 30.558
LA13	insertion	1187.58, 33.894	1178.00, 32.380	1218.65, 31.346	1161.00, 30.227
	swap	1215.77, 28.652	1152.00, 27.808	1202.34, 31.804	1150.00, 27.344

Table 10. The comparison between two methods

	GA		Hybrid method	
	Average	Best solution	Average	Best solution
ORB4	1092.11, 30.933	1087.00, 28.936	1164.09, 32.643	1155.63, 33.341
ORB5	954.28, 29.301	944.52, 28.300	1000.45, 28.366	982.22, 26.875
ORB6	1119.87, 31.984	1106.40, 30.750	1160.01, 32.025	1143.55, 33.103
ORB7	454.32, 28.468	444.975, 27.536	467.35, 29.998	458.925, 29.835
ORB8	1002.84, 33.056	995.06, 29.278	1033.67, 32.468	1026.74, 30.642
ORB9	1021.44, 31.133	980.87, 29.322	1053.37, 31.455	1037.00, 30.265
ORB10	1041.32, 31.356	999.062, 31.469	1053.45, 30.789	1031.7, 28.977

Table 10. *(continued)*

LA20	975.92, 29.871	957.02, 29.051	1006.22, 29.447	995.64, 24.971
LA19	914.35, 29.467	897.14, 27.471	975.28, 30.477	965.08, 28.305
LA15	1014.20, 27.485	997.95, 26.129	1038.05, 29.384	1026.46, 29.958
LA14	1296.18, 31.257	1292.00, 31.467	1300.25, 31.280	1294.00, 30.245
ABZ5	1334.98, 30.568	1315.77, 26.442	1360.87, 31.235	1348.67, 30.958
ABZ6	1014.20, 27.485	997.95, 26.129	1038.05, 29.384	1026.46, 29.957
FT10	1035.56, 27.813	1019.00, 29.958	1048.80, 30.608	1030.00, 30.145

From Table 10, it can be concluded that the proposed GA significantly outperform the hybrid method. The GA performs better than the hybrid method on 14 problems. For LA14, the hybrid method produces the similar solutions with the proposed GA; for other problem, the expected makespan of the hybrid method is always bigger than the corresponding value of the proposed GA. The proposed GA has promising advantage in SJSSP. The low performance of the hybrid method results from the limited optimization ability of SA, in which only one movement is used to produce neighborhood solutions.

6 Conclusions

The GA-based scheduling algorithm frequently uses the integer string to represent the solution of FJSSP. This chapter presents a random key genetic algorithm, which is based on random key representation, a new decoding procedure, elite strategy, binary tournament selection, TPX or DX and swap mutation. RKGA is tested and compared with SMGA and computational results show the good performance of RKGA.

Local search is often combined with the scheduling algorithms to intensity the optimization ability of the latter. RKGA also can be directly merged with local search such as 2-opt and 3-opt. We will consider the merging of the RKGA and local search, apply RKGA to other production scheduling problems such as flexible job shop scheduling in the near future.

Many meta-heuristics have been applied to FJSSP and fJSSP; however, these algorithms are seldom used to solve fJSSP in the fuzzy context. The main contribution of this chapter is to provide an effective path to the problem by GA. fJSSP with other fuzzy constraints should be investigated in the near future.

The application of meta-heuristics to stochastic job shop is seldom investigated in previous research. In this chapter, we proposed an effective approach to solve the problem with stochastic processing time. We will focus on the application of meta-heuristics to JSSP with other stochastic elements such as random breakdown in the near future.

With respect to multi-objective scheduling, most of the published papers considered the deterministic problem. Few papers addressed fuzzy scheduling problem and stochastic scheduling problem. Since most of the real-life scheduling problems involve uncertainty and multiple objectives, future researches on multi-objective scheduling with fuzzy or stochastic processing conditions are desirable and attractive.

Acknowledgments. This chapter is supported by China Hubei Provincial Science and Technology Department under grant Science Foundation Project (2007ABA332).

Appendix

Table A1. Trapezoid due-date

problem	Job1	Job2	Job3	Job4	Job5	Job6
1	(94,100, 112,121)	(65,76, 82,91)	(30,40, 49,60)	(78,85, 97,102)	(65,77, 83,89)	(35,44, 54,59)
2	(65,70, 81,89)	(50,58, 69,80)	(65,72, 84,92)	(35,43, 51,60)	(72,80, 90,96)	(58,65, 75,78)
3	(25,33, 43,50)	(75,86, 96,102)	(74,83, 93,103)	(58,65, 71,75)	(33,42, 49,54)	(42,52, 62,70)
4	(18,25, 30,40)	(18,27, 35,40)	(13,15, 20,28)	(19,26, 32,40)	(15,25, 30,35)	(23,30, 40,45)

Table A2. Trapezoid due-date of problem 5,6,7 and 8

	Job1	Job2	Job3	Job4	Job5	Job6	Job7	Job8	Job9	Job10
5	(115, 169, 184, 195)	(115, 123, 134, 145)	(90, 100, 110, 120)	(90, 102, 105, 115)	(110, 121, 136, 146)	(150, 167, 174, 185)	(110, 120, 130, 140)	(150, 163, 176, 185)	(70, 79, 94, 105)	(150, 160, 163, 170)
6	(120, 130, 151, 156)	(120, 130, 154, 157)	(90, 100, 106, 117)	(100, 115, 123, 138)	(70, 80, 85, 88)	(70, 80, 86, 94)	(100, 110, 120, 135)	(118, 130, 149, 158)	(92, 108, 117, 124)	(110, 121, 142, 148)
7	(100, 108, 124, 128)	(60, 72, 81, 95)	(70, 83, 92, 99)	(70, 83, 91, 103)	(90, 101, 109, 115)	(85, 92, 102, 107)	(96, 108, 118, 128)	(145, 159, 170, 178)	(55, 66, 75, 86)	(78, 86, 94, 107)
8	(24, 34, 45, 60)	(30, 40, 50, 60)	(35, 45, 50, 65)	(30, 40, 50, 65)	(30, 40, 50, 65)	(25, 35, 45, 60)	(25, 36, 45, 60)	(30, 41, 50, 60)	(25, 36, 45, 60)	(30, 40, 50, 60)

Table A3. Problem 9

	processing time and processing sequences				
Job1	{2,4,7}3	{9,13,16}4	{5,8,11}6	{8,11,15}10	{10,15,20}5
	{7,11,14}7	{9,14,17}1	{7,11,15}9	{10,14,17}2	{6,9,12}8
Job2	{8,11,15}4	{7,10,12}3	{4,6,8}1	{11,15,20}2	{12,17,21}10
	{9,11,15}9	{8,12,16}7	{5,6,9}6	{8,10,13}5	{7,11,15}8
Job3	{7,9,12}2	{6,7,9}1	{9,13,17}4	{10,15,20}5	{5,8,12}7
	{11,17,19}10	{8,12,16}9	{7,9,13}6	{8,13,17}3	{9,14,18}8
Job4	{10,14,18}5	{11,15,21}3	{9,13,17}9	{8,12,16}6	{7,11,12}4
	{11,15,19}8	{10,14,18}2	{8,13,17}7	{9,14,18}10	{3,5,8}1
Job5	{7,10,13}9	{7,11,15}10	{8,12,15}3	{9,13,16}5	{10,14,17}4
	{9,12,16}1	{10,15,17}8	{10,13,15}7	{11,14,17}2	{9,12,16}6
Job6	{11,15,21}9	{9,15,18}8	{8,12,16}7	{10,13,16}10	{7,11,14}3
	{9,13,17}2	{8,12,15}6	{10,14,17}5	{8,12,18}1	{12,17,20}4
Job7	{6,9,12}5	{7,10,13}6	{8,11,15}10	{9,12,16}4	{7,11,14}1
	{8,10,14}9	{10,14,16}7	{7,11,15}8	{7,10,11}3	{5,8,11}2
Job8	{9,12,16}6	{7,10,13}5	{8,11,14}3	{6,9,13}7	{4,7,9}2
	{9,13,17}8	{8,10,13}1	{7,8,11}4	{8,9,12}10	{6,8,10}9
Job9	{5,8,11}2	{7,12,14}6	{8,10,13}1	{6,7,8}4	{4,5,8}3
	{7,9,11}8	{8,11,13}9	{9,10,14}7	{7,9,12}10	{6,8,12}5
Job10	{4,5,8}3	{7,10,12}6	{8,12,16}7	{5,8,11}10	{3,5,8}2
	{4,7,9}4	{6,9,12}9	{7,10,12}1	{5,7,9}8	{10,14,17}5
Job11	{7,8,11}2	{8,9,12}5	{3,5,8}1	{5,7,10}3	{6,9,11}10
	{8,10,13}9	{7,10,11}6	{4,5,7}4	{7,11,12}8	{9,13,17}7
Job12	{6,8,11}6	{4,7,10}10	{5,6,9}1	{6,9,12}5	{5,8,10}7
	{3,5,9}4	{4,6,9}3	{5,8,12}2	{6,9,12}9	{4,7,10}8
Job13	{3,5,9}6	{7,10,12}10	{5,7,9}9	{6,9,11}8	{4,6,9}5
	{8,10,13}7	{9,11,15}4	{7,11,13}1	{5,8,9}2	{7,8,10}3
Job14	{5,8,11}2	{5,7,8}9	{4,5,8}1	{7,11,14}3	{6,9,12}10
	{5,9,10}4	{4,5,6}6	{8,11,14}7	{6,9,13}5	{5,8,11}8
Job15	{8,11,15}5	{7,10,12}4	{6,9,10}7	{5,9,10}6	{7,9,12}3
	{8,10,13}9	{4,6,9}2	{4,7,10}10	{6,7,11}8	{3,5,8}1

Table A4. Problem 10

	processing time and processing sequences				
Job1	{5,7,8}10	{9,10,13}6	{4,5,8}5	{7,8,11}3	{8,9,11}8
	{5,8,9}4	{6,7,10}2	{4,6,9}1	{8,11,14}9	{4,7,9}7
Job2	{3,5,8}4	{6,8,10}3	{7,10,12}5	{4,5,8}2	{2,4,6}10
	{5,8,11}1	{6,9,12}7	{4,5,7}6	{3,4,6}8	{6,8,11}9
Job3	{6,9,10}9	{3,5,7}8	{2,4,6}3	{4,6,8}1	{5,7,9}10
	{7,10,11}6	{5,6,9}7	{4,5,8}4	{1,3,5}2	{7,11,13}5
Job4	{7,11,14}4	{8,12,16}3	{6,9,11}7	{5,8,10}5	{9,13,17}8
	{10,14,18}9	{4,7,9}6	{8,10,13}10	{3,5,7}1	{6,8,10}2

Table A4. (*continued*)

Job5	{5,6,8} 5	{7,9,10}7	{8,11,16}2	{3,4,5}3	{10,14,18}8
	{7,10,13}1	{7,11,14}9	{6,8,9}6	{7,8,10}4	{2,3,6}10
Job6	{8,12,16}7	{6,9,12}1	{7,10,12}5	{5,8,11}4	{4,6,9}8
	{3,5,8}9	{6,9,12}2	{4,5,8}6	{8,12,16}3	{9,13,19}10
Job7	{10,14,17}4	{9,13,15}10	{11,15,19}7	{7,11,14}6	{9,14,18}1
	{6,10,11}9	{7,11,12}5	{12,18,21}3	{8,10,13}8	{10,12,14}2
Job8	{7,11,14}5	{3,5,8}2	{6,9,12}9	{4,6,9}1	{10,14,18}8
	{5,8,11}7	{6,7,9}6	{4,7,10}4	{4,5,8}10	{9,13,17}3
Job9	{13,17,22}10	{11,14,17}2	{9,11,16}5	{7,10,12}4	{6,9,13}9
	{7,11,14}3	{5,7,10}7	{8,12,16}1	{11,17,21}8	{1,2,3}6
Job10	{8,12,15}4	{7,10,13}3	{9,12,16}7	{10,13,15}10	{8,10,13}8
	{7,9,11}1	{6,8,10}5	{7,8,10}6	{6,7,10}2	{5,6,8}9
Job11	{7,11,14}2	{9,13,17}5	{6,9,12}1	{8,10,14}3	{10,14,17}10
	{8,11,15}7	{5,8,11}8	{8,12,16}9	{6,9,12}6	{4,6,9}4
Job12	{4,6,9}2	{5,7,10}4	{6,8,11}1	{5,9,12}3	{4,7,10}10
	{9,10,14}8	{7,9,12}9	{10,14,17}5	{7,9,11}7	{5,6,9}6
Job13	{3,5,8}6	{5,7,9}4	{6,8,10}7	{5,8,11}2	{4,6,9}1
	{3,4,6}8	{8,10,13}9	{4,6,8}10	{6,8,10}3	{8,10,13}5
Job14	{2,3,5}2	{4,5,8}1	{6,7,10}8	{5,6,9}5	{3,6,9}4
	{7,10,11}6	{6,9,12}10	{4,8,9}9	{5,6,8}7	{7,8,9}3
Job15	{5,8,11}5	{7,10,11}9	{6,8,9}3	{5,7,8}4	{4,6,7}2
	{8,12,16}7	{5,8,12}8	{9,13,17}10	{4,5,7}6	{3,6,9}1

Table A5. Fuzzy due-date of problem 9 and 10

part	Problem 9		Problem 10	
	doublet	Trapezoid	doublet	Trapezoid
Job1	217,251	171,208,217,251	152,179	92,110,152,179
Job2	223,258	164,194,223,258	127,157	81,101,127,157
Job3	233,269	180,217,233,269	120,147	76,98,120,147
Job4	250,288	194,234,250,288	181,225	115,146,181,225
Job5	247,278	183,219,247,278	161,193	99,121,161,193
Job6	264,302	204,246,264,302	173,221	113,142,173,221
Job7	211,242	159,191,211,242	233,283	144,183,233,283
Job8	200,231	143,168,200,231	164,207	106,133,164,207
Job9	183,210	128,150,183,210	212,263	134,166,212,263
Job10	173,200	132,160,173,200	184,217	111,133,184,217
Job11	176,201	125,148,176,201	198,249	127,159,198,249
Job12	152,183	119,144,152,183	167,205	105,128,167,205
Job13	171,196	124,148,171,196	139,169	87,107,139,169
Job14	162,187	124,151,162,187	129,155	80,99,129,155
Job15	168,195	125,150,168,195	153,189	97,124,153,189

Table A6. Fuzzy processing time of fJSSP instance 1*

J1 1	(5,8,11) (4,7,9) (10,13,17) (4,6,8) (6,9,11)(5,7,10) (6,9,12)(4,6,9) (8,10,13)(5,8,11)
2	(6,9,12) (4,7,10) (3,6,9)(3,5,8) (6,7,9)(5,6,8) (9,13,16) (7,10,12) (4,7,10)(5,7,10)
3	(9,11,14)(3,5,7)(5,7,10)(3,5,7)(4,7,9)(5,8,10)(5,7,10)(11,15,18)(8,10,13)(6,8,10)
4	(5,8,11)(9,12,15)(8,11,15)(6,9,11)(7,10,13)(13,15,18)(15,19,22)(7,9,13)(9,13,17) (7,9,13)
J21	(10,14,17)(4,7,10)(4,8,11)(5,6,9)(6,9,11)(5,8,11)(5,8,10)(7,10,12)(7,9,11)(5,8,10)
2	(9,11,15)(5,8,9)(6,9,10)(7,10,12)(5,7,9)(5,8,11)(7,9,12)(5,7,9)(8,11,13)(9,12,15)
3	(5,8,9)(4,7,9)(6,8,11)(7,8,10)(7,9,11)(4,8,10)(5,7,10)(6,8,12)(7,8,10)(8,9,10)
4	(7,8,10)(9,11,14)(8,10,13)(11,14,17)(13,17,20)(7,10,12)(8,11,12)(6,9,11)(5,8,12) (6,10,13)
J31	(3,4,5)(4,5,6)(2,3,6)(6,7,9)(7,8,10)(7,9,10)(4,5,7)(4,5,6)(5,7,8)(6,8,9)
2	(3,5,6)(7,9,12)(6,9,11)(7,8,11)(8,10,13)(5,6,8)(7,10,13)(7,9,12)(6,9,11)(5,9,12)
3	(10,14,17)(5,7,10)(10,13,17)(9,13,17)(8,11,15)(6,9,12)(5,8,11)(6,9,12)(10,12,14) (7,9,13)
4	(4,7,10)(3,5,9)(5,9,12)(6,8,12)(9,11,14)(5,9,12)(6,10,13)(19,24,28)(5,8,10)(7,10,12)
J41	(3,5,6)(4,7,10)(5,8,10)(5,7,10)(6,9,11)(7,9,11)(4,7,10)(3,5,8)(4,7,9)(10,11,13)
2	(3,4,5)(4,7,8)(7,9,12)(5,8,10)(6,8,11)(3,5,8)(4,7,8)(5,8,9)(11,13,16)(5,7,9)
3	(2,4,6)(7,9,12)(4,5,7)(5,8,10)(3,5,8)(4,5,7)(9,12,15)(7,9,13)(6,8,11)(8,11,15)
4	(5,8,11)(9,12,14)(8,11,13)(6,9,12)(5,8,11)(7,10,13)(6,9,11)(5,8,11)(7,9,12)(5,7,10)
J51	(3,6,8)(4,5,7)(8,9,11)(7,10,14)(4,6,9)(3,6,8)(5,8,10)(9,12,14)(5,6,8)(7,9,13)
2	(1,3,4)(5,6,8)(7,9,10)(3,5,8)(5,8,10)(5,7,9)(7,9,12)(8,10,13)(4,6,9)(5,6,8)
3	(8,11,14)(7,10,12)(6,7,8)(5,8,12)(4,7,10)(6,9,11)(8,11,15)(6,9,13)(6,8,9)(5,8,12)
4	(8,10,13)(7,9,12)(8,10,12)(6,9,12)(11,14,18)(5,8,10)(4,7,10)(6,8,11)(8,10,13)(5,8,9)
J6 1	(8,9,10)(5,9,12)(2,3,5)(7,9,10)(8,11,15)(4,6,9)(3,5,8)(7,8,10)(8,9,10)(4,5,7)
2	(6,9,12)(7,10,12)(8,11,13)(5,7,10)(8,11,13)(9,12,14)(6,8,10)(5,7,9)(5,7,9)(5,8,9)
3	(2,3,4)(4,7,8)(5,8,10)(3,5,6)(4,7,8)(6,9,11)(7,10,12)(5,8,10)(6,8,11)(4,5,7)
4	(3,4,5)(10,13,17)(6,8,11)(7,10,13)(4,7,8)(5,8,10)(3,6,8)(3,4,5)(10,14,17)(3,5,7)
J7 1	(2,4,6)(3,5,8)(4,6,8)(8,10,13)(4,6,9)(3,5,8)(7,8,10)(6,8,11)(1,2,4)(5,6,7)
2	(9,11,14)(6,8,9)(7,9,10)(8,12,15)(9,13,17)(5,9,13)(6,8,12)(5,8,11)(7,9,10)(7,10,12)
3	(5,8,10)(4,7,8)(10,12,15)(6,9,11)(6,9,11)(5,7,11)(6,9,12)(7,10,13)(6,8,10)(15,19,23)
4	(4,7,10)(5,8,10)(6,9,12)(4,7,9)(5,8,11)(4,7,9)(9,12,16)(8,11,15)(5,7,10)(4,7,10)

Table A6. (*continued*)

J8 1	(9,12,15)6,8,11)(4,7,10)(5,8,11)(10,13,15)(9,13,16)(8,11,15)(5,8,10)(6,8,11)(4,7,10)
2	(5,6,8)(2,3,5)(3,5,8)(6,8,11)(7,10,13)(4,7,8)(8,11,14)(6,9,13)(3,5,8)(15,19,24)
3	(5,8,10)(9,13,16)(7,10,14)(6,10,13)(7,10,12)(8,11,14)(8,11,14)(7,9,13)(10,13,15) (7,9,12)
4	(3,4,5)(8,11,13)(5,7,10)(7,9,11)(8,9,11)(5,7,10)(10,12,15)(3,5,6)(5,6,8)(5,8,10)
J9 1	(7,9,11)(10,14,17)(9,12,16)(8,10,12)(7,9,11)(8,11,14)(10,13,16)(8,11,15)(7,10,14) (5,7,9)
2	(4,6,7)(5,8,9)(7,8,10)(4,7,9)(35,39,44)(4,7,9)(7,10,13)(8,11,14)(7,9,13)(10,12,14)
3	(8,10,13)(7,8,9)(6,8,11)(9,12,14)(5,6,8)(7,9,12)(8,11,15)(6,8,11)(6,8,9)(7,10,12)
4	(2,4,5)(7,10,13)(8,10,12)(6,9,12)(3,5,8)(6,8,11)(6,7,9)(13,17,20)(6,7,9)(2,4,5)
J101	(3,4,6)(5,7,9)(5,7,9)(8,12,15)(7,10,12)(7,9,12)(6,9,12)(7,8,10)(19,23,26)(4,5,8)
2	(9,12,17)(6,8,10)(5,8,11)(4,7,9)(5,8,11)(6,9,12)(7,10,13)(6,7,9)(4,5,6)(10,13,17)
3	(6,8,9)(7,8,10)(8,10,11)(9,11,12)(10,13,15)(6,8,9)(11,15,18)(10,15,19)(7,8,10) (9,12,14)
4	(10,14,17)(5,8,10)(9,12,13)(6,8,9)(7,9,10)(5,6,8)(8,11,13)(5,6,8)(7,8,11)(9,10,12)

Table A7. Fuzzy processing time of fJSSP instance 2

J11	(7,10,14)(6,9,11)(10,13,17)(7,9,12)(8,11,15)(5,8,11)(8,11,15)(9,12,16)(9,12,16) (8,12,16)
2	(16,20,25)(14,19,23)(13,17,21)(12,15,19)(16,19,23)(15,16,19)(9,15,19)(10,15,19) (14,18,22)(14,18,22)
3	(10,16,17)(8,10,13)(10,12,16)(8,11,13)(9,12,14)(10,13,15)(9,12,14)(15,20,23) (13,15,18)(10,13,15)
4	(8,12,15)(13,16,19)(12,15,19)(10,13,14)(11,14,17)(16,19,23)(18,22,26)(10,13,14) (12,17,21)(11,13,17)
J21	(11,15,18)(5,8,10)(5,9,13)(6,7,10)(7,10,12)(6,9,12)(6,9,12)(8,11,13)(8,10,12)(6,9,11)
2	(10,12,14)(5,8,9)(5,9,10)(7,10,13)(5,7,9)(6,8,11)(7,10,13)(5,7,9)(8,11,14)(8,12,14)
3	(8,9,10)(10,13,14)(10,12,15)(11,13,16)(10,11,13)(8,12,15)(8,11,13)(6,9,12)(6,8,12) (8,9,10)
4	(6,9,11)(9,11,14)(8,10,13)(11,14,17)(7,10,12)(8,11,12)(13,17,20)(5,8,12)(6,10,13) (7,8,10)
J31	(6,7,9)(4,5,6)(6,9,12)(7,8,10)(6,7,9)(8,10,13)(5,7,10)(9,11,13)(9,12,14)(4,5,6)
2	(9,12,15)(6,8,12)(8,10,13)(7,8,11)(8,10,13)(5,6,8)(7,10,13)(5,9,12)(6,9,11)(5,9,12)
3	(5,6,8)(8,10,13)(10,14,19)(7,11,14)(6,8,11)(6,9,13)(9,13,17)(8,10,13)(7,11,14)(6,9,12)
4	(10,14,18)(11,15,20)(14,19,24)(13,17,20)(9,15,21)(9,14,18)(10,14,18)(15,19,26) (23,28,33)(8,12,17)
J4 1	(9,12,16)(9,13,17)(11,15,20)(8,13,19)(12,17,24)(10,15,19)(10,15,19)(13,16,20) (10,16,21)(12,16,23)

Table A7. (*continued*)

2	(11,15,20)(10,13,17)(13,18,23)(11,17,23)(11,16,23)(13,18,24)(11,15,20)(10,12,16)(11,17,24)(14, 19,23)
3	(15,21,26)(10,11,17)(9,13,17)(12,18,23)(10,11,17)(11,16,20)(13,17,21)(11,15,20)(13,19,24)(10,15,19)
4	(3,4,5)(6,7,10)(8,9,11)(9,11,14)(5,8,11)(8,11,16)(7,10,15)(8,12,16)(6,7,10)(9,12,16)
J5 1	(8,11,15)(7,10,12)(7,9,11)(6,9,12)(10,13,17)(7,10,12)(9,11,15)(7,9,12)(11,14,18)(5,8,11)
2	(6,8,9)(12,15,19)(7,10,13)(6,10,12)(7,9,12)(10,13,18)(7,11,15)(6,10,12)(8,11,15)(8,12,17)
3	(5,6,8)(6,9,11)(9,10,13)(7,9,10)(5,8,11)(11,14,18)(8,10,13)(9,12,16)(7,8,10)(7,8,10)
4	(4,7,9)(9,10,13)(5,8,12)(7,10,14)(6,8,11)(7,9,12)(6,8,11)(9,10,12)(5,9,12)(6,8,11)
J6 1	(10,13,17)(6,8,9)(7,9,11)(8,10,13)(8,10,13)(8,11,15)(6,10,13)(7,9,12)(9,11,13)(7,9,11)
2	(10,14,19)(11,15,20)(12,16,22)(9,13,17)(8,13,17)(10,13,18)(11,15,21)(12,19,25)(8,13,17)(10,14,17)
3	(7,10,12)(12,17,23)(12,19,25)(10,15,20)(9,16,27)(11,15,19)(13,18,24)(11,16,23)(10,15,20)(10,15,19)
4	(3,4,5)(2,3,5)(7,8,10)(5,8,10)(3,5,8)(6,8,10)(7,10,11)(6,8,10)(4,7,9)(11,14,18)
J7 1	(8,11,13)(7,9,11)(9,13,17)(10,13,17)(10,14,19)(7,10,13)(9,13,17)(8,12,16)(9,11,15)(8,10,14)
2	(6,9,11)(10,13,18)(7,8,10)(8,9,11)(8,9,11)(11,14,18)(10,13,16)(7,10,13)(10,12,14)(7,9,12)
3	(5,6,8)(4,5,7)(6,8,11)(8,10,13)(7,10,12)(5,8,10)(8,10,12)(9,11,15)(5,8,10)(11,15,20)
4	(11,14,18)(10,13,17)(8,12,17)(7,10,14)(9,14,20)(8,12,17)(10,15,21)(9,14,19)(10,14,19)(8,10,13)
J8 1	(7,10,12)(8,11,15)(9,14,18)(10,15,19)(11,14,18)(9,15,21)(8,12,17)(7,10,12)(9,13,18)(8,12,17)
2	(10,14,19)(9,12,15)(12,17,21)(13,18,23)(10,13,18)(11,15,20)(11,16,19)(11,16,19)(10,13,18)(12,16,21)
3	(4,6,8)(8,11,15)(7,11,14)(8,10,13)(6,8,10)(12,17,23)(7,10,12)(10,13,17)(9,14,18)(6,8,10)
4	(7,8,10)(4,5,7)(3,5,8)(7,9,11)(6,8,10)(5,7,9)(6,8,11)(7,10,12)(7,10,12)(8,10,13)
J9 1	(5,8,10)(10,13,18)(4,7,9)(8,10,13)(6,9,12)(5,7,10)(9,12,16)(7,9,13)(6,8,12)(5,7,10)
2	(2,3,5)(4,6,7)(8,10,11)(7,9,10)(5,6,8)(6,7,9)(9,12,14)(8,11,13)(4,6,7)(5,7,10)
3	(10,14,17)(6,9,11)(7,11,14)(5,9,12)(5,8,11)(8,10,14)(12,16,20)(6,8,11)(7,10,13)(10,13,15)
4	(9,11,15)(8,11,14)(7,10,12)(10,13,17)(8,10,13)(11,16,21)(7,9,11)(13,17,22)(10,14,17)(8,9,12)
J101	(5,8,11)(7,9,13)(6,9,11)(10,13,17)(7,10,13)(8,11,15)(9,12,16)(9,14,17)(10,14,18)(7,11,14)
2	(13,17,22)(7,10,12)(8,10,13)(6,10,13)(6,9,11)(9,13,16)(7,9,10)(8,11,15)(7,10,12)(9,13,17)
3	(5,8,10)(11,15,19)(7,9,11)(6,8,11)(8,10,13)(9,11,15)(7,9,11)(12,17,21)(5,9,12)(10,13,17)
4	(3,5,8)(4,5,7)(9,10,13)(5,8,10)(6,8,12)(7,9,12)(8,11,14)(10,15,19)(4,7,9)(11,16,20)

Table A8. Fuzzy processing time of fJSSP instance 3

J11	(3,4,6)(7,9,12)(5,7,10)(8,10,13)(9,11,14)(5,8,11)(10,14,18)(6,9,12)(7,9,10)(8,11,15)
2	(5,7,9)(8,10,13)(7,8,10)(6,8,9)(8,10,11)(9,11,14)(6,7,10)(5,7,10)(7,9,10)(10,14,18)
3	(7,9,11)(8,10,13)(7,8,11)(6,8,9)(5,7,8)(11,14,17)(5,8,11)(7,9,12)(6,8,11)(5,7,8)
4	(8,10,13)(5,8,10)(9,13,17)(6,8,11)(6,9,12)(7,10,13)(5,9,12)(6,9,12)(12,17,21)(8,12,16)
J21	(3,4,5)(6,8,9)(9,10,13)(5,7,8)(6,9,11)(7,10,12)(10,13,15)(5,7,10)(10,13,17)(6,9,12)
2	(10,13,17)(8,11,14)(7,10,12)(6,9,11)(6,9,11)(9,12,16)(8,11,15)(7,11,14)(6,10,13)(15,19,24)
3	(4,5,7)(8,10,13)(6,8,10)(5,8,9)(6,9,11)(6,8,9)(7,10,12)(11,14,18)(6,9,12)(9,11,14)
4	(6,9,11)(5,7,8)(4,6,7)(3,5,8)(6,8,9)(7,10,12)(8,10,13)(5,8,10)(6,9,11)(7,9,12)
5	(5,6,7)(9,10,14)(8,10,13)(7,9,10)(4,5,7)(6,8,10)(5,8,10)(6,8,9)(9,11,15)(3,5,8)
6	(8,11,15)(7,10,13)(6,9,11)(5,8,10)(7,9,12)(6,8,11)(8,10,13)(10,13,17)(5,9,13)(7,8,10)
J31	(7,9,12)(6,7,9)(4,6,9)(8,11,14)(9,13,15)(5,7,10)(6,9,12)(7,8,11)(9,11,13)(5,6,7)
2	(5,7,9)(7,9,12)(4,7,9)(8,9,12)(7,8,10)(6,9,10)(5,6,8)(3,5,8)(8,10,11)(6,8,9)
3	(10,14,18)(9,13,17)(12,16,21)(8,11,15)(7,11,14)(15,19,24)(11,15,19)(16,23,28)(9,12,16)(8,11,14)
4	(6,9,11)(5,8,10)(8,9,11)(7,10,13)(9,11,15)(10,13,17)(11,14,17)(8,10,13)(11,15,19)(6,10,12)
J41	(8,11,14)(7,9,12)(6,8,11)(5,8,11)(9,11,15)(7, 10,13)(6,10,13)(9,12,15)(5,7,10)(11,16,20
2	(4,5,7)(3,4,6)(5,8,10)(4,6,9)(2,3,5)(7,9,11)(8,9,12)(5,7,10)(8,10,11)(7,8,9)
3	(8,10,14)(9,12,16)(10,14,18)(7,10,13)(6,9,12)(5,8,10)(6,8,11)(7,11,14)(8,9,10)(6,9,12)
4	(6,8,10)(5,7,9)(7,10,12)(8,10,13)(9,11,15)(6,9,11)(11,15,18)(10,13,18)(7,9,12)(5,8,11)
5	(13,17,21)(8,11,14)(7,11,14)(9,12,15)(10,14,18)(8,10,13)(7,10,13)(14,18,23)(8,11,15)(9,11,15)
6	(7,10,12)(6,8,11)(5,8,10)(8,10,13)(9,12,17)(7,9,13)(11,15,19)(8,11,15)(6,9,12)(13,17,20)
J51	(9,12,16)(7,10,13)(6,9,12)(8,11,14)(7,9,10)(8,10,11)(13,15,16)(10,13,15)(9,12,16)(8,10,11)
2	(3,4,6)(7,8,10)(8,10,11)(9,11,15)(5,7,8)(6,8,10)(7,10,12)(5,8,10)(9,12,14)(10,13,18)
3	(7,9,12)(6,9,11)(8,11,15)(9,12,16)(9,13,17)(10,14,16)(7,10,13)(8,10,13)(11,15,19)(6,9,11)
4	(6,9,11)(5,8,11)(4,8,11)(7,11,14)(8,11,15)(7,9,12)(9,12,16)(10,14,18)(7,10,13)(8,11,14)
5	(2,4,6)(1,2,4)(4,6,8)(7,8,9)(8,10,12)(5,8,10)(3,5,8)(4,6,9)(7,9,11)(6,9,11)
J61	(7,10,13)(5,8,11)(8,10,13)(6,8,11)(7,9,11)(8,11,14)(9,10,12)(6,8,10)(7,10,13)(10,13,16)
2	(6,8,9)(5,6,7)(4,6,8)(3,5,8)(8,10,13)(4,5,6)(6,8,10)(7,9,11)(5,8,9)(6,9,11)
3	(9,11,14)(10,14,18)(8,11,15)(7,10,13)(8,10,13)(9,12,15)(7,9,11)(6,9,12)(10,15,20)(8,11,15)
4	(4,7,9)(5,8,10)(6,9,12)(7,10,12)(8,11,15)(9,10,13)(6,8,11)(10,14,17)(8,10,14)(5,7,10)
5	(8,10,13)(7,9,11)(7,10,13)(8,11,15)(6,9,12)(5,7,9)(6,8,10)(10,13,15)(8,10,13)(11,14,17)
J71	(6,8,9)(7,10,12)(8,11,13)(9,13,17)(5,8,11)(8,11,14)(7,9,12)(6,9,13)(5,9,12)(6,10,13)

Table A8. *(continued)*

2	(4,5,7)(5,6,7)(6,7,9)(7,9,11)(8,10,11)(9,11,13)(7,10,12)(5,7,10)(4,6,8)(8, 11,13)
3	(8,11,14)(7,8,10)(9,11,12)(10,13,14)(6,9,10)(7,9,10)(8,10,13)(11,14,18)(6,10,13) (13,17,21)
4	(11,13,17)(5,8,11)(4,7,10)(6,8,11)(7,10,12)(8,10,13)(9,11,14)(10,14,17)(8,11,15) (6,9,11)
5	(13,17,21)(7,10,13)(8,11,14)(9,11,13)(10,14,17)(14,19,22)(6,9,12)(5,8,12)(7,11,14) (8,11,15)
J81	(6,8,11)(5,8,12)(7,11,15)(8,10,13)(9,12,16)(10,14,18)(8,11,15)(6,9,12)(7,10,13)(8,9,11)
2	(15,19,24)(11,15,19)(9,10,13)(12,15,18)(8,12,17)(7,11,14)(17,21,26)(8,11,15) (10,14,17)(9,12,14)
3	(7,10,12)(6,9,11)(8,10,13)(9,12,14)(7,11,14)(10,13,17)(7,10,12)(11,15,19)(9,11,13) (8,10,13)
4	(8,10,13)(10,14,17)(6,9,12)(5,8,12)(7,10,13)(5,8,12)(12,15,19)(6,9,11)(7,11,14) (9,12,15)
5	(5,6,7)(6,8,11)(7,10,12)(5,8,12)(9,11,14)(10,14,18)(6,9,12)(7,11,14)(8,11,15)(5,7,10)
J9 1	(10,14,17)(9,12,15)(8,11,15)(7,11,14)(6,9,12)(7,10,13)(12,16,20)(13,18,23)(7,9,13) (8,10,13)
2	(7,9,11)(5,8,10)(7,10,13)(9,12,16)(6,9,12)(8,11,14)(10,14,16)(9,13,17)(6,10,13) (7,9,13)
3	(9,13,17)(6,9,12)(10,14,18)(8,11,15)(7,10,13)(6,9,11)(12,17,21)(8,10,13)(9,11,14) (11,14,17)
4	(4,5,7)(5,7,9)(6,8,9)(7,9,10)(8,10,13)(4,7,9)(5,6,8)(2,3,5)(6,9,11)(7,10,12)
5	(8,10,13)(9,11,14)(7,9,11)(6,9,13)(10,13,17)(5,7,10)(8,11,15)(9,12,16)(6,9,12) (7,10,12)
J101	(3,5,8)(7,9,11)(4,5,7)(5,8,11)(6,9,12)(8,10,13)(6,8,11)(10,14,17)(7,11,14)(8,10,11)
2	(8,11,15)(9,13,17)(7,11,14)(6,10,13)(5,9,12)(7,10,13)(6,9,13)(10,14,17)(8,10,13) (15,19,24)
3	(7,9,11)(13,17,22)(7,9,13)(8,10,13)(9,11,14)(8,11,15)(5,8,12)(6,9,13)(10,12,15) (7,10,12)
4	(6,9,12)(5,8,10)(8,10,13)(9,11,15)(7,9,12)(8,11,13)(5,9,12)(7,11,14)(10,14,17) (11,15,19)
5	(4,5,7)(7,8,9)(6,8,10)(3,5,8)(4,6,9)(5,7,10)(6,9,12)(7,10,12)(5,8,11)(8,10,13)

* In the first column of Tables 1, 2 and 3, numerals such as 1,2,3, 4 in the first column indicate the serial number of operations.

Table A9. The standard variance of the stochastic processing time

2.01126,	7.07227,	3.73974,	9.27866,	7.26508,
6.31886,	5.15262,	10.0637,	9.40556,	8.71944,
3.56697,	9.73049,	8.39451,	6.62181,	4.73595,
2.13486,	2.82263,	5.28007,	3.32582,	3.49309,
10.8967,	6.01123,	3.07175,	2.04202,	2.0802,
5.40092,	6.78497,	7.14066,	7.41588,	7.46449,
3.49611,	7.96741,	6.0571,	5.1691,	2.51335,
7.46916,	9.04987,	9.22346,	6.67895,	4.71755,
9.88375,	8.54009,	10.6031,	10.3315,	6.85418,
3.28104,	6.15873,	4.11795,	9.76016,	3.88641,
9.01691,	9.59288,	10.9712,	10.9973,	7.50349,
5.53194,	4.39592,	4.67553,	9.5613,	2.21369,
5.38279,	2.83361,	8.09485,	2.50594,	2.0791,
10.2691,	4.48299,	4.45607,	7.29118,	8.22065,
9.5385,	8.53844,	6.36445,	3.84823,	8.69362,
6.21613,	6.12165,	10.5424,	8.69994,	2.97452,
7.39143,	5.46712,	8.61507,	7.4807,	7.15165,
5.25205,	3.36399,	4.02594,	5.82638,	9.22593,
6.65395,	10.9099,	8.76394,	5.11005,	3.52083,
7.91577,	6.42708,	2.57186,	8.29783,	6.54326,

For 10×10 JSSP with stochastic processing time, the first ten real numbers are standard variance of the processing time of the operation $o_{11}, o_{12}, \cdots, o_{110}$ and the second ten real numbers correspond to operation $o_{21}, o_{22}, \cdots, o_{210}$ and so on. For 20×5 JSSP, the first five real numbers correspond to operation $o_{11}, o_{12}, \cdots, o_{15}$, the second five numbers correspond to operation $o_{21}, o_{22}, \cdots, o_{25}$ and so on.

References

[1] Rodammer, F.A., Preston, W.K.: A recent survey of production scheduling. IEEE Trans. Sys. Man Cyber. 188, 41–51 (1988)

[2] Kuroda, M., Wang, Z.: Fuzzy job shop scheduling. Int. J. Prod. Eco. 44, 45–51 (1996)

[3] Sakawa, M., Mori, T.: An efficient genetic algorithm for job shop scheduling problems with fuzzy processing time and fuzzy due date. Comput. Indus. Eng. 36, 325–341 (1999)

[4] Song, X.Y., Zhu, Y.L., Yin, C.W., Li, F.M.: Study on the combination of genetic algorithms and ant colony algorithms for solving fuzzy job shop scheduling problems. In: Proceedings of IMACS multi-conferences on computational engineering in systems applications, Beijing, pp. 1904–1909 (2006)

[5] Niu, Q., Jiao, B., Gu, X.S.: Particle swarm optimization combined with genetic operators for job shop scheduling problem with fuzzy processing time. Appl. Math.Comp. (in press)

[6] De, P., Ghosh, J.B., Wells, C.E.: On the minimization of the weighted number of tardy job with random processing times and deadline. Comp. Oper. Res. 18, 457–463 (1991)

[7] Soroush, H.M.: Optimal sequence in stochastic single machine shops. Comp. Oper. Res. 23, 705–721 (1996)

[8] Gourgand, M., Grangeon, N., Norre, S.: A contribution to the stochastic flow shop scheduling problem. Eur. J. Oper. Res. 151, 415–433 (2003)
[9] Luh, P.B., Cheng, D., Thakur, L.S.: An effective approach for job shop scheduling with uncertain processing requirements. IEEE Trans. Rob. Automat. 15, 328–339 (1999)
[10] Ginzburg, D.G., Gonik, A.: Optimal job-shop scheduling with random operations and cost objectives. Int. J. Pro. Eco. 76, 147–157 (2002)
[11] Tavakkoli-Moghaddam, R., Jolai, F., Vaziri, F., Ahmed, P.K., Azaron, A.: A hybrid method for solving stochastic job shop scheduling problem. App. Math. Comput. 170, 185–206 (2005)
[12] Lei, D.M., Xiong, H.J.: Job shop scheduling with stochastic processing time through genetic algorithm. In: Proceedings of International Conference on Machine Learning and Cybernetics, Kunming, China, pp. 941–946 (2008)
[13] Sakawa, M., Kubota, R.: Fuzzy programming for multi-objective job shop scheduling with fuzzy processing time and fuzzy due date through genetic algorithm. Euro. J. Oper. Res. 120, 393–407 (2000)
[14] Li, F.-M., Zhu, Y.-L., Yin, C.-W., Song, X.-Y.: Fuzzy programming for multi-objective fuzzy job shop scheduling with alternative machines through genetic algorithms. In: Wang, L., Chen, K., Ong, Y.S. (eds.) ICNC 2005. LNCS, vol. 3611, pp. 992–1004. Springer, Heidelberg (2005)
[15] Giffler, B., Thompson, G.L.: Algorithm for solving production scheduling problems. Oper. Res. 8, 487–503 (1960)
[16] Lei, D.M.: Pareto archive particle swarm optimization for multi-objective fuzzy job shop scheduling problems. Int. J. Adv. Manuf. Technol. 37, 157–165 (2008)
[17] Xing, Y.J., Wang, Z.Q., Sun, J., Meng, J.J.: A multi-objective fuzzy genetic algorithm for job-shop scheduling problems. In: 2006 International Conference on Computational Intelligence and Security, pp. 398–401 (2006)
[18] Ghrayeb, O.A.: A bi-criteria optimization: minimizing the integral value and spread of the fuzzy makespan of job shop scheduling problems. Appl. Soft. Comput. 2, 197–210 (2003)
[19] Javadi, B., Saidi-Mehrabad, M., Haji, A., et al.: No-wait flow shop scheduling using fuzzy multi-objective linear programming. Journal of the Franklin Institute (in press)
[20] Lei, D.M., Xiong, H.J.: An efficient evolutionary algorithm for multi-objective stochastic job shop scheduling. In: Sixth International Conference on Machine Learning and Cybernetics, pp. 19–22 (2007)
[21] Brindle, A.: Genetic Algorithms for Function Optimization, Doctoral dissertation, Univ. of Alberta, Canada (1981)
[22] Chakraborty, U.K., Deb, K., Chakraborty, M.: Analysis of selection algorithms: A Markov chain approach. Evolutionary Computation 4(2), 133–167 (1996)
[23] Bean, J.: Genetic algorithms and random keys for sequencing and optimization. ORSA J. Comput. 6, 154–160 (1994)
[24] Cheng, R.W., Gen, M., Tsujimura, Y.: A tutorial survey of job-shop scheduling problems using genetic algorithms: I. Representation. Compu. Indus. Eng. 30(4), 983–997 (1996)
[25] Rothlauf, F.: Representations for Genetic and Evolutionary Algorithms, 2nd edn. Springer, Heidelberg (2006)
[26] Chakraborty, U.K., Janikow, C.: An analysis of Gray versus binary encoding in genetic search. Information Sciences 156(3-4), 253–269 (2003)
[27] Bierwirth, C.: A generalized permutation approach for job shop scheduling with genetic algorithms. OR Spectrum, Special issue: Applied Local Search 17, 87–92 (1995)

[28] Bierwirth, C., Mattfeld, D., Kopfer, H.: On permutation representations for scheduling problems. In: Voigt, H.M. (ed.) Proceedings of Parallel Problem Solving from Nature IV, pp. 310–318. Springer, Berlin (1996)
[29] Mattfeld, D.C.: Evolutionary search and the job shop. In: Investigations on genetic algorithms and production scheduling.Springer, Berlin (1995)
[30] Pezzella, F., Morganti, G., Ciaschetti, G.: A genetic algorithm for the flexible job-shop scheduling problem. Comp. Oper. Res. 35(10), 3202–3212 (2008)
[31] Park, B.J., Choi, H.R., Kim, H.S.: A hybrid genetic algorithm for job shop scheduling problems. Comp. Ind. Eng. 45, 597–613 (2003)
[32] Zhang, C.Y., Li, P.G., Guan, Z.L., Rao, Y.Q.: A tabu search algorithm with a new neighbor structure for the job shop scheduling problem. Comp. Oper. Res. 34, 3229–3242 (2007)
[33] Sha, D.Y., Hsu, C.Y.: A hybrid particle swarm optimization for job shop scheduling problem. Comp. Ind. Eng. 51, 791–808 (2006)
[34] Goldberg, D., Lingle, R.: Alleles, loci and the traveling salesman problem. In: Proceedings of the First International Conference on Genetic Algorithms, pp. 154–159 (1985)
[35] Davis, L.: Applying adaptive algorithms to epistatic domains. In: Proceedings of the ninth International Joint Conference on Artificial Intelligence, pp. 162–164 (1985)
[36] Davis, L.: Job shop scheduling with genetic algorithms. In: Proceedings of the First International Conference on Genetic Algorithms, pp. 136–140 (1985)

Giffler and Thompson Procedure Based Genetic Algorithms for Scheduling Job Shops

S.G. Ponnambalam[1], N. Jawahar[2], and B.S. Girish[2]

[1] School of Engineering,
Monash University,
Malaysia
sgponnambalam@eng.monash.edu.my

[2] Department of Mechanical Engineering,
Thiagarajar College of Engineering,
Madurai, India, 625015
jawahartce@yahoo.co.uk, girishbs31@yahoo.co.in

Summary. This chapter addresses two well known job shop problems, namely the classical job shop scheduling problem (JSP) and the flexible job shop scheduling problem (FJSP). Both of them belong to the category of the toughest NP-hard problems. Genetic algorithm (GA) based heuristics that have adopted Giffler and Thompson (GT) procedure, an efficient active feasible schedule generation methodology for JSP, are discussed to solve the following job shop scheduling (JSS) models: JSP for single-objective criterion (minimization of makespan time), JSP for multi-objective criterion (minimization of weighted sum of makespan time, total tardiness and total idle time of all machine) and FJSP for makespan time criterion. The chromosome representation of the GAs proposed for the JSPs is the combination of priority dispatching rules '*pdrs*' (independent *pdrs* one each for one machine), which on decoding provides an active feasible schedule using GT procedure. The chromosome representation of the GA for FJSP consists of two strings of size equal to the total number of operations: one string for machine assignment that reduces the FJSP to a fixed route JSP and the other string is a permutation representation of priority numbers each corresponding to an operation that is used for resolving the conflict that arises while generating actives feasible schedules with GT procedure. The performance tests and validations of the proposed GAs are discussed along with future research directions.

1 Introduction

Scheduling involves the allocation of resources over a period of time to perform a collection of tasks (Baker 1974). It is a decision making process that exists in most manufacturing and production systems, transportation and distribution settings and in most information-processing environments (Pinedo 2005). Scheduling in the context of manufacturing systems refers to the determination of the sequence in which jobs are to be processed over the production stages, followed by the determination of the start-time and finish-time of processing of jobs (Conway et al. 1967). An effective schedule provides the basis for utilizing the plant effectively and attaining the strategic objectives of the firm as reflected in the production plan.

U.K. Chakraborty (Ed.): Comput. Intel. in Flow Shop and Job Shop Sched., SCI 230, pp. 229–259.
springerlink.com

The most common manufacturing system worldwide is the job shop. Job shops are associated with the production of small volumes/large variety products and operate in a make-to-order environment (Groover 2003). Hoitomt et al. (1993) mentions that approximately 50 to 75 % of all manufactured components fall into this category of low volume/high variety and due to the market trends this percentage is likely to increase. Even though flexible manufacturing systems are today's keywords that frequently appear in many research agendas, scheduling of job shops still receive ample attention from both researchers and practitioners due to the reason that job shop scheduling problems exist in many forms in most of the advanced manufacturing systems (Kutanoglu and Sabuncuoglu 1999). Besides, analysis of job shop scheduling problems provides important insights into the solution of the scheduling problems encountered in more realistic and complicated systems (Pinedo 2005). In this context, this chapter focuses on scheduling job shops which is an important task for manufacturing industry in terms of improving machine utilization or reducing lead time or adhering to due dates.

1.1 Job Shop Scheduling Problems

The classical job shop scheduling problem (JSP) is the most popular scheduling model in practice (French 1982, Brucker 1995, Pinedo 1995). It has attracted many researchers due to its wide applicability and inherent difficulty (Jain and Meeran 1999). The formulation of the JSP is based on the assumption that for each part type or production order (job) there is only one processing plan, which prescribes the sequence of operations and the machine on which each operation has to be performed. The n x m classical JSP involves n jobs and m machines. Each job is to be processed on each machine in a predefined sequence and each machine processing only one job at a time. It is also well known that JSP is NP-hard (Garey et al. 1976).

In practice, the shop-floor setup in a job shop typically consists of multiple copies of the most critical machines so that bottlenecks due to long operations or busy machines can be reduced (Ho et al. 2007). Therefore, an operation may be performed on more than one machine. Job shops also consists of multipurpose machines such as numerically controlled (NC) machines that are loaded with tool magazines and are capable of performing several different types of operations (Vaikartarakis and Cai 2003). Due to the overlapping capabilities of these machines, a given operation can be performed by more than one machine. However, in real life it has been a practice that machining operations are assigned to a certain machine tool during the process planning stage and the assignment of machine tools over time to different operations is performed during the scheduling stage. Recently, researchers considered the integration of process planning with scheduling by allowing alternative machine tool routings for operations at the scheduling stage (Hankins et al. 1984, Chryssoulouris and Chan 1985, Wilhelm and Shin 1985).

Unlike the JSP, the research on jobs shop scheduling associated with multiple routings is rather very limited even though it has more practical applications and advantages than the JSP. Two different scheduling models of job shop associated with multiple routings are addressed in the literature. The first model is referred as job shop scheduling with alternative machine tool routings, which was first addressed by Iwata et al. (1978). The same model was later addressed by Brandimarte (1993) as flexible job shop scheduling problem (FJSP). The second model is usually referred as job shop scheduling with multi-purpose machines (MPM-JSP), which was first addressed by Brucker and Schlie (1990). Dauzere-Peres and Paulli (1997) addressed the MPM-JSP as multiprocessor job shop scheduling problem (MJS). The difference between the two models (FJSP and MPM-JSP/MJS) is that, in the first model the processing time for each operation on its alternative routes differs with machine features, whereas in the second model the processing time is same for all the alternative machines of a particular operation. Since the FJSP can be represented as a generalized model of MPM-JSP/MJS, therefore, many recently published research articles refer both the models as FJSP. However, the introduction of alternative routing option adds an additional decision of machine allocation during scheduling that increases the complexity of the problem. Therefore, scheduling job shops that are associated with multiple routings are much more complex than the JSP.

1.2 Modeling and Solution Approaches for Scheduling Problems

A large number of approaches to the modeling and solution for job shop scheduling problems have been reported in the OR literature, with varying degrees of success. These approaches revolve around a series of technological advances that have occurred over that last four decades. These include optimization approaches such as mathematical programming, enumerative techniques, etc. and approximation approaches such as dispatching rules, artificial intelligence (AI) techniques, local search methods and metaheuristics (Brucker 1995).

Optimization algorithms provide optimal or near-optimal results if the problems to be solved are not too large and are restricted to low-dimensional over-simplified problems. With the growing uncertainty and complexity in manufacturing environment, most scheduling problems have been proven to be NP-hard, that is, the computational time requirements grow exponentially as a function of the problem size. This degrades the performance of conventional optimization techniques and hence optimization approaches are ruled out in practice. The approximation algorithms are capable of guaranteeing the solution to be within a fixed percentage of the actual optimum and are considered urgent and useful tools for solving discrete optimization problems. The performance of heuristics is satisfactory as long as the operating characteristics and objectives of the system remain the same. Heuristics yield good solutions, but are robust to the system. Local search based heuristics are known to produce excellent results in short run times, but they are susceptible of getting stuck in local entrapments.

Evolutionary programming, which belongs to the random search process, is regarded better than simulation in the sense that it guarantees near optimal solutions in actual cases. Also, by changing the evolution parameter of the genetic search process, the solutions can be obtained for other suitable objectives and can be made more flexible. These are useful to address the dynamic situations. The above discussion indicates that heuristics, local search algorithms, evolutionary search algorithms are useful tools for scheduling job shops.

1.3 Genetic Algorithm Based Heuristics for Scheduling: A Literature Review

In recent years, genetic algorithm (GA) is much used in job shop scheduling applications. The following work indicates the applications of GA in JSP and FJSP. Kopfer and Mattfield (1997) proposed a hybrid GA for the JSP and showed that the results are encouraging. Schultz and Mertens (1997) compared the GA with an expert system approach and priority rules. They indicated that the GA generally produces satisfactory schedules, and its performance depends on run time (i.e. population size and number of generations). Biegel and Davern (1990) showed the method of applying genetic concepts to scheduling problems. An elementary n-task, one-processor problem is provided to demonstrate the GA methodology for the job shop scheduling problem. Dorndorf and Pesch (1993) proposed a GA based on the idea of using a chain of priority rules which fits the needs of a particular problem. Within the GA each gene represents a priority rule from the set of priority rules. While decoding a chromosome, to generate a feasible schedule, the ith rule is applied for scheduling the ith conflict in the schedule generation procedure. In their GA process, they employed a Giffler and Thompson algorithm (Giffler and Thompson 1960) to generate an active feasible schedule and used the makespan time of the schedule as the fitness parameter. Jawahar et al. (1998) proposed a GA for scheduling flexible manufacturing systems. The proposed GA evolves a priority dispatching rule for each machine to resolve the conflicts that arise while generating active feasible schedules using Giffler and Thompson schedule generation procedure. Ponnambalam et al. (2001) proposed a multiobjective GA (MOGA) for the job shop scheduling problem for minimization of weighted sum of makespan, total tardiness of all jobs and total idle time of all machines. They used the chromosome representation proposed by Jawahar et al. (1998) in their proposed MOGA and showed the effectiveness of their approach by testing with various benchmark instances from literature.

Mesghouni et al. (1998) were the first to model GA for FJSP. They proposed a chromosomal representation known as parallel job representation in which a chromosome is represented by a matrix where each row consists of a set of ordered operations of each job. Due to the complexity of decoding the representation, their algorithm incurs significant computational cost. Hussain and Joshi (1998) proposed a two pass GA to solve job shop problem with alternative routing with the objective of minimizing the sum of squared weighted due date deviation for every job. The first pass picks the alternatives using a genetic algorithm and the second pass provides the order and start time of jobs on the selected alternatives by solving a non-linear program. Chen et al. (1999) proposed a GA that uses an A-B string representation to solve FJSP for minimum makespan time criterion. A string contains a list of all operations of all jobs and the machines selected for the corresponding operations while B string contains a list of operations that are processed on each machine. Moon and Lee (2000) developed a mixed integer linear programming (MILP) model and proposed a genetic algorithm (GA) for the job shop scheduling problem with alternative routings. The objective they considered is to minimize the mean flow time. The chromosome representation in their proposed GA consists of two strings, one for machine assignment and the other for schedule generation. Ho and Tay (2004) proposed a GA based tool, namely GENACE, for solving the FJSP for minimum makespan time criterion. The chromosome representation consists of two components, one component for machine selection and the other for operation sequence. Their methodology first generates an initial population using composite dispatching rules. A cultural evolution is then applied to preserve knowledge of schemata and resource allocations learned over each generation. The knowledge or belief spaces in turn influence mutation and selection of individuals.

Ho et al. (2007) proposed an architecture for learning and evolving of flexible job shop schedules for minimum makespan criterion called learnable genetic architecture (LEGA), a generalization of their previous approach GENACE (Ho and Tay 2004). The population generator module generates a set of feasible schedules equal to the population size using composite dispatching rules and then encodes it into chromosomes of initial population for subsequent evolution in the EA module. During genetic evolution, the SL module modifies the offspring schedules to improve solution quality and to preserve feasibility based on a memory of conserved schemas resolved from sampled schedules sent dynamically from EA module. Tay and Ho (2008) proposed a genetic programming (GP) based approach for evolving effective composite dispatching rules for solving the multi-objective FJSP. The objective they considered is to minimize the weighted sum of makespan time, mean flow time and mean tardiness. They proposed a GP framework in which an individual is composed of terminals (like job release dates, due date, processing time, current time, remaining time, etc.) and algebraic functions. Their GP solves a specific problem by carefully selecting the terminals and functions and generating a composite dispatching rule that satisfies the requirements of that particular problem. They generated five composite dispatching rules using a large training set and compared the results with other popular rules like FIFO, SPT, etc. The coding schemes adopted in the most of the above GAs for FJSP requires repair mechanisms to maintain solution feasibility. Most of the GAs proposed for FJSP, therefore, have chances of missing the best optimal solution even under extensive searches for larger size problems. Girish and Jawahar (2008) proposed a GA for the FJSP for minimum makespan time criterion. The chromosome representation of their proposed GA consists of two strings: one string for machine assignment and the other string for sequencing the operations on the assigned machines using Giffler and Thompson schedule generation procedure (Giffler and Thompson 1960). The chromosomal representation of their proposed GA does not require a repair mechanism and is capable to rummage through the entire search space.

In this chapter, genetic algorithm based heuristics that adopt Giffler and Thompson schedule generation procedure, which is a proven method to generate active feasible schedules for JSP, are presented to evolve optimal or near optimal schedules to the well known JSP and FJSP formulations. The rest of the chapter is organized as follows: section 2 describes the job shop scheduling models considered in this chapter; the description with numerical illustration and performance analysis of the proposed GAs for the single-objective JSP, multi-objective JSP and single-objective FJSP are presented in sections 3, 4 and 5, respectively; section 6 concludes with directions for future research.

2 Description of Job Shop Scheduling Models

2.1 Model 1: Scheduling Job Shop for Makespan Time Criterion

2.1.1 Environment

- There are n jobs to be processed in one or more of m machines.
- Each job i require J_i precedence-constrained operations to be performed.
- Each operation O_{ij} can be processed only on one machine and its processing time is t_{ij}.

2.1.2 Assumptions

- Jobs are independent and no priorities are assigned to any job type.
- Each machine can process only one job at a time.
- The revisit of jobs for another operation to a same machine is not allowed.
- Job pre-emption or cancellation is not allowed.
- Set up and inspection times are included in the processing time.
- All jobs are simultaneously available at time zero.
- After a job is processed on a machine it is transported to the next machine immediately and the transportation time is negligible or included in the operation time.
- Breakdowns are not considered.

2.1.3 Objective

The objective is to complete all operations at the earliest possible time, which is known as minimum makespan time. This objective would distribute the workload evenly among all processing stations or work centers and all the processing stations would be freed at the makespan time for planning another set of jobs of next planning horizon.

2.1.4 Problem Formulation

The mathematical formulation for the problem under discussion with the objective of minimizing makespan time is presented below:

Objective:

$$\text{Minimize}\,[\text{Max}(C_{1J_1}, C_{2J_2} \ldots, C_{nJn})] \tag{1}$$

Subject to:

$$C_{ij} - S_{ij} - t_{ij} = 0 \qquad \forall \quad i, j \tag{2}$$

$$\begin{aligned} &C_{i'j'} - C_{ij} + H(1 - Y_{iji'j'}) \geq t_{i'j'}, \\ &\forall (i, j), (i', j') : O_{ij} \in N_k, O_{i'j'} \in N_k \end{aligned} \tag{3}$$

$$\begin{aligned} &C_{ij} - C_{i'j'} + H(Y_{iji'j'}) \geq t_{ij}, \\ &\forall (i, j), (i', j') : O_{ij} \in N_k, O_{i'j'} \in N_k \end{aligned} \tag{4}$$

$$S_{ij} \geq 0, \quad \forall i, j \tag{5}$$

$$S_{ij+1} - C_{ij} \geq 0, \quad \forall\, i,\, j = 1, \ldots, J_i - 1 \tag{6}$$

$$Y_{iji'j'} = \begin{cases} 1, & \text{if operation } O_{ij} \text{ precedes } O_{i'j'} \\ 0, & \text{otherwise} \end{cases} \tag{7}$$

where, S_{ij} and C_{ij} are the start time and completion time of job i, H is a very large positive integer, N_k is the set of operations $\{O_{ij}\}$ that can be loaded on machine k and $Y_{iji'j'}$ is a decision variable that generates a sequence between the operations O_{ij} and $O_{i'j'}$.

The constraint set (2) imposes that the difference between the completion time and the starting time of an operation is equal to its processing time. This constraint satisfies the assumption that once an operation has started, it cannot be pre-empted until its completion. Constraint sets (3) and (4) ensure that no two operations can be processed simultaneously on the same machine. The disjunctive constraint (3) becomes inactive when $Y_{iji'j'}=0$ and the disjunctive constraint (4) becomes inactive when $Y_{iji'j'}=1$. Constraint set (5) ensures that the start time of an operation is always positive. Constraint set (6) represents the precedence relationship among various operations of a job.

2.2 Model 2: Scheduling Job Shop for Multiobjective Criteria

The problem environment and assumptions for this model are the same as that of the job shop scheduling model described in section 2.1. Each job in this model is additionally subjected to job deadlines (due date) that are assumed between 1 to 5 times that of its total processing time. Besides, the objective is to minimize the weighted sum of makespan time, total tardiness and total idle time of machines and is given below.

$$\text{Minimize } w_1 \times \{\max [C_{iJi}]\} + w_2 \times \sum_{i=1}^{n} \max[0, C_{iJ_i} - d_i] + w_3 \times \sum_{k=1}^{m} I_k \tag{8}$$

Where,

$$\text{Makespan} = \max [C_{iJi}], \tag{9}$$

$$\text{Total tardiness} = \sum_{i=1}^{n} \max[0, C_{iJ_i} - d_i], \tag{10}$$

$$\text{Total idle time} = \sum_{k=1}^{m} I_k , \tag{11}$$

C_{iJi} is the completion time of job i, d_i is the due-date of job i and I_k is the Idle time of machine k. The constraints for this model are the same as single-objective job shop problem described in section 2.1.4.

2.3 Model 3: Scheduling Flexible Job Shop for Makespan Time Criterion

2.3.1 Environment

- There are m machines in the system and n jobs to be processed.
- Each job i require J_i precedence-constrained operations to be performed.
- Each operation O_{ij} can be processed on a number of alternative (non-identical) machines and the processing time t_{ijk} differs with machine features. This addresses the multiple routings for jobs. An alternative routing could be used if one machine tool is temporarily overloaded while another is idle. The alternative routing is useful where capacity problem arises.

- The objective is to complete all operations at the earliest possible time, which is known as minimum makespan time.

2.3.2 Assumptions

- Jobs are independent and no priorities are assigned to any job type.
- Job pre-emption or cancellation is not allowed.
- Set up and inspection times are included in the processing time.
- All jobs are simultaneously available at time zero.
- After a job is processed on a machine it is transported to the next machine immediately and the transportation time is negligible or included in the operation time.
- Breakdowns are not considered.

2.3.3 Problem Formulation

The mathematical formulation for the problem under discussion with the objective of minimizing makespan time is presented below:

Objective:

$$\text{Minimize}\,[\text{Max}(C_{1J_1}, C_{2J_2} \ldots, C_{nJn})] \tag{12}$$

Subject to:

$$C_{ij} - S_{ij} - \sum_{\{k:Oij \in N_k\}} (t_{ijk} \,.\, X_{ijk}) = 0 \qquad \forall \quad i, j \tag{13}$$

$$\begin{aligned} C_{i'j'} - C_{ij} + H(1 - Y_{iji'j'k}) + H(1 - X_{ijk}) + H(1 - X_{i'j'k}) \geq t_{i'j'k}, \\ \forall \quad k, (i, j), (i', j') : O_{ij} \in N_k, O_{i'j'} \in N_k \end{aligned} \tag{14}$$

$$\begin{aligned} C_{ij} - C_{i'j'} + H(Y_{iji'j'k}) + H(1 - X_{ijk}) + H(1 - X_{i'j'k}) \geq t_{ijk}, \\ \forall \quad k, (i, j), (i', j') : O_{ij} \in N_k, O_{i'j'} \in N_k \end{aligned} \tag{15}$$

$$S_{ij} \geq 0, \quad \forall \quad i, j \tag{16}$$

$$S_{ij+1} - C_{ij} \geq 0, \quad \forall\, i,\, j = 1, \ldots, J_i - 1 \tag{17}$$

$$\sum_{k:\, Oij \in N_k} X_{ijk} = 1, \quad \forall\, i, j \tag{18}$$

$$X_{ijk} = \begin{cases} 1, & \text{if operation } O_{ij} \text{ is assigned to machine } k \\ 0, & \text{otherwise} \end{cases} \tag{19}$$

$$Y_{iji'j'k} = \begin{cases} 1, & \text{if operation } O_{ij} \text{ precedes } O_{i'j'} \text{ on machine } k \\ 0, & \text{otherwise} \end{cases} \tag{20}$$

where, S_{ij} and C_{ij} is the start time and completion time of job i, H is a very large positive integer, N_k is the set of operations $\{O_{ij}\}$ that can be loaded on machine k, X_{ijk} is a decision variable for machine selection for operation O_{ij} and $Y_{iji'j'k}$ is a decision variable that generates a sequence between the operations O_{ij} and $O_{i'j'}$ for loading on machine k. The constraint set (13) imposes that the difference between the completion time and the starting time of an operation is equal to its processing time on the machine to which it is assigned. This constraint satisfies the assumption that once an operation has started, it cannot be pre-empted until its completion. Constraint set (14) and (15) ensures that no two operations can be processed simultaneously on the same machine. This disjunctive constraint (14) becomes inactive when $Y_{iji'j'k}$=0 and the disjunctive constraint (15) becomes inactive when $Y_{iji'j'k}$=1. Constraint set (16) ensures that the start time of an operation is always positive. Constraint set (17) represents the precedence relationship among various operations of a job. Constraint set (18) imposes that an operation can only be assigned to one machine.

3 GA for Single Objective JSP

3.1 Description of the Proposed GA

The different modules of the GA that is proposed to evolve optimal schedule to the job shop problem for minimum makespan time criterion is outlined as flow chart given in fig. 1.

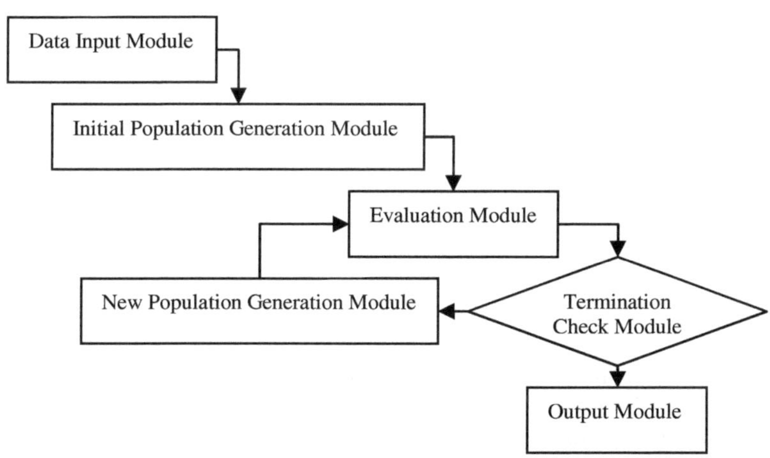

Fig. 1. Procedure of the proposed GA for JSP

Data input module: The following data pertaining to the problem are given as input: number of jobs (n), number of machines in the shop (m), number of operations J_i of each job i ($\forall i$), the machine number K_{ij} corresponding to the operation j of job i along with its processing time t_{ij} ($\forall i$, $\forall j$) and the job due date d_i.

Initial population generation module: The genetic search process starts with a randomly generated set of chromosomes called the initial population. The size of the population *(pop_size)* depends on the solution space. Each gene (*g*) in a chromosome of the proposed GA represents a priority dispatching rule (*pdr*) code (0, 1, 2 and 3), one each for one machine. The description of the *pdr* codes is given in the table 1. Floating-point encoding has been used to identify the *pdr* code. The chromosome *c*, the length of which is equal to the number of machines in the system, represents a *machine-wise-pdr* set and is representative of a feasible solution. The position of the gene in a chromosome indicates the machine number and the *pdr* code in that position identifies the *pdr* for conflict resolution by that machine.

Table 1. Priority dispatching rules and the respective codes

Priority dispatching rule	Symbol	*pdr* code
Shortest total processing time (min. of t_{ij})	SPT	0
Longest total processing time (max. Of t_{ij})	LPT	1
Earliest due time (min. of d_i)	EDT	2
Minimum Slack time (min. of (d_i- t_{ij}-t))	MINSLK	3

The possible number of combinations of *machine-wise-pdr* sets is 4^m where m is the number of machines in the system. Hence, the population size is related to the number of machines in the system and has been assumed to be equal to the number of machines in the system. The *machine-wise-pdr* set of a chromosome is applied in the Giffler and Thompson (GT) procedure to give a feasible schedule. This produces a timetable with the start and end of the processing period, and the makespan time. The fitness parameter (*fit*(*c*)) is the makespan time. It is found through the schedule generated using the *machine-wise-pdr* set and is represented by the chromosome *c*.

$$fit(c) = \text{makespan time corresponding to chromosome } c. \tag{21}$$

Each chromosome in the current population is updated as the global best chromosome, if its fitness value is less than or equal to the global best solution.

Termination Check Module: A specified number of generations (*no_iter*) are used to terminate the GA. On satisfactory termination, the output module prints the global best solution.

New population generation module: Roulette wheel selection procedure (Michalewicz 1996, Chakraborthy et al. 1996) is adopted to select chromosomes for the next generation. The process of selecting the chromosomes has the following steps:

1. Conversion of the fitness parameter value to a new fitness value (*new_fit*(*c*)), a parameter suitable for minimization objective.

$$new\_fit(c) = 1 - \frac{fit(c)}{F} \tag{22}$$

where F is a sum of the fitness parameter of all chromosomes

$$F = \sum_{c=1}^{pop\_size} fit(c) \tag{23}$$

2. Conversion of new fitness parameter to an expected frequency of selection ($p(c)$).

$$p(c) = \frac{new\_fit(c)}{\sum_{c=1}^{pop\_size} new\_fit(c)} \tag{24}$$

3. Calculation of the cumulative probability of survival ($cp(c)$)

$$cp(c) = \sum_{c=1}^{c=c} p(c) \tag{25}$$

A random selection procedure, which is explained below, generates the next population of the same size. A random number *rand*() between 0 and 1 is obtained and a chromosome *c* is selected which satisfies the following condition:

$$cp(c-1) < rand() \leq cp(c) \tag{26}$$

This selection process is repeated a number of times equal to the population size. The method used here is more reliable in that it guarantees that the most fit individuals will be selected, and that the actual number of times each is selected will be its expected frequency ±1. This procedure enables the fittest chromosome to have multiple copies and the worst to die off.

The next step is to carry out the crossover operation, which is a reproduction method. This involves two steps:

1. Selection of chromosome for crossover.
2. Crossover operation.

The probability of crossover *(p_cross)* is the one vital parameter that needs attention at this juncture. The value for *p_cross* has been assumed to be 0.3, so that at least 30% of the chromosomes selected for the new population will undergo the crossover operation and produce offspring. The procedure for this selection is as follows. Random numbers between 0 and 1 are generated for all chromosomes and those chromosomes that obtain a random number less than the *p_cross* value are the chromosomes selected for crossover. If the number of selected chromosomes is odd, then the above procedure is repeated until one more chromosome gets selected and the number of selected chromosomes becomes even. The genetic literature addresses many crossover operators (Michalewicz 1996). Notable among them are: partially mapped crossover, ordinal mapped crossover and edge crossover. They use either single-point crossover or two-point crossover. This program uses the edge crossover method because of its simplicity in operation and because the chromosome is short. This splits the parent chromosomes into two parts with a random number generated with the range 1...(*m*- 1) and interchanges the genes from that crossover position.

The purpose of mutation is to introduce new genetic material or to recreate good genes that were lost by chance through a poor selection of mates. To do this effectively, the effect of mutation must be a major one. At the same time, the valuable gene pool must be protected from wanton destruction. Thus, the probability of mutation would be small (Masters 1993). On the above grounds, the value of the probability of mutation *(p_mut)* has been assumed to be 0.05. The repetition of the whole process (iteration) of evaluation, selection, reproduction and mutation depends on the size of the problem. The number of iterations is related to the number of jobs n to be scheduled, and has been fixed as $4 \times n$, subject to a maximum of 100.

Output Module: This module prints the schedule corresponding to the global best solution for minimum makespan criterion.

3.2 Numerical Illustration of the Proposed GA

The input job data of 10 jobs that requires processing on 6 machines is given in table 2.

For each machine [1...*j*...*m*] generate an integer random number [0...3] and put as a string. The position number represents the machine number and the number in that position is the *pdr* code to be followed by that machine. Similarly generate *m* strings. Each string represents one chromosome and table 3 gives the entire population.

Table 2. Data for the illustration problem

Job i	No. of operations J_i	Machine No. (Processing time) K_{ij} (t_{ij})						Due time d_i
		j=1	j=2	j=3	j=4	j=5	j=6	
1	5	2(24)	1(16)	3(20)	5(10)	6(10)	--	280
2	4	3(35)	2(30)	1(40)	6(15)	--	--	360
3	6	2(20)	1(25)	3(15)	4(10)	5(5)	6(5)	160
4	6	1(25)	3(35)	2(45)	5(15)	6(20)	4(10)	750
5	5	2(30)	1(20)	3(40)	4(10)	6(10)	--	660
6	6	2(20)	1(20)	3(30)	6(15)	4(10)	5(5)	450
7	4	3(15)	1(15)	4(20)	6(10)	--	--	240
8	4	1(40)	2(10)	6(15)	5(25)	--	--	270
9	3	2(12)	4(23)	6(15)	--	--	--	100
10	4	3(35)	2(45)	5(30)	4(10)	--	--	360

Table 3. Initial population of the *pdr* codes

Chromosome No. c	Machine No. k					
	1	2	3	4	5	6
1	1	1	3	1	2	0
2	3	1	1	1	3	2
3	1	0	1	0	1	1
4	0	1	3	0	3	2
5	1	0	3	2	1	3
6	2	1	0	2	0	1

The makespan time of the schedules obtained using *machine-wise-pdr* set of all the chromosomes (c = 1 to *pop_size*) is given in table 4.

Table 4. Fitness value of the initial population

Chromosome No. c	1	2	3	4	5	6
Makespan time $fit(c)$	425	489	309	353	346	384

The best schedule corresponds to chromosome c = 3 and the makespan time is 309. Total value of the evaluation function of the population

$$F = \sum_{c=1}^{pop\_size} fit(c) = 2306.$$

The probabilities of selection of chromosomes and their respective cumulative probabilities, which have been calculated using the parameter *new_fit*(c), are given in table 5. The random numbers generated and chromosomes selected for the next generation are given in table 6.

Table 5. Probability of selection of chromosomes

c	1	2	3	4	5	6
$p(c)$	0.1631	0.1576	0.1732	0.1694	0.1700	0.1667
$cp(c)$	0.1631	0.3207	0.4939	0.6633	0.8333	1.0000

Table 6. Population to represent next generation

New chromosome c'	1'	2'	3'	4'	5'	6'
rand()	0.6309	0.2538	0.1627	0.8413	0.7572	0.4409
Old chromosome c	4	2	1	6	5	3

Table 7. Chromosomes selected for crossover

rand()	0.4409	0.3507	0.0079	0.4224	0.5220	0.7023
rand() less than *p_cross*	no	no	yes	no	no	no
Selected	--	--	3'(1)	--	--	--

The chromosomes selected with *a p_cross* of 0.3 for crossover from the new set are shown in table 7. Since only one chromosome is selected (i.e. 3': 1 1 3 1 2 0), a null chromosome O' (0 0 0 0 0 0) is added to make the number of chromosomes selected even, and they become the parents (3" and O") and undergo crossover. The parents, and their respective offspring (3''' and O'''), produced with a crossover point 3, are given in table 8. The crossed Y'' replaces the 3' and becomes 3" in the new population. No element has been selected for mutation. The new population obtained after crossover and mutation is given in table 9.

Table 8. Parents and offspring

Parents	Chromosome 3"(3')	1	1	3	1	2	0
	Null chromosome O"(O')	0	0	0	0	0	0
Offspring	Crossed chromosome 3'''	1	1	0	0	0	0
	Crossed chromosome O'''	0	0	3	1	2	0

Table 9. New population

c'	*k*=1	*k*=2	*k*=3	*k*=4	*k*=5	*k*=6
1'(4)	0	1	3	0	3	2
2'(2)	3	1	1	1	3	2
3'(3")	1	1	0	0	0	0
4'(6)	2	1	0	2	0	1
5'(5)	1	0	3	2	1	3
6'(3)	1	0	1	0	1	1

Repeat the steps of generation and evaluation of the new population for *no_iter* iterations.

Best makespan time : 299
Solution at (it_no) : 18th iteration
Optimal *machine-wise-pdr* : 1-2-2-0-1-1
Schedule : Table 10

3.3 Performance Analysis of the Proposed GA

Varied comments on the feasibility of the application of the proposed methodology to this scheduling problem are discussed in this section. The problems considered address a typical range of problems for short-term planning. Many data sets have been experimented with, and the results obtained compared with the extended B-B technique proposed by Jawahar et al (1996) and the direct application of *pdrs.* The makespan time of the schedules and the computational time of a sample of twenty problems (randomly generated) obtained with all the methodologies are given in table 11.

The extended B-B methodology takes much more time and the computational complexity is also high. The direct application of *pdrs* for resolving conflict does not guarantee optimal or near optimal solutions and no generalization is possible. The computational time is less than for the other methods; but the weakness of this method is that most of the time it provides poor solutions.

The application of a genetic algorithm (with classical genetic operators) to this problem is useful as the values of the objective function are optimal, or very close to the optimal. The values obtained are comparable to the best solution obtainable with the extended B-B technique. Also a near optimal solution can be obtained with reasonable computational time.

Table 10. Schedule for the illustration problem

Job i	Operation j	Machine $K_{ij}(k)$	Start time S_{ij}	Completion time C_{ij}
1	1	2	50	74
1	2	1	90	106
1	3	3	106	126
1	4	5	149	159
1	5	6	159	169
2	1	3	50	75
2	2	2	119	149
2	3	1	149	189
2	4	6	189	204
3	1	2	12	32
3	2	1	65	90
3	3	3	90	105
3	4	4	105	115
3	5	5	159	164
3	6	6	169	174
4	1	1	40	65
4	2	3	126	161
4	3	2	199	244
4	4	5	244	259
4	5	6	259	279
4	6	4	289	299
5	1	2	169	199
5	2	1	209	229
5	3	3	239	279
5	4	4	279	289
5	5	6	289	299
6	1	2	149	169
6	2	1	189	209
6	3	3	209	239
6	4	6	239	254
6	5	4	254	264
6	6	5	264	269
7	1	3	0	15
7	2	1	106	121
7	3	4	159	179
7	4	6	204	214
8	1	1	0	40
8	2	2	40	50
8	3	6	50	65
8	4	5	65	90
9	1	2	0	12
9	2	4	12	35
9	3	6	35	50
10	1	3	15	50
10	2	2	74	119
10	3	5	119	149
10	4	4	149	159

Table 11. Makespan time of schedules generated with different methods

Problem No.# n=10 m=6	Makespan time of schedules generated with Direct application of *pdr*				TIEs resolved in Branch and Bound method with *pdr* (Jawahar et al. 1996)				*Machine-wise-pdr* evolved through GA
	SPT	LPT	MINSLK	EDT	SPT	LPT	MINSLK	EDT	
1	361	524	353	332	291**	316	276*	298	299
2	280	270	255	285	300	230*	255	295	255**
3	195	210	235	225	160**	172	155*	165	185
4	295	365	372	368	275	295	245*	275	270**
5	300	350	325	315	280*	335	285**	295	295
6	219	265	243	212	218	217	205**	209	203*
7	270	200	185	210	200	180*	195	195	I85**
8	260	345	290	285	230*	230*	230*	230*	235**
9	430	455	535	515	380**	490	380	370*	395
10	300	270	275	240**	244	280	230*	250	240**
11	248	272	263	253	244	237	225**	292	196*
12	287	274	230	256	268	221**	259	292	200*
13	231	222	293	220	188*	197	244	192	191**
14	218	279	236	338	191*	204	240	240	194**
15	231	277	268	215	204	152*	264	239	196**
16	210	236	243	234	191	232	185*	201	190**
17	303	314	323	330	367	225*	244**	--	249
18	294	287	318	286	272**	296	--	300	239*
19	294	303	360	340	256	340	244*	279	245**
20	292	300	331	359	378	276*	522	269	279**
Average computational time (s)	0.0782	0.0921	0.0642	0.1093	45.2345	54.2340	46.3245	53.2341	3.7834

data set of the example problems are given in the Jawahar et al. (1998)
*indicates the best solution;
**indicates the second best solution.

4 Multiobjective GA for JSP

4.1 Description of the Proposed GA

The different modules of the multiobjective GA (MOGA) that is proposed to evolve schedule to the job shop problem for minimization of weighted sum of makespan time, total tardiness of all jobs and total idle time of all machines is same as given in fig. 1.

The description of the different modules of the proposed MOGA is as follows:

Data input module: The data as described in the input module of proposed GA for single objective JSP (section 3.1) is given as input for this module.

Initial population generation module: The genetic search process starts with a randomly generated set of chromosomes (*machine-wise-pdr* sets) called the initial population. The size of the population (*pop_size*) depends on the solution space. The chromosome representation of the MOGA is the same as that of the GA for single objective JSP described in section 3.1. The possible number of combinations of *machine-wise–pdr* sets in MOGA is 7^m, where m is the number of machines and 7 *pdrs*. Hence, the population size is related to the number of machines in the system and has been assumed to be equal to the number of machines in the shop. The description of the *pdr* codes used in MOGA is given in the table 12.

Table 12. Value of gene and their corresponding *pdr*

Value representing the gene	Corresponding pdr
0	SPT–Shortest processing time
1	LPT–Longest processing time
2	EDT–Earliest due date
3	MINSLK–Least slack
4	SPO–Smallest ratio of slack per operation
5	JSR–job slack ratio
6	CR–Smallest critical ratio

The *machine-wise–pdr* set is used to generate schedule using GT procedure and from that the makespan, total tardiness and the total idle time of all the machines are calculated. The fitness parameter *fit*(c) is the weighed sum of makespan, total tardiness and the total idle time of machines.

$$Fit(c) = \{ w_1 \, fit1(c) + w_2 \, fit2(c) + w_3 \, fit3(c) \} \tag{27}$$

The randomly generated weights are arranged in such a way that $w_1 > w_2 > w_3$ and $fit1 > fit2 > fit3$ to avoid entrapment in local minima. This may happen when assigning a very high weight to an objective, whose value is nearer to zero, and very low weights to the other two objectives whose value being comparatively higher. This in turn leads this weighed sum to be an optimal solution, which is actually not an optimum one. The weights w_1, w_2 and w_3 are assigned randomly by generating three random numbers. In general, the value of each weight can be randomly determined. For a multi-objective optimization problem with n objective function ($n \geq 2$), a random real number can be assigned to each weight as follows (Muarata et al. 1996).

$$w_i = \frac{rand_i}{\sum_{j=1}^{n} rand_j} \tag{28}$$

where, $rand_i$ and $rand_j$ are non-negative random integers (or non-negative random real numbers). From the above equation it can be seen that n random real numbers are generated for the weights w_is to calculate the weighed sum in equation (27) when evaluating the chromosomes. Since only three objectives are considered here, it is enough to generate three random numbers to find the three weights w_1, w_2 and w_3. The weights assigned to the multiple objective functions are not constant. If we use the weighed sum in equation (27) with the constant weight w_is, the search direction in genetic algorithms is also constant. The idea is to realize various search directions.

The termination criterion module and new population generation module for the proposed MOGA are the same as that of the GA for single-objective GA for JSP described in section 3.1. The parameter set for the proposed MOGA is given in table 13.

Table 13. MOGA parameters

Initial population	Randomly generated
Population size	Equal to the number of machines
Length of the chromosome	Equal to no. of machines
Crossover operator	Edge crossover (single point)
Mutation operator	Random
Crossover probability	0.3
Mutation probability	0.01
Selection procedure	Rowlette wheel method
Fitness parameters	Weighed sum of makespan, total tardiness, and total machine idle time
Assignment of weights	Random
Termination condition	When no. of iterations is equal to 100

4.2 Numerical Illustration for the Proposed MOGA

The working of proposed MOGA is explained by considering a problem instance ft06. The input data for the example problem is given in table 14.

Table 14. Data for the illustration problem (Problem ID: ft06, size 6 x 6)

Job i	No. of operations J_i	Machine No. (Processing time) K_{ij} (T_{ij})						Due time d_i
		j=1	j=2	j=3	j=4	j=5	j=6	
1	6	3 (1)	1 (3)	2 (6)	4 (7)	6 (3)	5 (6)	52
2	6	2 (8)	3 (5)	5 (10)	6 (10)	1 (10)	4 (4)	94
3	6	3 (5)	4 (4)	6 (8)	1 (9)	2 (1)	5 (7)	68
4	6	2 (5)	1 (5)	3 (5)	4 (3)	5 (8)	6 (9)	70
5	6	3 (9)	2 (3)	5 (5)	6 (4)	1 (3)	4 (1)	25
6	6	2 (3)	4 (3)	6 (9)	1 (10)	5 (4)	3 (1)	45

For each machine [1...*j*...*m*] generate an integer random number [0…6] and put as one string. The position represents the machine number and the number in that position is the *pdr* code to be followed by that machine. Similarly, generate *m* number of strings. Each string represents one chromosome and table 15 gives the entire population.

Table 15. Initial population of the *pdr* codes

Chromosome No. c	Machine No. k					
	1	2	3	4	5	6
1	2	1	2	5	5	3
2	5	4	0	3	1	3
3	0	4	3	2	4	2
4	5	3	2	0	0	5
5	0	3	5	3	1	0
6	4	5	5	4	3	3

The weighed sum of makespan, total tardiness and total machine idle time of the schedules obtained using *machine-wise pdr* set of all the chromosomes is given in table 16.

Table 16. Fitness value of the initial population

c	Makespan	Total machine idle time	Total tardiness	w_1	w_2	w_3	*Fit*(c)*
1	105	433	181	0.4908	0.4422	0.0669	299**
2	98	391	61	0.8200	0.1779	0.0020	338
3	**96**	**379**	**32**	**0.4172**	**0.3504**	**0.2324**	**199**
4	98	391	84	0.5013	0.4812	0.0175	244
5	85	313	66	0.7541	0.2410	0.0050	256
6	101	409	31	0.7338	0.1588	0.1074	319

*$fit(c) = fit1 * w_1 + fit2 * w_2 + fit3 * w_3$

**$fit(c1) = 433 * 0.4908 + 181 * 0.4422 + 105 * 0.0669 = 299$.

The process of termination check and new population generation (includes selection, crossover and mutation) is performed with the initial population given in table 15 in the same way as illustrated in section 3.2. Results obtained with MOGA for the example problem (ft06) is given below:

Best fitness value: 137
Solution at: Second iteration
Optimal *machine-wise-pdr*: 1-2-5-2-4-5
Schedule given in: job-wise schedule in table 17
makespan: 76
Total Tardiness: 31
Total Idle Time: 259
Fitness value: 137

Table 17. Schedule for the example problem

Job i	Operation j	Machine $K_{ij}(k)$	Start time S_{ij}	Completion time C_{ij}	Job completion time C_{iJi}	Due time d_i	Tardiness	Earliness
1	1	3	32	33	58	52	6	0
1	2	1	33	36				
1	3	2	36	42				
1	4	4	42	49				
1	5	6	49	52				
1	6	5	52	58				
2	1	2	0	8	76	94	0	18
2	2	3	27	32				
2	3	5	37	47				
2	4	6	52	62				
2	5	1	62	72				
2	6	4	72	76				
3	1	3	22	27	73	68	5	0
3	2	4	27	31				
3	3	6	34	42				
3	4	1	44	53				
3	5	2	53	54				
3	6	5	66	73				
4	1	2	25	30	75	70	5	0
4	2	1	39	44				
4	3	3	44	49				
4	4	4	49	52				
4	5	5	58	66				
4	6	6	66	75				
5	1	3	13	22	40	25	15	0
5	2	2	22	25				
5	3	5	25	30				
5	4	6	30	34				
5	5	1	36	39				
5	6	4	39	40				
6	1	2	8	11	38	45	0	7
6	2	4	11	14				
6	3	6	14	23				
6	4	1	23	33				
6	5	5	33	37				
6	6	3	37	38				

4.3 Performance Analysis of the Proposed MOGA

Twenty-eight problems available in the open literature are used for the evaluation of the three objectives. The first 23 benchmark problems available in the OR library are available at internet site http://mscmga.ms.ic.ac.uk/ and the next five proposed by Jawahar et al. (1998) are used for the evaluation purpose. The consolidated results of 28 problems are tabulated in table 18.

Table 18. Consolidated results

Problem No.	Problem ID instance	Problem size	Makespan	Total machine idle time	Total tardiness	Weighted sum of objectives
1	abz5	10 × 10	1587	1948	8097	4218
2	abz6	10 × 10	1369	1882	7744	4052
3	ft10	10 × 10	1496	3459	9851	5461
4	la16	10 × 10	1452	1127	9169	4378
5	la17	10 × 10	1172	1779	7044	3429
6	la19	10 × 10	1251	1581	7164	3372
7	la20	10 × 10	1419	1451	8745	4122
8	orb01	10 × 10	1704	3052	11631	5530
9	orb02	10 × 10	1284	1565	7585	3631
10	orb03	10 × 10	1643	4140	11138	6168
11	orb04	10 × 10	1543	4951	9802	5548
12	orb05	10 × 10	1323	2195	8322	4026
13	orb06	10 × 10	1645	2601	10836	5098
14	orb07	10 × 10	583	699	3423	1862
15	orb08	10 × 10	1340	3498	8840	4621
16	orb09	10 × 10	1462	2029	9439	4539
17	orb10	10 × 10	1382	1806	8271	3850
18	la01	10 × 5	1256	3324	3431	2863
19	la02	10 × 5	1066	2081	2687	2167
20	la03	10 × 5	821	1926	1722	1492
21	la04	10 × 5	861	3194	1798	2034
22	la05	10 × 5	893	1716	2182	1752
23	ft06	6 × 6	76	31	259	137
24	ex01	10 × 6	330	140	1030	530
25	ex02	10 × 6	230	625	625	542
26	ex03	10 × 6	185	130	598	315
27	ex04	10 × 6	305	532	1028	623
28	ex05	10 × 6	380	335	1495	750

5 GA for FJSP

5.1 Description of the Proposed GA

The different modules of the proposed GA that is proposed to evolve simultaneously the optimal route choice and schedule to the flexible job shop problem is same as given in fig. 1.

The description of the different modules is as follows:

Data Input Module: The following data pertaining to the problem are given as input: number of jobs (n), number of machines in the shop (m), number of operations J_i of each job i ($\forall i$), number of alternative machines (routes) R_{ij} for operation j of job i ($\forall i$, $\forall j$), the machine number K_{ijr} corresponding to the route r of operation j of job i along with its processing time T_{ijr} ($\forall i$, $\forall j$, $\forall r$).

Initial Population Generation Module: A set of chromosomes equal to the size of the population (*pop_size*) is randomly generated in this module. Each chromosome comprises of two parts. The genes of the first part of each chromosome represent the route choices for the operations of all jobs. This is divided into number of sets of genes equal to the number of jobs n; one set for one job such that 1st set corresponds to the 1st job, 2nd set corresponds to the 2nd job and so on. A gene of any set is the representation of route choice of an operation. So the number of genes in a set that corresponds to the job i becomes J_i and the total number of genes of 1st part is equal to the total number of operations of all the jobs (i.e. $\sum_{i=1}^{n} J_i$). The second part of the chromosome with as many number of genes equal to total number of operations, represents the priority of one operation over the other for loading on the machines. The sequence priority of the 1st operation of job i is represented at the 1st position of J_i number of genes allotted for job i, 2nd operations' sequence priority at 2nd position and so on.

Evaluation Module: An active feasible schedule with Giffler and Thompson (1960) procedure for each chromosome is found by reducing the alternate route choice problem to a fixed route problem using the first part of the chromosome and resolving the conflicts with the priorities in the second part of the chromosome in the reduced fixed route job shop problem. Giffler and Thompson method is used for generating active feasible schedules for the job shop problem. The procedure ensures that no subsequent left shifting is possible since as soon as a job completes its processing on one machine it becomes a contender for processing on the next machine as determined by the technological order restriction. If there are two or more contenders for the same machine, a conflict will occur which is resolved by choosing only one of the contenders to be processed next on the machine. The sequence priority string is used for resolving the conflicts that arise between the jobs during the schedule generation.

The makespan time of the schedules corresponding to the chromosome c thus becomes the objective function or fitness value ($fit(c)$) of it. In order to suit the probability of survival, the $fit(c)$ is modified with negative exponential function as:

$$new\_fit(c) = e^{-x fit(c)} \tag{29}$$

Each chromosome in the current population is updated as the global best chromosome, if its fitness value is less than or equal to the global best solution.

Termination Check Module: A specified number of generations (*no_iter*) are used to terminate the GA. On satisfactory termination, the output module prints the global best solution.

New Population Generation Module: A new population, size equal to *pop_ size*, is selected from the previous population based on the concept of probability of survival. Roulette wheel selection method has been adopted for generation of new population. The chromosomes for crossover are selected from the new population based on the probability of crossover (*p_cross*=0.6). Edge crossover is the crossover operator used for both route choice and schedule generation strings. Crossover is followed by mutation in which each gene of all the chromosomes is mutated with a probability of mutation (*p_mut*=0.05). Swap operator is used for mutating the route choice and schedule generation strings.

Output Module: This module prints the global best solution of the optimal route choices of all operations along with its schedule for minimum makespan criterion.

5.2 Numerical Illustration for the Proposed GA

Table 19 provides the process data of 3 jobs - 5 machines problem that is used for illustrating the proposed GA.

Table 19. Process data of the illustrative problem

Job i	Operation j	Number of route choices R_{ij}	Machine No. with Processing time K_{ijr} (T_{ijr}) corresponding to each route r	
			r =1	r =2
1	1	2	2 (3)	3 (7)
	2	2	1 (4)	4 (2)
	3	2	1 (1)	2 (2)
2	1	2	2 (5)	5 (2)
	2	2	2 (3)	3 (6)
	3	2	1 (3)	5 (7)
3	1	2	2 (4)	3 (5)
	2	2	1 (2)	4 (3)
	3	2	1 (2)	3 (3)

The above data is given as input in the input module. An initial population of size, *pop_size*=10 is randomly generated. Table 20 shows the information for the first chromosome of the initial population which is used to generate schedule for the operations and to determine the makespan time.

Table 20. Information of Genes corresponding to chromosome c=1

		Chromosome c=1								
Route choice	Gene No.	*g1*	*g2*	*g3*	*g4*	*g5*	*g6*	*g7*	*g8*	*g9*
	String	1	2	1	2	1	1	1	1	1
Sequence priority	String	9	4	2	6	1	3	7	5	8
	Gene No.	*g10*	*g11*	*g12*	*g13*	*g14*	*g15*	*g16*	*g17*	*g18*
Corresponds to Operation j of job i (i,j)		(1,1)	(1,2)	(1,3)	(2,1)	(2,2)	(2,3)	(3,1)	(3,2)	(3,3)

Table 21 shows the phenotype values of process data and priority number for all operations corresponding to the decoded information shown in table 20.

Table 22 illustrates the Giffler and Thompson schedule generation procedure for the information given in table 21. Any conflict if arises during the schedule generation, is resolved using the priority number from the sequence priority string. The operation with the higher priority number precedes the other conflicting operations. The makespan time for the above schedule is 13 and this becomes the evaluation of fitness parameter for the chromosome "121211111 942613758".

Table 21. Phenotype information of chromosome c=1

Job i	Operation j	Route Selected r	Machine No. K_{ijr}	Processing time T_{ijr}	Priority Number
1	1	1	2	3	9
	2	2	4	2	4
	3	1	1	1	2
2	1	2	5	2	6
	2	1	2	3	1
	3	1	1	3	3
3	1	1	2	4	7
	2	1	1	2	5

Table 22. Active feasible schedule generation

Machine No. k	Job i	Steps of schedule generation s								
		1	2	3	4	5	6	7	8	9
1	1				6*					
	2								13	13*
	3						9	10	10*	
2	1	3	3							
	2		5	6	6	6	10	10		
	3	4	4	7	7	7				
3	1									
	2									
	3									
4	1			5						
	2									
	3									
5	1									
	2	2								
	3									
Datum Time		2	3	5	6	6	9	10	10	13**
Conflict		-	I	-	-	II	-	-	III	

*Flow time of jobs

**Makespan time

Table 23. Illustration of different stages of new population generation

Initial Population Generation Module		Evaluation Module	Iteration 1: New Population Generation Module							Evaluation Module
			Selection (Roulette wheel method)				Crossover (*p_cross*=0.6)		Mutation (*p_mut* =0.05)	Evaluation Module
c	Random generation of Initial Population (*pop_size*=10)	Fitness *fit(c)*	*p(c)*	*cp(c)* = *Σp(c)*	Random Number *rand()*	Chromosome *c* Selected that satisfies *cp(c-1) ≤ rand() < cp(c)*	Order of chromosomes Selected (cut point)[#]	Chromosomes After Crossover (Edge Crossover operator)	Chromosomes after Mutation (swap operator) New Population	Fitness *fit(c)*
1	121211111 942613758	13*	0.126	0.126	0.734	221211121 354286791	x	221211121 354286791	221211121 354286791	13*
2	222222121 495162783	19	0.093	0.220	0.053	121211111 942613758	x	121211111 942613758	121211111 942613758	13
3	212121222 845762391	24	0.072	0.293	0.999	212221121 246158397	1 - (3)	21**2221121** 24**6158397**	212121121 246158397	18
4	121122222 453718926	24	0.072	0.366	0.418	111212111 169823457	6 - (3)	11**1211121** 53**4286791**	111211111 536284791	17
5	111212111 169823457	15	0.114	0.480	0.938	212221121 246158397	2 - (3)	21**2221121** 24**6158397**	212221121 246158397	21
6	121121221 216974853	14	0.120	0.601	0.900	222121111 183725964	3 - (5)	2221**12111** 1869**23457**	221121221 586913427	19
7	212122111 478539216	20	0.089	0.690	0.788	221211121 354286791	x	221211121 354286791	221211121 354286791	13
8	221211121 354286791	13	0.126	0.816	0.127	222222121 495162783	x	222222121 495162783	222222121 495162783	19
9	222121111 183725964	18	0.098	0.915	0.467	111212111 169823457	4 - (5)	1112**21111** 1378**25964**	121121111 137825964	16
10	212221121 246158397	21	0.084	1.000	0.728	221211121 354286791	5 - (3)	221**212111** 61**9823457**	221212111 619823457	14

*Generation best solution

x - Not selected for crossover

- Parental sets for crossover = {(3, 5), (6, 9), (10, 4)}

Table 23 illustrates the new population generation mechanism. The parameters used for the generation of new population are as follows:

Probability of survival $p(c)$ of chromosome c: $p(c) = e^{-xfit(c)} / \sum e^{-xfit(c)}$
Constant x value : 0.05

The cumulative probabilities of survival $cp(c)$ of all chromosomes are then found out using the equation $cp(c) = \sum_{c=1}^{c=C} p(c)$.

The chromosomes selected for the new generation is shown in the table 23. The selected chromosomes then undergo crossover and mutation. The following are the parameters used for crossover and mutation:

Probability of crossover (*p_cross*) : 0.6
Crossover operator : Edge Crossover
Probability of Mutation (*p_mut*) : 0.05
Mutation Operator : Swap operator

The best solution of this generation corresponds to the chromosome c=1, which replaces the global best if it is better than the previously stored global best solution. The process of evaluation and new population generation is repeated for 100 generations, which is the termination criterion for this problem. The best solution evolved is given in table 24.

Table 24. Optimal Solution (121211221 912348675).

Job *i*	Operation *j*	Machine Allotted	Start Time	Finish Time	Flow Time of Job	Makespan Time
1	1	2	0	3	6	10
	2	4	3	5		
	3	1	5	6		
2	1	5	0	2	9	
	2	2	3	6		
	3	1	6	9		
3	1	3	0	5	10	
	2	4	5	8		
	3	1	9	10		

5.3 Performance Analysis of the Proposed GA

The performance of the proposed GA is evaluated by comparing its solutions with the best known solutions (BKS) for a set of benchmark instances from literature. The first set of benchmark instances are from Thomalla (2001), in which all the problems are flexible job shop instances with total flexibility, i.e., all the operations in each of the problem instances can be performed on all the machines. The second set of benchmark instances are from Brandimarte (1993), in which all the problems are flexible job shop instances with partial flexibility. The results of the proposed GA are evolved with the programs coded in C language.

The termination criterion used for GA is the total number of iterations which is equal to 100 times the total number of operations of all jobs. The crossover probability and mutation probability considered for the analysis is 0.6 and 0.05, respectively. The parameter values for the proposed GA are obtained by fine tuning through trials. The three proposed algorithms are run 5 times for each problem and the best solution obtained has been taken for comparison. Table 25 shows the results of the proposed GA that are obtained with Pentium-IV 2.4 GHz processor.

Table 25. Result obtained by GA for the set of data from literature

Reference	Problem Name	Problem Size $n \times m$	Makespan time		
			Lower Bound*	BKS*	GA
Thomalla (2001)	EX1	3×3	--	117	117
	EX2	4×3	--	109	109
	EX3	6×10	--	316	348
Brandimarte (1993)	MK01	10×6	36	40	40
	MK02	10×6	24	26	29
	MK03	15×8	204	204	204
	MK04	15×8	48	60	71
	MK05	15×4	168	173	188
	MK06	10×15	33	58	81
	MK07	20×5	133	144	152
	MK08	20×10	523	523	523
	MK09	20×10	299	307	378
	MK10	20×15	165	198	265

*reported in Mastrolilli and Gamberdella(2000)

The comparison between the proposed GA and the best known solution (BKS) in the literature for the above benchmark problems reveals that proposed GA is competent with the existing methodologies. For five problems the solution obtained with GA is the same as the BKS. For the remaining problems the solution obtained with GA is closer to the BKS. For the last two problems (MK09 and MK10) the solution obtained with the proposed GA is considerably poor. To improve the performance, the various parameters and operators considered in the GA could be varied and fine tuned so that the above limitation could be overcome. Local search methods such as tabu search, simulated annealing, bottleneck shifting procedure, etc. could be incorporated to enhance the performance of the proposed GA.

6 Conclusion

In this chapter, genetic algorithm based heuristics are presented for the two well known job shop scheduling models, the JSP and the FJSP. The genetic algorithms adopt the Giffler and Thompson (GT) schedule generation procedure for active feasible schedule generation. The proposed GAs for the JSP derives optimal *machine-wise-pdr* set that is used for generating active feasible schedules with GT algorithm. The performance of the proposed GAs for JSP is analyzed for both single objective and multiple objective criteria and the results obtained reveals that the optimal *machine-wise-pdr* obtained with the proposed GA is efficient in providing optimal solutions for the JSP in reasonable computational time. The chromosome encoding scheme used in the proposed GA for FJSP makes it capable to rummage through the entire solution space and provide all possible instances that an enumerative search can and therefore is capable of finding the optimal or near-optimal solutions under extensive searches. The performance of the proposed GA for FJSP is analyzed with various benchmark instances for makespan time criterion which reveals that the proposed GA is competent with the existing approaches. The performance of the proposed GA for FJSP can be improved by incorporating local search methods, such as simulated annealing algorithm, tabu search, etc. The proposed GAs can be extended to solve more complex job shop models like the assembly job shop problem.

Nomenclature

c, c', c'', c'''	Index for chromosome ($c = 1,\ldots, pop\_size$)
C_{ij}	Completion time of operation O_{ij}
$cp(c)$	Cumulative probability of survival of chromosome c
d_i	Due date of job i
$fit(c)$	Fitness value of chromosome c
FJSP	Flexible job shop scheduling problem
GA	Genetic Algorithm
GT	Giffler and Thompson schedule generation procedure
gbest	Global best solution
H	A very large positive integer
i, i'	Index for job ($i = 1,\ldots, n$)
j, j'	Index for operations on job ($j = 1,\ldots, J_i$)
J_i	Number of operations required to complete job i
JSP	Classical job shop scheduling problem
k	Index for machine ($k = 1,\ldots, m$)
K_{ij}	Machine number for operation O_{ij} in JSP
K_{ijr}	Machine number for operation O_{ij} in its route r in FJSP
m	Number of machines in the shop
$new\_fit(c)$	Modified fitness value of chromosome c
n	Number of jobs
N_k	Set of operations $\{O_{ij}\}$ that can be loaded on machine k
no_iter	Number of iterations
O_{ij}	Operation j of job i
$p(c)$	Probability of survival of chromosome c
p_cross	Probability of crossover
p_mut	Probability of mutation
pop_size	Population size

r	Index for route choice ($r = 1,\ldots, R_{ij}$) in FJSP
$rand()$, $rand_i$	Random number between 0 and 0.999
R_{ij}	Number of alternate routes for operation O_{ij} in FJSP
S_{ij}	Start time of operation O_{ij}
t_{ij}	Processing time of operation O_{ij} in JSP
t_{ijk}	Processing time of operation O_{ij} on machine k in FJSP
T_{ijr}	Processing time of operation O_{ij} in its route r in FJSP
tn	Iteration identifier ($tn = 1, \ldots, no\_iter$)
w_i	Weight assigned to the objective function i
x	Scaling parameter
X_{ijk}	Decision variable for machine selection for operation O_{ij} in FJSP
$Y_{iji'j'}$	Decision variable for generating a sequence between the operations O_{ij} and $O_{i'j'}$ in JSP
$Y_{iji'j'k}$	Decision variable for generating a sequence between the operations O_{ij} and $O_{i'j'}$ for loading on machine k in FJSP

References

1. Baker, K.R.: Introduction to Sequencing and Scheduling. Wiley, New York (1974)
2. Biegel, J.E., Davern, J.J.: Genetic algorithms and job shop scheduling. Computers and Industrial Engineering 19(1-4), 81–90 (1990)
3. Bierwirth, C., Mattfeld, D.C.: Production scheduling and rescheduling with genetic algorithms. Evolutionary computation 7, 1–17 (1999)
4. Blackstone, J.H., Phillips, D.T., Hogg, G.L.: A state of the art survey of dispatching rules for manufacturing job shop operations. International Journal of Production Research 20, 27–45 (1982)
5. Blazewicz, J., Ecker, K., Schmidt, G., Wegalrz, J.: Scheduling in computer and manufacturing systems. Springer, Heidelberg (1993)
6. Brandimarte, P.: Routing and scheduling in a flexible job shop by tabu Search. Annals of Operations research 41, 157–183 (1993)
7. Brucker, P.: Scheduling algorithms. Springer, Heidelberg (1995)
8. Brucker, P., Schlie, R.: Job shop scheduling with multi-purpose machines. Computing 45, 369–375 (1990)
9. Chakraborthy, U.K., Deb, K., Chakraborthy, M.: Analysis of selection algorithms: A Markov chain approach. Evolutionary computation 4(2), 133–167 (1996)
10. Chen, H., Ihlow, J., Lehmann, C.: A genetic algorithm for flexible job-shop scheduling. In: Proceedings of the IEEE International Conference on Robotics and Automation, vol. 2, pp. 1120–1125 (1999)
11. Choi, I.C., Choi, D.S.: A local search algorithm for job shop scheduling problems with alternative operations and sequence-dependent setups. Computers and Industrial Engineering 42, 43–58 (2002)
12. Chryssolouris, G., Chan, S.: An integrated approach to process planning and scheduling. Annals of CIRP 34, 413–417 (1985)
13. Conway, R.W., Maxwell, W.L., Miller, L.M.: Theory of Scheduling. Addison-Wesley, Reading (1967)
14. Dauzere-Peres, S., Paulli, J.: An integrated approach for modeling and solving the general multiprocessor job-shop scheduling problem with tabu search. Annals of Operations Research 70, 281–306 (1997)

15. Dorndorf, U., Pesche, E.: Combining genetic and local search for solving the job shop scheduling problem. In: Proceedings of APMOD 1993, Budapest, vol. 1, pp. 142–149 (1993)
16. Dorndorf, U., Pesche, E.: Evolution based learning in a job shop environment. Computers and Operations Research 22, 25–40 (1995)
17. French, S.: Sequencing and Scheduling: An introduction to the mathematics of Job-Shop. Ellis Horwood Limited, Chichester (1982)
18. Garey, M.R., Johnson, D.S., Sethi, R.: The complexity of flowshop and jobshop scheduling. Mathematics of Operations Research 1, 117–129 (1976)
19. Giffler, B., Thompson, G.L.: Algorithms for solving production scheduling problems. Operations Research 8, 487–503 (1960)
20. Girish, B.S., Jawahar, N.: Scheduling job shops associated with multiple routings with genetic and ant colony heuristics. International journal of production research (2008) (in print) (available online) DOI:10.1080/00207540701824845
21. Groover, M.P.: Automation, production systems, and computer integrated manufacturing. Prentice Hall of India Pvt. Ltd, New Delhi (2003)
22. Hankins, S.L., Wysk, R.A., Fox, K.R.: Using a CATS database for alternative machine loading. Journal of Manufacturing Systems 3, 115–120 (1984)
23. Ho, N.B., Tay, J.C.: GENACE: An Efficient Cultural Algorithm for Solving the Flexible Job-Shop Problem. In: Proceedings of the IEEE Congress on Evolutionary Computation, vol. 1, pp. 1759–1766 (2004)
24. Ho, N.B., Tay, J.C.: Evolving dispatching rules for solving the flexible job-shop problem. In: Proceedings of the 2005 IEEE Congress on Evolutionary Computation, vol. 3, pp. 2848–2855 (2005)
25. Ho, N.B., Tay, J.C., Lai, E.M.K.: An effective architecture for learning and evolving flexible job-shop schedules. European Journal of Operational Research 179, 316–333 (2007)
26. Hoitomt, D.J., Luh, P.B., Pattipati, K.R.: A practical approach to job shop scheduling problems. IEEE transactions on Robotics and Automation 9(1), 1–13 (1993)
27. Hussain, M.F., Joshi, S.B.: A Genetic Algorithm for Job Shop Scheduling problems with Alternate Routing. In: Proceedings of IEEE International conference on systems, man and cybernetics, vol. 3, pp. 2225–2230 (1998)
28. Hutchison, J.: Current and future issues concerning FMS scheduling. International journal of management sciences 19(6), 529–537 (1991)
29. Iwata, K., Murotsu, Y., Oba, F., Uemura, T.: Optimization of selection of machine tools, loading sequence of parts and machining conditions in job-shop type machining systems. Annals of the CIRP 27, 447–451 (1978)
30. Jain, A.S., Meeran, S.: Deterministic job-shop scheduling: Past, present and future. European Journal of Operational Research 113(2), 390–434 (1999)
31. Jawahar, N., Aravindan, P., Ponnambalam, S.G., Arvindkarthikeyan, A.: Branch bound technique in combination with priority dispatching rules for scheduling FMS. In: Proceedings of the International Conference on CAD/CAM, Automation, Robotics and Factories of the future (INCARF 1996), New Delhi, vol. 1, pp. 143–150 (1996)
32. Jawahar, N., Aravindan, P., Ponnambalam, S.G.: A Genetic Algorithm for Scheduling flexible manufacturing systems. International journal of Advanced Manufacturing Technology 14, 588–607 (1998)
33. King, J.R.: Production Planning and control. Pergamon press, Oxford (1975)
34. Kim, Y.K., Park, K., Ko, J.: A symbiotic evolutionary algorithm for the integration of process planning and job shop scheduling. Computers and operations research 30, 1151–1171 (2003)

35. Kopfer, H., Mattfield, D.C.: A hybrid search algorithm for the job shop problem. In: Proceedings of the First International Conference on Operations and Quantitative Management, vol. 2, pp. 498–505 (1997)
36. Kutanoglu, E., Sabuncuoglu, I.: An analysis of heuristics in a dynamic job shop with weighted tardiness objectives. International Journal of Production Research 37(1), 165–187 (1999)
37. Masters, T.: Practical Neural Network Recipes in C++. Academic Press, USA (1993)
38. Mastrolilli, M., Gamberdella, L.M.: Effective neighbourhood for the flexible job shop problem. Journal of Scheduling 3(1), 3–20 (2000)
39. Mesghouni, K., Hammadi, S., Borne, P.: Evolution programs for job shop scheduling. In: Proceedings of the IEEE international conference on computational cybernetics and simulation, vol. 1, pp. 720–725 (1998)
40. Michalewicz, Z.: Genetic Algorithms + Data Structures = Evolution Programs, 3rd edn. Springer, Heidelberg (1996)
41. Moon, J., Lee, J.: Genetic Algorithm Application to the Job Shop Scheduling problem with Alternative Routing. Technical report-Brain Korea 21 logistics Team, Pusan National University (2000)
42. Muarata, T., Ishibuchi, H., Tanaka, H.: Multiobjective genetic algorithm and its applications to flow shop scheduling. Computers Industrial Engineering 30(4), 957–968 (1996)
43. Pinedo, M.L.: Scheduling theory: theory algorithms and systems. Englewoodcliffs, New Jersey (1995)
44. Pinedo, M.L.: Planning and scheduling in manufacturing and services. Springer, New York (2005)
45. Ponnambalam, S.G., Ramkumar, V., Jawahar, N.: A multiobjective genetic algorithm for job shop scheduling. Production Planning and control 12(8), 764–774 (2001)
46. Schultz, J., Mertens, P.: A comparison between an expert system, a GA and priority for production scheduling. In: Proceedings of the First International Conference on Operations and Quantitative Management, vol. 2, pp. 505–513 (1997)
47. Tay, J.C., Ho, N.B.: Evolving dispatching rules using genetic programming for solving multi-objective flexible job-shop problems. Computers and Industrial Engineering 54, 453–473 (2008)
48. Tay, J.C., Wibowo, D.: An Effective Chromosome Representation for Evolving Flexible Job Shop Schedules. In: Deb, K., et al. (eds.) GECCO 2004. LNCS, vol. 3103, pp. 210–221. Springer, Heidelberg (2004)
49. Thomalla, C.S.: Job shop scheduling with alternative process plans. International Journal of Production Economics 74(1-3), 125–134 (2001)
50. Vairaktarakis, G.L., Cai, X.: The value of processing flexibility in multipurpose machines. IIE transactions 35, 763–774 (2003)
51. Wilhelm, W., Shin, H.: Effectiveness of alternative operations in a flexible manufacturing system. International Journal of Production Research 23, 65–79 (1985)

Scheduling Practice and Recent Developments in Flow Shop and Job Shop Scheduling

Betul Yagmahan[1] and Mehmet Mutlu Yenisey[2]

[1] Uludag University, School of Engineering and Architecture
Department of Industrial Engineering
Gorukle Campus, 16059 Bursa Turkey
betul@uludag.edu.tr
[2] Istanbul Technical University, School of Management
Department of Industrial Engineering
34367 Macka, Istanbul Turkey
yenisey@itu.edu.tr

Summary. Each plant and/or service provider performs several tasks to satisfy customer demand. Every task consumes several resources in order to be completed. Scheduling deals with the allocation of limited resources to tasks over time. Because the resources used in manufacturing activities are very limited, scheduling becomes a very important concept in managerial decision-making. This importance draws the attention of both practitioners and academicians to scheduling.

Scheduling problems usually lie in the NP-hard problem class. Difficulty especially increases as the number of jobs or machines involved increases. As the problem size increases, exact solution techniques become insufficient. This chapter provides an overview of recent developments in computational intelligence approaches to flow shop and job shop scheduling.

1 Scheduling Theory and Problems

Sequencing and scheduling are important research and application pitches in both manufacturing and service systems. It is always important to meet customer-demanded shipping dates for customer satisfaction. Furthermore, better schedules in terms of the performance measure(s) used as objective will improve the system's performance. Hence, a manufacturer or a service provider can maintain lower costs in order to strengthen his power against the intense competition in today's global environment.

The terms "sequencing" and "scheduling" are usually used interchangeably. However, distinguishing them is useful. Convey et al. (1967) claim that whenever there is a choice as to the order in which a number of tasks can be performed, there will be a sequencing problem. Baker (1974) discusses the sequencing problem as a specialized scheduling problem in which the ordering of jobs completely determines a schedule. Pinedo (2002) defines a sequence as usually corresponding to a permutation of n jobs or the order in which jobs are to be processed on a given machine. It is clear that, considering the above arguments, if the studied problem is only to order the task, then the problem falls into the category of sequencing problems.

U.K. Chakraborty (Ed.): Comput. Intel. in Flow Shop and Job Shop Sched., SCI 230, pp. 261–300.
springerlink.com

The term of "scheduling" has two meanings in the literature. The first definition relates to function while the second definition relates to theory. In the scheduling function, managers seek the answers to these questions: What product or service is produced? What will be the production scale? Which resources will be used? The planning function of an enterprise finds the answers to these questions. Several models can be used to find the answers required. The oldest – and probably the best known – model is the Gantt chart. The Gantt chart consists of horizontal bars, which represent jobs. The lengths of bars show the duration of jobs. The bars are arranged by the resources they use.

Scheduling Theory mainly focuses on the mathematics, models, and solution techniques for the scheduling function. All models and solution techniques for scheduling aim to find the answers to these two questions: Which resources will be allocated to perform each task? When will each task be performed? The first question involves allocation decisions while the second question pertains to sequencing problems.

Morton and Pentico (1993) define scheduling more broadly. They claim that scheduling is the process of organizing, choosing, and timing resources to perform all the activities required to meet customer demand. From their viewpoint, scheduling is strategic.

All definitions that made for scheduling lead us to one result. Scheduling is one of the most important decision-making processes in the management of enterprises as it forms an important basis for planning activities. Moreover, it has a wide area of application, covering project planning, shop management, timetabling, routing of transportation vehicles, etc.

French (1982) classifies the scheduling problem into four categories based on the dichotomies of static vs. dynamic and deterministic vs. stochastic, which are founded on the job-arrival discipline and uncertainty, respectively.

In scheduling problems, the objective function is defined in terms of several performance measures. These measures can be flow time, makespan, earliness, lateness, tardiness, number of tardy jobs, etc. It is necessary to define preliminaries of scheduling before discussing the details.

The following parameters are the bases for computation and are given:

Processing time (t_i) : Length of time required for job i to be completed

Ready time (r_i) : Time point at which job i is ready to be processed

Due date (d_i) : Time point at which job i should be completed no later than

The following parameter is found after the complete schedule was determined:

Completion time (C_i) : Time point at which job i is completed

The following parameters are basic quantitative measures, based on completion time, used to evaluate the schedule:

Flow time (F_i) : Length of time job i spends in the system

Lateness (L_i) : Length of time that the completion of job i exceeds its due date

These parameters are calculated as follows:

$$F_i = C_i - r_i \tag{1.1}$$

$$L_i = C_i - d_i \tag{1.2}$$

Lateness may be negative, zero, or positive. Non-negative values show good performance of the schedule. However, negative values stand for bad performance. Negative lateness points out earliness for that job. Usually, there is no reward for early jobs, but late jobs incur several costs. Therefore, tardiness is defined for absolute late jobs as follows:

$$T_i = \max\{0, L_i\} \tag{1.3}$$

Similarly earliness can be defined:

$$E_i = \max\{0, -L_i\} \tag{1.4}$$

Schedules are evaluated using several performance measures. These measures are usually based on completion times. Assume that we have n jobs scheduled. The most common measures can be defined as follows:

Mean flow time: $$\overline{F} = \frac{\sum_{i=1}^{n} F_i}{n}. \tag{1.5}$$

Mean tardiness: $$\overline{T} = \frac{\sum_{i=1}^{n} T_i}{n}. \tag{1.6}$$

Maximum flow time: $$F_{\max} = \max_{1 \le i \le n}\{F_i\}. \tag{1.7}$$

Maximum lateness: $$L_{\max} = \max_{1 \le i \le n}\{L_i\}. \tag{1.8}$$

Maximum tardiness: $$T_{\max} = \max_{1 \le i \le n}\{T_i\}. \tag{1.9}$$

Makespan: $$C_{\max} = \max_{1 \le i \le n}\{C_i\}. \tag{1.10}$$

Number of tardy jobs: $$N_T = \sum_{i=}^{n} \delta(T_i), \text{ where } \begin{cases} \delta(x) = 1, \text{if } x > 0 \\ \delta(x) = 0, \text{otherwise} \end{cases}. \tag{1.11}$$

Performance measures can be divided into two categories: measures based on completion times and measures based on due dates. The mean flow time, mean completion time, maximum flow time, and makespan are in the first category while mean lateness, maximum lateness, mean tardiness, maximum tardiness, and number of tardy jobs fall into the second category. Moreover, jobs can be weighted according to their importance and these weights can be added into the measures.

Although some problems generally deal with only one objective, problems that aim to achieve more than one objective are also gaining increasing interest and importance. T'kindt and Billaut (2002) discuss multicriteria scheduling problems in detail.

Brucker and Knust (2006) present models and algorithms for complex scheduling problems. They discuss both project and machine scheduling and summarize the well known exact solution and heuristic methods.

2 Scheduling Problem Types

Scheduling has a very wide area of application. Almost every service provider and manufacturer experiences a kind of scheduling problem. For example, airports have landing and take-off sequencing problems; airline operators have timetabling and routing problems; a university must decide on class and exam schedules; a manufacturer experiences several shop problems in order to meet customer demand. The variety of these problems leads both researchers and practitioners to study a wide range of scheduling problems and solution techniques. In this section, we summarize the taxonomy of scheduling problems.

2.1 Project Scheduling

Project Scheduling mainly deals with the sequencing of activities subject to precedence constraints and allocation of resources to these activities in a project. Pinedo (2005) claims that the project scheduling problem is similar to parallel machine problem that has an infinite number of machines. The objective is to minimize the makespan. In another words, project scheduling and planning are the longest path problems in terms of Graph Theory.

The well known methods used in project scheduling are CPM (Critical Path Method) and PERT (Program Evaluation and Review Technique). CPM is used for projects with deterministic activity durations while PERT is used for projects with probabilistic activities.

2.2 Single Machine Scheduling

Although the Single Machine Scheduling Problem is the simplest formulation in scheduling, it constitutes the foundation of Scheduling Theory. All other problems arise from the single machine scheduling formulation. Therefore, it plays a crucial role in both theory and application.

Basically, the single machine scheduling problem is concerned with the sequencing of multiple jobs on a single machine. Examples of single machine problems are the running of processes on one CPU computer or landing and take-off scheduling in a one-runway airport. The characteristics of jobs are process time, ready time, and due date. The objectives can be related to throughput measures, like total flow time, mean flow time, weighted flow time, or waiting times, or to measures related to the due date, like total tardiness, weighted tardiness, or number of tardy jobs. It is clear that makespan is independent to the schedule in a single machine environment.

The primary rules for solving single machine scheduling problems are SPT (Shortest Processing Time), WSPT (Weighted Shortest Processing Time), EDD (Earliest Due Date), and MST (Minimum Slack Time). Additionally, several techniques, like Hodgson's Algorithm, Wilkerson-Irwin Algorithm, Dynamic Programming Approach, or Branch-and-Bound Approach, can be used to solve these problems.

2.3 Parallel Machines Scheduling

The generalization of the single machine scheduling problem leads us to multiple machine problems. If we are to extend single machine scheduling, the first problem area is the Parallel Machine Scheduling Problem.

Assigning customers to bank teller windows in a bank or computing on a multiprocessor computer are examples of parallel scheduling problems.

Regarding multi-machine problems, the performance measure of makespan becomes meaningful and objective. Other performance measures for parallel machine problems, besides makespan, are mean flow time, weighted mean flow time, maximum lateness, and number of tardy jobs.

Brucker (2004) categorizes parallel machine problems into three classes according to machine types.

- Identical machines: All machines have the same specifications. Thus, there is no difference in the processing of jobs among machines.
- Uniform machines: The machines have different speeds (s_j). In this problem category, each job has a processing requirement (p_i). The processing of job i on machine j requires p_i/s_j time units. If s_j is set equal to 1 for all machines, then a parallel identical machines problem presents itself.
- Unrelated machines: Each job has different processing times on different machines. This model is the generalization of the uniform parallel machine problem.

Another important point is that jobs may be independent or have precedence constraints.

Baker (1974) constructs an integer programming formulation for problems of parallel identical processors with independent jobs, as given below:

$$\text{Minimize } y \tag{2.1}$$

Subject to:

$$y - \sum_{i=1}^{n} t_i x_{ij} \geq 0, \qquad 1 \leq j \leq m \tag{2.2}$$

$$\sum_{j=1}^{m} x_{ij} = 1, \qquad 1 \le i \le n \tag{2.3}$$

$$x_{ij} \ge 0, \text{ and integer}$$

In this formulation, y stands for makespan, x_{ij} is decision variable which is equal to 1 if job i is assigned to machine j and t_i represents the processing time of job i.

2.4 Shop Scheduling

So far, jobs have been part of a single operation, and we have been interested in one resource. Even in the case of the parallel machine problem, we have actually dealt with a single resource of similar machines.

Brucker (2004) defines general shop scheduling as being composed of problems having n jobs (i=1,…,n), and m machines (M_1,…,M_m). However, each job i consists of a set of operations O_{ij} (j=1,…,n_i). The processing times of these operations are t_{ij} and each operation must be processed on a machine $\mu_{ij} \in \{M_1,\ldots,M_m\}$. Moreover, there may be precedence relationships among the operations. Furthermore, each job can be processed only by one machine at a time while a machine can process only one job at a time. The objective is to find out a feasible schedule that minimizes a performance measure. This performance measure usually a function of completion time. Additionally, the defined problem may aim to satisfy more than one objective.

The shop scheduling problem is divided into several categories according to processing of the shop, flow of jobs on the shop-floor, and routing of production. The following sections discuss the types of shop scheduling problems and the differences among them.

2.4.1 Flow Shop Scheduling

The flow shop scheduling problem will be discussed in Section 3 in detail. However, it will be useful to give a brief introduction and basic derivation of flow shop scheduling in order to achieve consistency throughout this chapter.

As explained above, there are m machines and each job has m operations in a shop environment. The main characteristic of a flow shop is that the flow of work is unidirectional. Machines have a natural order in the flow shop according to work progress. Hence, the machines can be numbered 1,2,…,m and the operations of a job i have corresponding numbers (i,1), (i,2),…,(i,m). If all jobs require one operation on each machine, then it is called a pure flow shop. Jobs require fewer than m operations in the general flow shop.

A number of variants can be defined for the flow shop, like in a skip-shop or a re-entrant flow shop. Jobs may skip some machines in a skip-shop while some machines may be visited more than once in a re-entrant flow shop.

The objective of the flow shop scheduling problem is to find a job order of π_j ($\pi_j = \{(i,1), (i,2),\ldots,(i,m) | i=1,\ldots,m\}$ for each machine j in order to minimize a performance measure based on completion time, like makespan.

If the solution is limited to job sequences $\pi_1, \pi_2, \ldots, \pi_m$ where $\pi_1 = \pi_2 = \ldots = \pi_m$, then this is called a permutation flow shop.

2.4.2 Job Shop Scheduling

The job shop problem generalizes the flow shop problem. There are n jobs $i=1,\ldots,n$ and m machines $M_1,\ldots,M_m$. Job i is made of a sequence of n_i operations; $O_{i1}, O_{i2}, \ldots, O_{in_i}$. The precedence constraints are defined between the operations of each job like $O_{ij} \rightarrow O_{ij+1}$ ($j=1,\ldots,n_{i-1}$). A processing time t_{ij} is associated with each operation O_{ij} to be processed on machine $\mu_{ij} \in \{M_1,\ldots,M_m\}$. The objective is to find a feasible schedule that minimizes some performance measure depending on the completion time C_i of the last operation O_{in_i} of each job. It is assumed that $\mu_{ij} \neq \mu_{ij+1}$ for $j=1,\ldots,n_{i-1}$ if otherwise not stated.

The main difference between a flow shop and a job shop is that the job shop does not have a unidirectional work flow. Therefore, it is necessary to consider the machine number in the route of the jobs on the shop floor. For this purpose, the third subscript indicator is used in order to express which operation of a job should be processed on which machine.

Conway et al. (1967) give an integer programming formulation for the job shop problem following Manne's (1960) model. Moreover, they discuss the modification of the objective according to several performance measures.

Baker (1974), Morton and Pentico (1993), Błażewicz et al. (2001), Pinedo (2002), and Pinedo (2005) give a number of examples for the integer programming model based on a disjunctive constraint formulation.

Disjunctive constraint formulation is based on graph theory. A directed graph is developed to represent the routes of operations for each job. Two kinds of arcs are used in this graph. The conjunctive arcs represent the routes while disjunctive arcs stand for the operations of different jobs to be processed on the same machine. The nodes correspond to the operations to be performed for particular jobs. Pinedo's (2005) formulation is given below for the reader's information.

$$\text{Minimize } C_{\max} \tag{2.4}$$

Subject to

$$y_{hj} - y_{ij} \geq t_{ij} \qquad \text{for all } (i,j) \rightarrow (h,j) \in A, \tag{2.5}$$

$$C_{\max} - y_{ij} \geq t_{ij} \qquad \text{for all } (i,j) \in N, \tag{2.6}$$

$$y_{ij} - y_{ik} \geq t_{ik} \text{ or } y_{ik} - y_{ij} \geq t_{ij} \qquad \text{for all } (i,k) \text{ and } (i,j),\ i=1,\ldots,m, \tag{2.7}$$

$$y_{ij} \geq 0 \qquad \text{for all } (i,j) \in N, \tag{2.8}$$

where y_{ij} denotes the starting time of operation (i,j), N is the set of all operations (i,j), A stands for the set of precedence constraints $(i,j) \rightarrow (h,j)$ and $C_{\max}$ represents the makespan. In this formulation, (i,j) and (h,j) denotes two consecutive operations of job j. The first constraint set guarantees the precedence relationship between the operations of each job while the third set of constraints ensures the order of the operations of different jobs to be processed on the same machine. These constraints are called disjunctive constraints and are why this formulation is called disjunctive programming.

2.4.3 Open Shop Scheduling

An open shop problem is a special case in which there is no precedence relationship between the operations of jobs. In another words, it is a generalization of the flow shop problem. In this problem, each job i consists of m operations O_{ij} $(j=1,\ldots,m)$. The operation O_{ij} must be processed on machine M_j. The objective is to find job sequences (orders of the operations of the same job) and machine sequences (orders of the operations to be performed on the same machine).

2.5 Other Examples

The application of scheduling is not only limited to machine scheduling in manufacturing systems in the manner of single or multiple processors. In practice, there are numerous interesting applications to service systems. A few examples are (Pinedo 2005):

- Reservation systems in car-rental agencies
- Exam scheduling
- Classroom assignments
- Scheduling and timetabling for sports tournaments
- Scheduling network television programs
- Conference presentation scheduling
- Transportation scheduling and timetabling
- Workforce scheduling
- Computer resource scheduling

3 Solution Techniques in Scheduling

3.1 Basic Descriptions

Recently, flow shop production has been widely used in many industrial applications. For this reason, the flow shop scheduling problem has become an attentively studied problem over the last 50 years. The flow shop is characterized by a unidirectional flow of work, i.e., all jobs have the same processing order through the machines. A flow shop contains a natural machine order. Thus, it is possible to number the machines so that if the ith operation of any job precedes the jth operation, then the machine required by ith operation has a lower number than the machine required by the jth operation. The machines in a flow shop are numbered as $1,2,\ldots,m$, and the operations of job i are correspondingly numbered as $(i,1),(i,2),\ldots,(i,m)$. Figure 1 represents a pure flow shop. In this system, all jobs require one operation on each machine. Figure 2 represents a more general flow shop. In the second case, jobs may require fewer than m operations, and their operations may not always require adjacent machines in the numbered order. Additionally, the first and last operations may not always occur at machines 1 and *m*, respectively.

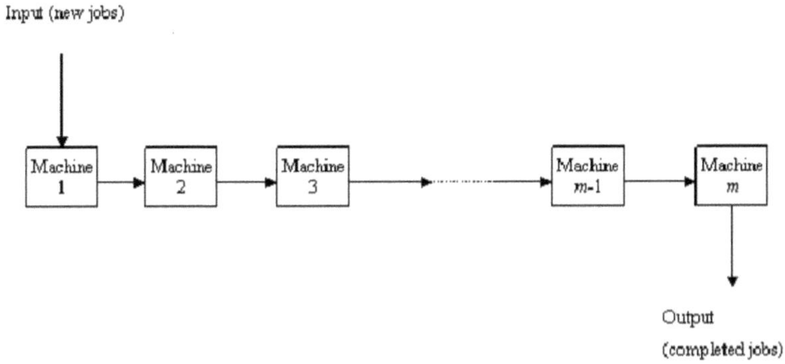

Fig. 1. The pure flow shop

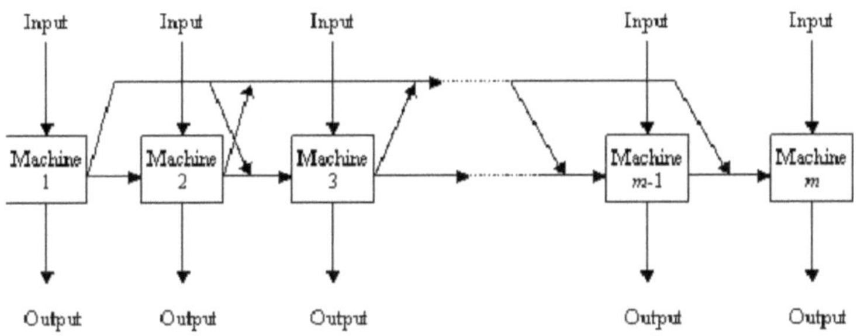

Fig. 2. The general flow shop

The flow shop scheduling problem has these main assumptions (Baker, 1974):

- A set of *n* multiple-operation jobs is available for processing at time zero.
- Setup times for the operations are sequence-independent and are included in processing times.
- Jobs descriptions are known in advance.
- *m* different machines are continuously available.
- Individual operations are not preemptable.

Most of the literature on flow shop scheduling is limited to a special case of the flow shop, the permutation flow shop, in which each machine processes jobs in the same order. Thus, in a permutation flow shop, once the job sequence on the first machine is fixed, the sequences will be kept on all remaining machines. The resulting schedule is called a permutation schedule (Błażewicz et al., 1996).

3.2 Objectives

The flow shop scheduling problem consists of scheduling n jobs with the same order, given processing times on m machines for a given objective. The objective of this problem is mostly to minimize the total completion time, i.e., makespan.

The Gantt chart example for the four-job five-machine permutation flow shop scheduling problem is given in Figure 3.

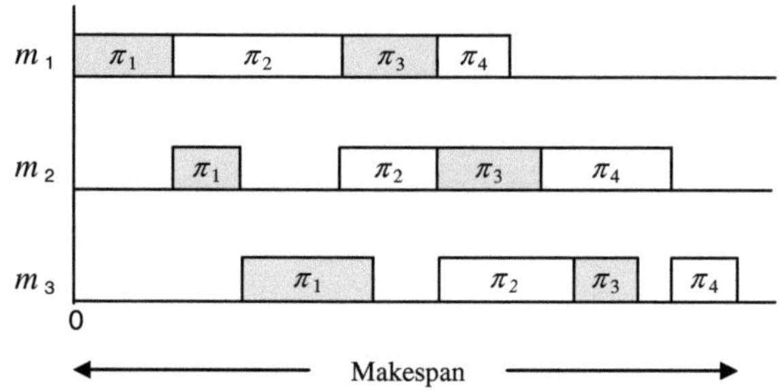

Fig. 3. The Gantt chart for the flow shop scheduling problem

The n-job, m-machine flow shop scheduling problem of minimizing makespan ($n/m/P/C_{max}$) is described as follows:

$t(i,j)$	: processing time for job i on machine j (i=1,2,...,n), (j=1,2,...,m)
n	: total number of jobs to be scheduled
m	: total number of machines in the process
$\{\pi_1, \pi_2, \ldots, \pi_n\}$	: permutation job set

The makespan can be formulated as follow:

Completion times $C(\pi_i, j)$:

$$C(\pi_1, 1) = t(\pi_1, 1), \qquad (3.1)$$

$$C(\pi_i, 1) = C(\pi_{i-1}, 1) + t(\pi_i, 1), \qquad i=2,\ldots,n \qquad (3.2)$$

$$C(\pi_1, j) = C(\pi_1, j-1) + t(\pi_1, j), \qquad j=2,\ldots,m \qquad (3.3)$$

$$C(\pi_i, j) = \max\{C(\pi_{i-1}, j), C(\pi_i, j-1)\} + t(\pi_i, j)\}, \qquad i=2,\ldots,n; j=2,\ldots,m \qquad (3.4)$$

Makespan is defined as:

$$C_{max}(\pi) = C(\pi_n, m). \qquad (3.5)$$

Moreover, different objectives, such as total flow time (*TFT*), mean flow time ($\overline{F}$), maximum tardiness (T_{max}), total tardiness (*TT*), and idletime (*IT*) can be considered as objectives in the flow shop scheduling problem.

These objectives are described as follows where $d(\pi_i)$ is due date for job *i*:

$$TFT = \sum_{i=1}^{n} C(\pi_i, m), \tag{3.6}$$

$$\overline{F} = (1/n) \cdot \sum_{i=1}^{n} C(\pi_i, m), \tag{3.7}$$

$$T_{max} = \max\{\max\{C(\pi_i, m) - d(\pi_i), 0\} | i = 1, \ldots, n\}, \tag{3.8}$$

$$TT = \sum_{i=1}^{n} \max\{C(\pi_i, m) - d(\pi_i), 0\}, \tag{3.9}$$

$$IT = \left\{ C(\pi_1, j-1) + \sum_{i=2}^{n} \{\max\{C(\pi_i, j-1) - C(\pi_{i-1}, j), 0\}\} | j = 2, \ldots, m \right\}, \tag{3.10}$$

3.3 Mathematical Model

A single objective model for the flow shop scheduling problem is given by following formulations (Błażewicz et al., 1996 and 2001). The decision variables are:

$$z_{ik} = \begin{cases} 1 & \text{if job } i \text{ is assigned to the } k\text{th position in the permutation} \\ 0 & \text{otherwise} \end{cases} \tag{3.11}$$

x_{kj} : Idle time (waiting time) on machine *j* before the start of the job in position *k* in the permutation of jobs

y_{kj} : Idle time (waiting time) of the job in the *k*-th position in the permutation, after finishing of processing on machine *j*, while waiting for machine *j*+1 to become available

$$\textit{Minimize} \quad C_{max} \tag{3.12}$$

$$\text{s.t.} \quad \sum_{k=1}^{n} z_{ik} = 1, \qquad i = 1, \ldots, n \tag{3.13}$$

$$\sum_{i=1}^{n} z_{ik} = 1, \qquad k = 1,...,n \tag{3.14}$$

$$\sum_{i=1}^{n} t_{ri} z_{i,k+1} + y_{k+1,r} + x_{k+1,r} = y_{k,r} + \sum_{i=1}^{n} t_{r+1,i} z_{ik} + x_{k+1,r+1},$$

$$k = 1,...,n-1; r = 1,...,m-1 \tag{3.15}$$

$$\sum_{k=1}^{n} \sum_{i=1}^{n} t_{mi} z_{ik} + \sum_{k=1}^{n} x_{km} = C_{\max}, \tag{3.16}$$

$$\sum_{r=1}^{j-1} \sum_{i=1}^{n} t_{ri} z_{i1} = x_{1j}, \qquad j = 2,...,m \tag{3.17}$$

$$y_{1j} = 0, \qquad j = 1,...,m-1 \tag{3.18}$$

Equations (3.13) and (3.14) assign jobs and permutation positions to each other. Equation (3.15) provides Gantt chart accounting between all adjacent pairs of machines in the m-machine flow shop. Equation (3.16) determines the makespan. Equation (3.17) accounts for the machine idletime of the second and subsequent machines while they wait for the arrival of the first job. Equation (3.18) ensures that the first job in the permutation always pass immediately to each successive machine.

3.4 Complexity

The flow shop scheduling problem of minimizing makespan is a classical combinatorial optimization problem for the NP-hard problem class (Garey et al., 1976; Gonzalez et al., 1978). Only a few particular cases are efficiently solvable (Błażewicz et al., 1996 and 2001):

- The two machine flow shop case is simple. In the same way, the case of three machines is a solvable problem in polynomial time under very restrictive requirements on the processing times of the intermediate machine.
- The two machine flow shop scheduling of Johnson can be applied to a case with three machines if the intermediate machine is not the bottleneck.
- The two machine flow shop can be solved using the graphical method.
- Johnson's algorithm solves the preemptive two machine flow shop.
- If the definition of precedence constraints $\pi_i < \pi_j$ specifies that job i must complete its processing on each machine before job j may start processing on that machine, then the two machine flow shop problem with three or series-parallel precedence constraints and makespan minimization is solvable in polynomial time.

3.5 The Flow Shop Scheduling Solution Algorithms

Initial research concerning flow shop scheduling problem was done by Johnson (1954). Johnson described an exact algorithm to minimize makespan for the n-jobs two-machine flow shop scheduling problem. Later, algorithms, such as branch-and-bound and beam search, that yield the exact solution for this problem were proposed. The flow shop scheduling problem that includes many jobs and machines is a combinatorial optimization problem for the NP-hard problem category. Therefore, near optimum solution techniques are preferred. Several heuristic approaches for the flow shop scheduling problem are developed. In recent years, metaheuristic approaches, such as simulated annealing, tabu search, and genetic algorithms, have become very desirable in solving combinatorial optimization problems because of their computational performance. The metaheuristic is a rather general algorithmic framework that can be applied to different optimization problems with minor modifications. By considering recent studies on the flow shop scheduling problem, it is obvious that solution methods based on metaheuristic approaches are frequently proposed.

3.5.1 Exact Solution Methods

Johnson's Rule

Consider the n-jobs two-machine flow shop problem of minimizing makespan. Each job has the same order on both machines. Johnson's rule is used for this type of general two-machine scheduling problem. These measures must be optimized by job sequence:

- Minimization of finishing time
- Minimization of average waiting time of jobs
- Minimization of average idle time of machines

Figure 4 represents Johnson's rule for the two-machine flow shop problem of minimizing makespan (Baker, 1974; Johnson, 1954).

t_{ij} : processing job i on machine j

Step1: Schedule the group of jobs U that are shorter on the first machine than the second. $U = \{i | t_{i1} < t_{i2}\}$ as the first priority group. Schedule the group of jobs V that are shorter on the second machine than the first. $V = \{i | t_{i2} \le t_{i1}\}$ as the second priority group.

Step2: Schedule within U by Shortest Processing Time (SPT) on the first machine. Schedule within V by Longest Processing Time (LPT) on the second machine.

Step3: An optimal sequence is the ordered U followed by the ordered V.

Fig. 4. Johnson's rule for the two-machine flow shop problem

Extension of Johnson's Rule

For the case in which $m = 3$, exact results have not been obtained yet. However, exact results are possible in certain cases by extending Johnson's rule. This extension can be applied to problems in which the second machine has uniformly shorter processing times than the first machine (or the third machine). If $\min_i\{t_{i1}\} \geq \max_i\{t_{i2}\}$ or if $\min_i\{t_{i3}\} \geq \max_i\{t_{i2}\}$, then the problem may be solved with Johnson's rule as a two-machine problem with defined times $T_{i1} = t_{i1} + t_{i2}$ and $T_{i2} = t_{i2} + t_{i3}$.

Branch-and-Bound Algorithm (The Ignall-Schrage Algorithm)

The basic branch-and-bound procedure for the m-machine flow shop problem of minimizing makespan was developed by Ignall and Schrage (1965). The problem is constructed as a tree. Each node in the tree represents a partial solution. The first node corresponds to the initial state in which no jobs are scheduled. From this node, there are n branches corresponding to the possible n jobs that can be assigned to the first position in the sequence. Each of these nodes has n-1 branches corresponding to the n-1 jobs available to be placed in the second position, and so on. (Ignall and Schrage, 1965; Baker, 1974).

For each node on the tree, a lower bound for the makespan associated with any completion of the corresponding partial sequence is obtained by considering the work that remains unscheduled on each machine.

To illustrate the bounds for $m = 3$, let σ' denote the set of jobs that are not contained in the partial sequence σ.

For a given beginning partial sequence σ and remainder set σ':

q_1 : The latest completion time on machine 1 among jobs in σ.
q_2 : The latest completion time on machine 2 among jobs in σ.
q_3 : The latest completion time on machine 3 among jobs in σ.

The amount of processing time still required on the first machine is:

$$\sum_{i \in \sigma'} t_{i1} . \tag{3.19}$$

Moreover, there must be a particular job k that is the last job on machine 1. After it is completed on machine 1, job k must be completed on machines 2 and 3, which takes at least ($t_{k2} + t_{k3}$). The most favorable situation that could occur is:

- There is no idle time in assigning jobs on machine 1.
- There is no idle time in assigning any jobs of the operations of the last job k.
- Job k has the minimal sum ($t_{k2} + t_{k3}$) among the jobs in σ'.

Thus, one lower bound on the makespan is:

$$b_1 = q_1 + \sum_{i \in \sigma'} t_{i1} + \min_{i \in \sigma'} \{t_{i2} + t_{i3}\} . \quad (3.20)$$

Similarly, a lower bound on machine 2 is:

$$b_2 = q_2 + \sum_{i \in \sigma'} t_{i2} + \min_{i \in \sigma'} \{t_{i3}\} . \quad (3.21)$$

Finally, a lower bound on machine 3 is:

$$b_3 = q_3 + \sum_{i \in \sigma'} t_{i3} . \quad (3.22)$$

A lower bound at a node is:

$$B = \max\{b_1, b_2, b_3\} . \quad (3.23)$$

3.5.2 Heuristic Solution Methods

Palmer's Heuristic

For the m-machine flow shop problem of minimizing makespan, Palmer (1965) proposed a slope index s_i to specify job priority:

$$s_i = -(m-1)t_{i1} - (m-3)t_{i2} - (m-5)t_{i3} + \ldots + (m-3)t_{i,m-1} + (m-1)t_{im} \quad (3.24)$$

Job priorities are determined so that jobs with processing times that tend to increase from one machine to another should be given higher priority than jobs with processing times that tend to decrease from one machine to another.

A permutation schedule is constructed using the job index with respect to decreasing s_i. That is:

$$s_{[1]} \geq s_{[2]} \geq \ldots \geq s_{[n]} \quad (3.25)$$

CDS Heuristic

Campbell et al. (1970) proposed a heuristic that is the most accurate extension of Johnson's rule for the m-machine flow shop problem of minimizing makespan. CDS creates several schedules from which a best schedule can be chosen. In this approach, Johnson's rule is applied to the sum of the first two and last two processing times.

In general, at iteration k, the sum of times for job i on the first j machine T_{i1} and the sum for the last j machine T_{i2} is calculated as follows:

$$T_{i1} = \sum_{j=1,k} t_{ij} \ . \tag{3.26}$$

$$T_{i2} = \sum_{j=1,k} t_{i,m-j+1} \ . \tag{3.27}$$

For each iteration, we apply Johnson's rule and a job sequence and makespan M_k are obtained. Finally, the makespan is taken as $M = \min\{M_k\}$.

Gupta's Heuristic

Gupta (1972) proposed a priority rule in the form of Palmer's heuristic so that it would produce good schedules.

The priority index s_i for job i is defined as follows:

$$s_i = \frac{e_i}{\min\limits_{1 \le k \le m-1} \{t_{ik} + t_{i,k+1}\}} \ , \tag{3.28}$$

where:

$$e_i = \begin{cases} 1 & \text{if} \quad t_{i1} < t_{im} \\ -1 & \text{if} \quad t_{i1} \ge t_{im} \end{cases} \tag{3.29}$$

Then a permutation schedule is constructed using the job index with respect to decreasing s_i . That is:

$$s_{[1]} \ge s_{[2]} \ge \ldots \ge s_{[n]} \tag{3.30}$$

NEH Heuristic

The NEH heuristic was proposed by Nawaz et al (1983) to solve the m-machine flow shop problem of minimizing makespan.

The heuristic is based on the assumption that a job with more processing time on all machines will be given higher priority while a job with less processing time on all machines will receive lower priority. Accordingly, the two jobs with highest processing times are determined from the n-jobs problem. The best partial sequence for these two jobs is found by considering the two possible partial schedules. The relative positions of these two jobs with respect to each other are fixed in the remaining steps of the heuristic. Next, the job with the third highest processing time is determined and three partial sequences are tested in which this job is placed at the beginning, middle, and end of the partial sequence found before. The best partial sequence fixes the relative positions of these three jobs in the remaining steps of the heuristic. This procedure is repeated until all jobs are fixed and scheduled.

3.6 Other Studies

In this section, we continue to present other studies concerning the flow shop scheduling problem. All these reviews and evaluations are mainly focused on the most recent heuristics and metaheuristics approaches. A summary of studies on minimizing makespan in the flow shop scheduling problem in the literature is given in Table 1. A summary of studies in the literature for the flow shop scheduling problem for objectives other than makespan is given Table 2.

These methods and many other less known heuristics are well-reviewed in Framinan et al. (2005a). Ruiz and Maroto (2005) give an updated and comprehensive review of flow shop heuristics and metaheuristics. Another recent review is given by Reza Hejazi and Saghafian (2005). The literature in which the flow shop scheduling problem is modeled as a traveling salesman problem (TSP) is reviewed by Bagchi et al. (2006). Gupta and Stafford (2006) provide the developments in flow shop scheduling over the last 50 years.

Table 1. Flow Shop Scheduling Studies on Minimizing Makespan

Solution Approach	References
Exact Solution Methods	Johnson, 1954;
(branch-and-bound, elimination methods, mixed binary integer programming)	Ignall and Schrage, 1965;
	McMahon and Burton, 1967;
	Ashour, 1970;
	Szwarc, 1973;
	Baker, 1975;
	Haouari and Ladhari, 2003;
	Ladhari and Haouari, 2005;
	Šeda, 2007;
	Ziaee and Sadjadi, 2007

Table 1. (*continued*)

Solution Approach	References
Heuristic	Page, 1961;
	Palmer, 1965;
	Smith and Dubek, 1967;
	Gupta, 1971a;
	Gupta, 1971b;
	Campbell et al., 1970;
	Dannenbring, 1977;
	Stinson and Smith, 1982;
	Nawaz et al., 1983;
	Hundal and Rajgopal, 1988;
	Widmer and Hertz, 1989;
	Werner, 1993;
	Moccellin, 1995;
	Lai, 1996;
	Lourenço, 1996;
	Davoud Pour, 2001;
	Nagano and Moccellin, 2002;
	Agarwal et al., 2006;
	Chakraborty and Laha, 2007;
	Jin et al., 2007;
	Laha and Chakraborty, 2007;
	Ruiz and Stützle, 2007;
	Dong et al., 2008;
	Kalczynski and Kamburowski, 2008;
	Vallada and Ruiz, 2008;
	Rad et al. 2009.
Simulating annealing	Osman and Potts, 1989;
	Ogbu and Smith, 1991;
	Ishibuchi et al., 1995;
	Zegordi et al., 1995;
	Low et al., 2004;
	Nearchou, 2004a;
	Nearchou, 2004b.

Table 1. (*continued*)

Solution Approach	References
Tabu search	Taillard, 1990;
	Nowichi and Smutnicki, 1996;
	Ben-Daya and Al-Fawzan, 1998;
	Grabowski and Pempera, 2001;
	Grabowski and Wodecki, 2004;
	Solimanpur et al., 2004;
	Ekşioğlu et al., 2008.
Genetic algorithm	Chen et al., 1995;
	Reeves, 1995;
	Murata et al., 1996;
	Cotta and Troya, 1998;
	Reeves and Yamada, 1998;
	Wang et al., 2003;
	Wang and Zheng, 2003;
	Iyer and Saxena, 2004;
	Wang et al., 2004;
	Ruiz et al. 2006;
	Wang and Zhang, 2006;
	Zhang et al., 2006;
	Cheng and Chang, 2007;
	Nagano et al., 2008.
Ant colony optimization	Stützle, 1998a;
	Rajendran and Ziegler, 2004;
	Ying and Liao, 2004.
Particle swarm optimization	Lian et al., 2006;
	Liao et al., 2007;
	Tasgetiren et al., 2007;
	Jarboui et al., 2008;
	Lian et al., 2008;
	Zhang et al., (2008).

Table 1. (*continued*)

Solution Approach	References
Scatter Search Algorithm	Nowichi and Smutnicki, 2006; Haq et al., 2007; Saravanan et al., 2008.
Differential evolution algorithm	Tasgetiren et al., 2004; Onwubolu and Davendra, 2006; Pan et al, 2008; Qian et al., 2008.
Artificial immune system	Gao and Liu, 2007.
Greedy randomized adaptive search procedure (GRASP)	Prabhaharan et al., 2006.
Iterated Local Search	Stützle, 1998b.

Table 2. Single-objective Flow shop Scheduling Studies on Different Objectives

Objective	Solution Approach	References
Total flow time	Heuristic	Rajendran and Chaudhuri, 1991; Rajendran, 1993; Ho, 1995; Wang et al., 1997; Woo and Yim, 1998; Liu and Reeves, 2001; Allahverdi and Aldowaisan, 2002; Tang and Liu, 2002; Framinan and Leisten, 2003; Framinan et al., 2005b; Laha and Chakraborty, 2008; Pan et al., 2008;

Table 2. (*continued*)

	Branch-and-bound	Ahmadi and Bargchi, 1990.
	Genetic local search algorithm	Yamada and Reeves, 1998.
	Ant colony optimization	Rajendran and Ziegler, 2004.
Total weighted flow time	Heuristic	Rajendran and Ziegler, 1997.
Mean flow time	Differential evolution algorithm	Onwubolu and Davendra, 2006.
Total tardiness	Heuristic	Ow, 1985.
	Genetic algorithm	Kim, 1995; Onwubolu and Mutingi, 1999; Yong and Sannomiya, 2001.
	Tabu search	Armentano and Ronconi, 1999.
	Simulated annealing	Hasija and Rajendran, 2004.
	Differential evolution algorithm	Onwubolu and Davendra, 2006.
Mean tardiness	Heuristic	Kim, 1993.
Weighted tardiness	Heuristic	Gelders and Samdandam, 1978.
	Genetic algorithm	Neppalli et al., 1994.

4 Selected Recent Literature on Flow Shop Scheduling

In this section, we concentrate on some implemented metaheuristics for flow shop scheduling. This review will focus on the recent studies and developments on the flow shop permutation problem using makespan as the measure of performance. There are many algorithms that have been implemented in the flow shop scheduling problem, like simulated annealing, tabu search, genetic algorithms, ant colony optimization, particle swarm optimization, differential evolution, artificial immune systems, and explorative local search methods. Additionally, hybrid algorithms combining some of these methods have been developed in many studies.

4.1 Simulated Annealing

The simulated annealing algorithm inspired by the Metropolis algorithm for statistical mechanics has been successfully applied to many complex combinatorial optimization problems. The fundamental idea comes from the field of metallurgy, in which a solid is first melted and then is slowly chilled. The SA algorithm allows for movements that result in a better solution than the current solution (uphill movements) in order to escape local minima. The probability of making such a movement decreases during the search. The SA algorithm is summarized in Figure 5. The algorithm begins by generating an initial solution x either randomly or heuristically and by initializing the so-called temperature parameter. A candidate solution x' is randomly generated from the current solution x in each iteration and is compared to the two solutions. The candidate solution is accepted as depending on objective functions $f(x), f(x')$ and temperature T. If $f(x') < f(x)$, then the SA algorithm accepts the candidate solution by replacing x with x'. If $f(x') \geq f(x)$, then the candidate solution is accepted with a probability that is a function of $f(x), f(x')$ and T. The temperature T is decreased during the search process according to cooling schedule. The algorithm runs until a stopping condition is met. Several stopping conditions used for the SA algorithm, such as number of iterations, or zero or near-zero temperature (Pinedo, 2002; Blum and Roli, 2003; Nearchou, 2004b).

Step 1:

Generate an initial solution x
Select an initial temperature T_0
$T = T_0$

Step 2:

Generate candidate solution x' from current solution x
if $f(x') < f(x)$, then set $x = x'$
if $f(x') \geq f(x)$, then set $x = x'$ with a certain probability

Step 3:

Update T according to cooling schedule
If stopping condition is met, then STOP; otherwise, go to Step 2.

Fig. 5. Simulated Annealing Algorithm

The simulated annealing algorithm for solving the flow shop scheduling problem has been pointed out in the works of several researchers. First, Osman and Potts (1989) and Ogbu and Smith (1991) have reported high-quality results using the basic simulated annealing algorithm.

Ishibushi et al. (1995) proposed two simulated annealing algorithms with a modified generation mechanism. Several neighbors of a current solution are evaluated and the move to the best of these neighbors is examined using this mechanism.

Zegordi et al. (1995) presented a simulated annealing algorithm with problem-specific information, which yielded a form of index in a "move desirability for jobs" table.

Low et al. (2004) proposed a modified simulated annealing searching procedure consisting of the "restarting solution mechanism" and some additional termination conditions to assure the solution's quality and efficiency.

Finally, Nearchou (2004a) presented a new hybrid simulated annealing algorithm which integrated the basic structure of a simulated annealing algorithm with features borrowed from the fields of genetic algorithms and local search techniques. The algorithm works from a population of candidate schedules and generates new populations of neighbor schedules by applying suitable small perturbation schemes. During the annealing process, an iterated hill climbing procedure is stochastically applied to the population of schedules in order to achieve a desertion from possible local minima and to improve the algorithm's performance. Nearchou (2004b) proposed another algorithm, that is similar to the previous one, which combines the canonical characteristics of simulated procedure with the features of genetic algorithm's population of individuals.

4.2 Tabu Search

The Tabu Search algorithm was first proposed by Glover (1989, 1990). The TS algorithm is dependent on the following parameters: initial solution, moves, neighborhood, searching strategy, tabu list, aspiration criterion, and stopping criteria. The basic idea of this method consists in starting from an initial solution and then moving successively among neighborhood solutions. At each iteration, a move is made to the best solution in the neighborhood of the current solution, which may not be an improving solution. Tabus are used to prevent cycling when moving away from local optima through non-improving moves. Tabus are stored in the tabu list. At every iteration of TS, a move will be assigned to the tabu list when the move is chosen to lead the search from the current solution to its neighborhood solution. This move will then not be chosen for a number of immediately succeeding iterations. The size of the tabu list is bound by tabu list size. The size of the tabu list could be fixed or variable. A candidate solution x' is accepted if it is not on the tabu list or if an aspiration criterion is satisfied. An aspiration criterion could allow a tabu move when the neighborhood has an objective function value better than the best objective encountered so far. If x^* is a better solution, the objective function transforms $f(x)$ into $f(x^*)$. The search is terminated when some stopping condition is satisfied. The structure of the TS algorithm is shown in Figure 6 (Ben-Daya and Al-Fawzan, 1998; Gupta et al., 1999; Pinedo, 2002; Glover and Kochenberger, 2003; Ekşioğlu et al., 2008).

Several TS algorithms have been proposed for the flow shop scheduling problem. Taillard (1990) presented a tabu search technique that obtained better solutions than the NEH. Later Nowichi and Smutnicki (1996) proposed a tabu search technique with a specific neighborhood definition employing block properties to reduce the neighborhood structure.

Ben-Daya and Al-Fawzan (1998) proposed implementation of the tabu search approach that suggested simple techniques for generating neighborhoods of a given sequence and combined a scheme for intensification and diversification that had not been considered before.

Grabowski and Pempera (2001) presented and discussed some new properties of blocks in the flow shop problem. These properties allow reductions in the neighborhood size in the tabu search and direction of the search trajectory into a promising region of the solution space.

Grabowski and Wodecki (2004) also presented and discussed some new properties of the problem associated with the blocks. In order to decrease the computational effort of the search in tabu search, they proposed calculation of the lower bounds on the makespans instead of computing makespans explicitly for the best solution.

Solimanpur et al. (2004) developed a neural network-based tabu search method to solve the flow shop scheduling problem. This algorithm exploits a neuro-dynamical structure to iteratively improve the initial permutation. The proposed algorithm is different from the other tabu search methods, as it reduces the tabu effect exponentially.

Recently, Ekşioğlu et al. (2008) investigated a tabu search procedure for the flow shop scheduling problem with the makespan minimization criterion. It is different from other tabu search procedures. The neighborhood of a solution is generated using a combination of three different exchange mechanisms (adjacent exchange, random exchange, and insertion). This resulted in a well-diversified search procedure.

Step 1:

Generate an initial solution x

$x^* = x$

Initialize the tabu list

Step 2:

While set of candidate solutions X' is not complete

Generate candidate solution x' from current solution x

Add x' to X' if x' is not on the tabu list T or satisfy the aspiration criterion

Step 3:

Find the best candidate solution x^* in X'

if $f(x) < f(x^*)$, then set $x^* = x$

Update the tabu list and aspiration conditions

Step 4:

If stopping condition is met then stop; otherwise, go to Step 2.

Fig. 6. Tabu Search Algorithm

4.3 Genetic Algorithm

The genetic algorithm is a population-based method that is based on the mechanics of natural selection and natural genetics. The GA maintains a population of individuals $P(t)$ for generation t. Each individual represents a solution to the problem. These solutions are encoded into chromosomes. Every individual in the population is evaluated and assigned a fitness value. Then the population undergoes genetic operations to form new individuals. During a number of iterations, this population evolves until some stopping criterion is satisfied. Figure 7 shows the general framework of a genetic algorithm. The selection operator picks from the population some individuals according to the assigned fitness value in such a way that the fittest individuals have a greater chance of being selected. The crossover operator creates new individuals by combining the good properties of different individuals. The mutation operator creates new individuals by making changes to a single individual (Gen and Cheng, 2000; Ruiz et al., 2006).

Step 1:

  $t=0$

  Form the initial population $P(t)$

Step 2:

  $P'(t) \leftarrow$ crossover($P(t)$)

  $P''(t) \leftarrow$ mutation($P'(t)$)

  Evaluate $(P''(t))$

  $P(t+1) \leftarrow$ Select($P''(t) \cup P(t)$)

  $t = t+1$

Step 3:

  If the stopping criteria are met (number of generation), then stop; Otherwise, go to step 2.

Fig. 7. Genetic algorithm

The application of genetic algorithms to the flow shop scheduling problem has been widely studied. Chen et al. (1995) developed one of the earliest genetic algorithms for the flow shop scheduling problem with the makespan minimization criterion.

Reeves (1995) also described the concept of genetic algorithms and applied it solving the flow shop scheduling problem with makespan as a criterion.

Murata et al. (1996) examined the performance of genetic algorithms in order to specify some genetic operators and parameters for the flow shop scheduling problem. They then proposed two hybrid genetic algorithms to improve the performance of the genetic algorithm. One is the genetic local search algorithm and the other is a genetic simulated annealing algorithm. They also introduced some modifications of search mechanisms in these hybrid genetic algorithms.

Cotta and Troya (1998) studied different representations for the flow shop scheduling problem using forma analysis. They proposed some new operators that run on these representations.

Reeves and Yamada (1998) re-considered the implementation of a genetic algorithm for the flow shop scheduling problem using the representative neighborhood and path re-linking.

Wang et al. (2003) presented a class of order-based genetic algorithms for the flow shop scheduling problem. This algorithm borrows from the idea of ordinal optimization to ensure the quality of the solution found with a reduced computation effort. It is applied to evolutionary search mechanisms and learning capabilities of genetic algorithms to effectively perform exploration and exploitation.

Wang and Zheng (2003) proposed an effective hybrid heuristic for the flow shop scheduling problem. They incorporated the NEH heuristic into the random initialization of a genetic algorithm, used multicrossover operators acting on the divided subpopulations, and replaced mutation by the simulated annealing metropolis sample process with multiple neighbor state generators.

Iyer and Saxena (2004) improved the standard implementation of the genetic algorithm by tailoring the various genetic algorithm operators to suit the structure of the problem.

Wang et al. (2004) first formulated the determination of optimal genetic control parameters. Then the ordinal optimization and the optimal computing budget allocation techniques are applied to determine the best genetic control parameters among all the alternative parameter combinations.

Ruiz et al. (2006) proposed a robust genetic algorithm and a rapid hybrid implementation for solving the permutation flow shop scheduling problem. These algorithms use new genetic operators, advanced techniques like hybridization with local search, an efficient population initialization, and a new generational scheme.

Wang and Zhang (2006) presented a novel and systematic approach based on ordinal optimization and optimal computing budget allocation techniques to determine the optimal combinations of genetic operators for flow shop scheduling problems.

Zhang et al. (2006) proposed an adaptive genetic algorithm with multiple operators for the flow shop scheduling problem. This adaptive genetic algorithm uses multiple crossover and mutation operators in an adaptively hybrid sense, according to their contribution to the search process.

Cheng and Chang (2007) used genetic algorithms to solve the flow shop scheduling problem and adopted Taguchi's experimental design to effectively obtain optimal parameter design in the genetic algorithm.

Nagano et al. (2008) described the application of a constructive genetic algorithm that includes a population of dynamic sizes composed of schemata and structures, and the possibility of using heuristics in structure representation and in fitness function definitions.

4.4 Ant Colony Optimization

Ant colony optimization (ACO) is proposed as a new metaheuristic approach for solving difficult combinatorial optimization problems in the literature. The main idea of ACO metaheuristics is based on the behavior of real ants that use the pheromone trail for communication and cooperation. The first example of the ACO algorithm is the Ant System (AS) algorithm, proposed by Dorigo et al. (1991a, 1991b) for the Traveling Salesman Problem (TSP). Studies then tried to improve its performance and, consequently, various ACO algorithms were proposed. These extensions include Ant Colony System (ACS), Ant-Q, the Max–Min Ant System (MMAS), and Rank Based Ant System. The structure of ACO is given in Figure 8. At the initialization step, pheromone trails, heuristic information, and parameters are initialized. Then, in the iterative step, until a complete solution is constructed, each ant repeatedly selects the next solution component by applying a certain transition probability rule. Then, the updating rule is applied to increase pheromones between components of the best solution up to the current iteration. Thus, all ants will focus on a better solution. Finally, until reaching the stopping condition, the procedure is repeated (Dorigo and Stützle, 2004; Yagmahan and Yenisey, 2008).

Recently, attempts have been made to solve the flow shop scheduling problem by using ACO algorithms. Stützle (1998a) developed the first ant colony optimization algorithm that incorporated a new local search technique in MMAS.

Rajendran and Ziegler (2004) proposed the two ant-colony algorithms. The first algorithm incorporates the summation rule and a new local search technique in the max–min ant system. The second proposed ant-colony algorithm is based on a new technique for local search (job-index-based local search).

Ying and Liao (2004) presented an ant colony system algorithm. They revised the slope index of Palmer's method as the heuristic desirability.

Step 1:
  Pheromone trails, heuristic information, and parameters are initialized

Step 2:
  Construct a complete solution for each ant
  Apply local search process
  Update the pheromone trail

Step 3:
  If the stopping condition is realized, then STOP; otherwise go to step 2.

Fig. 8. Ant colony optimization algorithm

4.5 Particle Swarm Optimization

The particle swarm optimization algorithm is one of the latest population-based optimization methods. It is based on sociological behavior associated with bird flocking or fish schooling. PSO consists of a swarm of *m* particles, where each particle represents a solution to an optimization problem. Each particle moves at a position $X_i = \{x_{i1}, x_{i2}, \ldots, x_{in}\}$ in the multi-dimensional search space with a certain velocity $V_i = \{v_{i1}, v_{i2}, \ldots, v_{in}\}$, where $i = 1,2,\ldots,m$. Each particle moves towards its best previous position of the *i*th particle that gives the best objective function value

($lbest$) denoted by $P_i = \{p_{i1}, p_{i2}, \ldots, p_{in}\}$. On the other hand, each particle moves towards the best particle in the whole swarm that gives the best objective function value ($gbest$) denoted by $G = \{g_1, g_2, \ldots, g_n\}$. Each particle moves according to a function of its current position, velocity, *lbest*, and *gbest* in the search space along the iterations. Each particle adjusts its velocity in order to update the position of each particle. Velocity is added to the position coordinates of the particle.

The new velocity and particle position at t iteration are calculated using the following equations:

$$v_{ij}^{t} = v_{ij}^{t-1} + c_1 r_1 (p_{ij}^{t-1} - x_{ij}^{t-1}) + c_2 r_2 (g_j^{t-1} - x_{ij}^{t-1}), \tag{3.31}$$

$$x_{ij}^{t} = x_{ij}^{t-1} + v_{ij}^{t}, \tag{3.32}$$

where c_1 is the cognition learning factor, c_2 is the social learning factor, and r_1 and r_2 are random numbers uniformly distributed in [0, 1]. The general PSO algorithm is summarized in Figure 9 (Tasgetiren et al., 2007; Jarboui et al., 2008).

Step 1:

Randomly initialize particle positions and velocities
For each particle *i*:
Evaluate the objective function $f(x)$ at current position X_i

Step 2:

For each particle *i*:
If $f(X_i) < f(lbest)$ then $lbest = X_i$ and $P_i = X_i$
If $f(X_i) < f(gbest)$ then $gbest = X_i$ and $G_i = X_i$

Step 3:

For each particle *i*:
Update velocity V_i using equation (3.31)
Update position X_i using equation (3.32)

Step 4:

If the stopping condition is realized, then STOP; otherwise go to step 2.

Fig. 9. Particle swarm optimization algorithm

Currently, several papers have been published that solve the flow shop scheduling problem based on a PSO algorithm. Lian et al. (2006) first proposed a similar particle swarm optimization algorithm and applied it to the permutation flow shop scheduling problem of minimizing makespan. This algorithm investigates the effect of various operators (crossovers) under the framework of the problem.

Liao et al. (2007) proposed the discrete version of particle swarm optimization for the flow shop scheduling problem. In the algorithm, the particle is moved to the new sequence by applying an efficient approach to the construction of a sequence. A new neighborhood structure of particles is also designed.

Tasgetiren et al. (2007) presented a particle swarm optimization algorithm in order to solve the permutation flow shop sequencing problem. A heuristic rule, called the smallest position value, was developed in order to apply the continuous particle swarm optimization algorithm to all classes of sequencing problems. In addition, they applied a local search procedure based on variable neighborhood search in order to obtain good quality solutions.

Jarboui et al. (2008) described a combinatorial particle swarm optimization. Furthermore, they added an improvement phase based on the simulated annealing approach.

Lian et al. (2008) presented a novel particle swarm optimization algorithm and successfully applied to the permutation flow shop scheduling problem to minimize makespan. They described some novel particle swarm optimization operators (crossovers and mutations) and investigated its effectiveness under the framework of the flow shop scheduling problem.

Recently, Zhang et al. (2008) proposed an improved particle swarm optimization algorithm to solve the flow shop scheduling problem with the objective of minimizing makespan. The particle swarm optimization algorithm effectively combined with genetic operators. When a particle is going to stagnate, the shift mutation operator is used to search its neighborhood.

4.6 Scatter Search Algorithm

Scatter search (SS) is a population-based optimization method that has been successfully applied to optimization problems. SS generates a trial set from using the seed solutions corresponding to feasible solutions to the problem under consideration. An improvement method is used to attempt to improve trial solutions and update the reference set. A reference set contains the best solutions found so far in terms of the objective function. A subset of solutions is produced by combining solutions in the reference set. These newly created subset solutions are improved and used to update the reference set. This search is terminated when the stopping criteria are satisfied. The SS algorithm is summarized in Figure 10 (Blum and Roli, 2003; Glover and Kochenberger, 2003; Saravanan et al., 2008).

Recently, the SS algorithm has been successfully applied to the flow-shop scheduling problem. Nowichi and Smutnicki (2006) provided a new view on the solution space and the search process. The new approximate algorithm uses some elements of scatter search as well as the path re-linking technique. This algorithm also offered unprecedented accuracy within a short computing time.

Haq et al. (2007) solved the flow shop scheduling problem using the generalized template created for evolutionary scatter search algorithms and compared results with a multilevel hybrid system based on scatter search, path re-linking, and tabu search.

Saravanan et al. (2008) applied a novel metaheuristic approach called scatter search for the flow shop scheduling problem. The algorithm compared the various existing metaheuristic and heuristic methods in the literature. The experiments verified the effectiveness and efficiency of the SS algorithm over other metaheuristics.

Step 1:
  Generate trial solutions from the seed solutions
  Apply the improvement method to produce one or more enhanced trial solutions
  Update the reference set
Step 2:
  Generate new subsets from the reference set
  Combine these subsets to obtain one or more new trial solutions
  Apply the improvement method to the trial solutions
Step3:
  Update the reference set
Step 4:
If the stopping condition is realized, then STOP; otherwise go to step 2.

Fig. 10. Scatter search algorithm

4.7 Differential Evolution Algorithm

Differential evolution (DE) is an evolutionary algorithm proposed by Price and Storn (1995). DE can be classified as an evolutionary optimization algorithm. In a DE algorithm, candidate solutions are represented by chromosomes based on floatingpoint numbers. DE works as follows: First, all individuals are randomly initialized and evaluated. At each generation, the mutation and crossover operators are applied to individuals to generate a new population. In the mutation process, the weighted difference between two randomly selected population members is added to a third member to generate a mutated solution. Then, a crossover operator follows to combine the mutated solution with the target solution to generate a trial solution. A selection operator is applied to compare the fitness function value of both competing solutions, namely, target and trial solutions to determine who can survive for the next generation. As long as the termination condition is not fulfilled, this process is executed. The basic algorithm of differential evolution is shown in Figure 11 (Pan et al., 2008; Qian et al., 2009).

First, Tasgetiren et al. (2004) reported the application of the differential evolution algorithm to the flow shop scheduling problem with makespan criterion. The smallest position value rule is used in differential evolution algorithms to convert a continuous parameter vector to a job permutation.

Onwubolu and Davendra (2006) described a novel differential evolution algorithm. The techniques for handling discrete variables are described as well as the techniques needed to handle boundary constraints. Other objective functions considered in this work include mean flow time and total tardiness.

Step 1:	
	Initialize population
Step 2:	
	Evaluate the objective values of all individuals, find the best individual
Step 3:	
	Mutation
	Crossover
	Selection
	Update the best individual
Step 4:	
	If a stopping criterion is satisfied, then STOP; otherwise go to Step 3.

Fig. 11. Differential evolution algorithm

Pan et al. (2008) presented a new and novel discrete differential evolution algorithm and the iterated greedy algorithm for the permutation flow shop scheduling problem with the makespan criterion. Furthermore, they proposed a new and novel referenced local search procedure hybridized with both algorithms to further improve the solution quality.

Qian et al. (2008) proposed a hybrid algorithm combining the differential evolution based search and local search. A largest-order-value rule is presented to convert the continuous values of individuals in differential evolution to job permutations. After the DE-based exploration, a simple but efficient local search is applied to emphasize exploitation.

4.8 Artificial Immune System

The artificial immune system algorithm is an adaptive system, inspired by theoretical immunology and observed immune functions, principles, and models, which is applied to solve problems.

The main aim of the immune system is to recognize disease-causing organisms, called *pathogens*, to defend against invasion and to eliminate malfunctioning cells. Pathogens are not directly recognized by the components of the immune system. *Antigens* are small portions of the pathogens molecules, which are recognized by the immune system. There are two types of antigens: *self* and *non-self*. Non-self antigens are disease-causing elements, whereas self antigens are harmless to the body. Two major groups of immune cells are *B-cells* and *T-cells*. B-cells can recognize the antigens free in solution, while T-cells require antigens to be presented by other assisting cells. Both B-cells and T-cells contain the surface receptors capable of recognizing antigens. Antigens are covered with molecules to be recognized by receptor molecules. An *antibody* is the B-cell receptor molecule. When an antigen is recognized by immune cell receptors, the immune system produces antibodies. Binding an antibody to antigens is a signal to remove disease-causing organisms. There are several selection mechanisms used in AIS algorithms. Negative selection, clonal selection, and immune network models are examples. Figure 12 depicts the negative selection principle (de Castro and Timmis, 2002; de Castro, 2002).

The AIS algorithm has recently been applied to scheduling problems such as job-shop and flow-shop. Gao and Liu (2007) presented a novel artificial immune system algorithm for the flow shop scheduling problem with makespan criterion. The algorithm was tested on flow shop problem benchmarks. Computational results show that artificial immune system algorithms give good results.

```
Step 1:
        Generate random candidate detectors (C)
Step 2:
        While detectors set (M) not produced do
            Compare the elements in C with the elements in self set P
            If an element of P is recognized by an element of C, then
            eliminate this element of C; else place this candidate detector
            of C in M
        End while
Step 3:
        Monitor a new set of self for any variation after M has been gener-
        ated. This means that if any element of M matches an element of the
        new self-set, then a non-self element was detected.
```

Fig. 12. Artificial immune system algorithm

4.9 Explorative Local Search Methods

4.9.1 GRASP

The greedy randomized adaptive search procedure is an iterative process. Basically, this metaheuristic consists of two phases: a construction phase and a local search phase. In the construction phase, a feasible solution is iteratively constructed, one new element at a time. In each iteration, all elements are ranked according to an adaptive greedy heuristic criterion that gives them a score as a function of the benefit if inserted in the current partial solution. The candidate list, called a restricted candidate list (RCL), is composed of the best α elements. One element is randomly selected from a restricted candidate list. The heuristic values are updated during each iteration of the construction phase to reflect the changes brought about by the selection of the previous elements. Figure 13 describes the construction phase. In the second phase, the solution is improved using a local search, which may be a basic local search algorithm such as iterative improvement, or a more advanced technique such as SA or TS. The best overall solution found is kept. The search finishes when a termination criterion is verified. The GRASP algorithm is given in Figure 14 (Blum and Roli, 2003; Glover and Kochenberger, 2003).

A few attempts have been made to solve flow shop scheduling problems using GRASP. Prabhaharan et al. (2006) implemented a greedy randomized adaptive search procedure to solve a flow shop scheduling problem. These computational experiments indicate that the GRASP algorithm outperforms the traditional NEH algorithm.

```
S= ∅
Determine candidate list length α
While solution is not complete do
        Build RCL_α
        Select from RCL_α an element x at random
        S = S ∪ { x }
        Update the greedy heuristic values
End while
```

Fig. 13. Greedy randomized solution construction

```
While termination conditions not met do
        Construct greedy randomized solution
        Apply local search
        Memorize best found solution
End while
```

Fig. 14. Greedy Randomized Adaptive Search algorithm

4.9.2 Iterated Local Search

Iterated local search is a very simple and powerful metaheuristic that consists of repeatedly applying a local search algorithm to modifications of previously visited local optimal solutions. The algorithm starts with an initial solution and applies a local search until a local optimum is found. Then, the algorithm perturbs the current solution and a different local optimum is obtained by performing local search. Finally, acceptance criteria depending on the search history are used to decide from which solution the search is continued in the next iteration. The ILS algorithm can be described using the pseudo-code shown in Figure 15 (Stützle, 1998b; Glover and Kochenberger, 2003).

Generate initial solution x_0.

x^* =LocalSearch(x_0).

repeat

  x' =Perturbation(x^* , *history*)

  x'' =LocalSearch(x').

  x^* =AcceptanceCriterion(x', x'' , *history*)

until termination condition met

Fig. 15. Iterated local search algorithm

Stützle (1998b) applied an iterated local search algorithm to the permutation flow shop scheduling problem. The iterated local search algorithm is based on a straightforward local search implementation. Computational results show that iterated local search approach also performs well compared to other approaches proposed for the flow shop scheduling problem.

5 Conclusion

In this chapter, scheduling problems are discussed and several examples of recent developments in the scheduling literature are given. Clearly, scheduling is a very important and developing research area. It has very interesting uses in both theory and application. Manufacturing with the lowest cost becomes very important in today's global competitive environment. All manufacturers, in both goods and services, seek ways to lower costs. Moreover, they not only focus on costs but also production and service speeds. Thus, scheduling theory and its applications are becoming crucial in manufacturing.

However, scheduling is a complex and difficult problem. Conventional optimization methods are insufficient for large problems in terms of solution time. Different techniques have been developed in order to solve scheduling problems. These techniques are generally based on heuristic approaches. However, although these techniques provide solutions in an appropriate amount of time, they do not guarantee the optimum result. They find the near-optimum solutions that are satisfactory for large and complex problems. At the least, an acceptable solution can be obtained for problems that are technically unsolvable.

Scheduling has various areas which could be improved based upon recent literature. One such development area involves the objectives; recently, multi-objective applications have become widespread. The second progressing area is development of solution techniques. Researchers are working on both improving the performance of existing algorithms and creating new techniques to solve scheduling problems. Some researchers are attempting to combine several techniques in order to provide a better algorithm..

Another important point is that recent studies mainly focus on metaheuristic algorithms. When recent articles and papers discuss scientific meetings, it can be easily claimed that the studies on metaheuristic algorithms have been rapidly increasing. These metaheuristic algorithms are affected by events in nature and are inspired from the behavior of animals like ants or swarms, biological entities like neurons or genes, or some physical event like annealing.

References

Agarwal, A., Colak, S., Eryarsoy, E.: Improvement heuristic for the flow-shop scheduling problem: An adaptive-learning approach. Eur. J. Oper. Res. 169, 801–815 (2006)

Ahmadi, R., Bargchi, U.: Improved lower bound for minimizing the sum of flowtimes of n jobs over m machines in a flow shop. Eur. J. Oper. Res. 44, 331–336 (1990)

Allahverdi, A., Aldowaisan, T.: New heuristics to minimize total completion time in m-machine flowshops. Int. J. Prod. Econ. 7, 71–83 (2002)

Armentano, V.A., Ronconi, D.P.: Tabu search for total tardiness minimization in flowshop scheduling problems. Comput. Oper. Res. 26(3), 219–235 (1999)

Ashour, S.: An experimental investigation and comparative evolution of flowshop sequencing techniquess. Oper. Res. 18, 541–549 (1970)

Bagchi, T.P., Gupta, J.N.D., Sriskandarajah, C.: A review of TSP based approaches for flowshop scheduling. Eur. J. Oper. Res. 169, 816–854 (2006)

Baker, K.R.: Introduction to sequencing and scheduling. John Wiley & Sons Inc., New York (1974)

Baker, K.R.: A comparative study of flowshop algorithms. Oper. Res. 23, 62–73 (1975)

Ben-Daya, M., Al-Fawzan, M.: A tabu search approach for the flow shop scheduling problem. Eur. J. Oper. Res. 109(1), 88–95 (1998)

Błażewicz, J., Ecker, K.H., Pesch, E., Schmidt, G., Węglarz, J.: Scheduling computer and manufacturing processes. Springer, Berlin (1996)

Błażewicz, J., Ecker, K.H., Pesch, E., Schmidt, G., Węglarz, J.: Scheduling computer and manufacturing processes, 2nd edn. Springer, Berlin (2001)

Blum, C., Roli, A.: Metaheuristics in combinatorial optimization: overview and conceptual comparison. ACM Comput. Surv. 35(3), 268–308 (2003)

Brucker, P.: Scheduling algorithms, 4th edn. Springer, Berlin (2004)

Brucker, P., Knust, S.: Complex scheduling. Springer, Berlin (2006)

Campbell, H.G., Dudek, R.A., Smith, M.L.: A heuristic algorithm for the n job m machine sequencing problem. Manage. Sci. 16, 630–637 (1970)

Chakraborty, U.K., Laha, D.: An improved heuristic for permutation flowshop scheduling. Int. J. Inf. Commun. T. 1(1), 89–97 (2007)

Chen, C.L., Vempati, V.S., Aljaber, N.: An application of genetic algorithms for flow shop problems. Eur. J. Oper. Res. 80, 389–396 (1995)

Cheng, B.W., Chang, C.L.: A study on flowshop scheduling problem combining Taguchi experimental design and genetic algorithm. Expert. Syst. Appl. 32, 415–421 (2007)

Conway, R.W., Maxwell, W.L., Miller, L.W.: Theory of scheduling. Addison Wesley Publishing Company, Massachusetts (1967)

Cotta, C., Troya, J.M.: Genetic forma recombination in permutation flowshop problems. Evol. Comput. 6(1), 25–44 (1998)

Dannenbring, D.G.: An evaluation of flowshop sequencing heuristics. Manage. Sci. 23, 1174–1182 (1977)

Davoud Pour, H.: A new heuristic for the n-job, m-machine flowshop problem. Prod. Plan. Control. 12(7), 648–653 (2001)

De Castro, L.N.: Immune, swarm and evolutionary algorithms Part I: basic models. In: Proceedings of the ICONIP Conference (International Conference on Neural Information Processing), Singapura, pp. 1464–1468 (2002)

De Castro, L.N., Timmis, J.: Artificial immune systems: a novel paradigm to pattern recognition. In: Corchado, J.M., Alonso, L., Fyfe, C. (eds.) Artificial Neural Networks in Pattern Recognition, pp. 67–84. University of Paisley, UK (2002)

De Castro, L.N., Timmis, J.I.: Artificial immune systems as a novel soft computing paradigm. Soft. Comput. 7(7), 526–544 (2003)

Dong, X., Huang, H., Chen, P.: An improved NEH-based heuristic for the permutation flowshop problem. Comput. Oper. Res. 35, 3962–3968 (2008)

Dorigo, M., Maniezzo, V., Colorni, A.: Positive feedback as a search strategy. Technical Report, No. 91-016, Politecnico di Milano, Italy (1991a)

Dorigo, M., Maniezzo, V., Colorni, A.: The ant system: An autocatalytic optimizing process. Technical Report, No. 91-016 (Revised), Politec-nico di Milano, Italy (1991b)

Dorigo, M., Stützle, T.: Ant colony optimization. MIT Press, Cambridge (2004)
Ekşioğlu, B., Ekşioğlu, S.D., Jain, P.: A tabu search algorithm for the flowshop scheduling problem with changing neighborhoods. Comput. Ind. Eng. 54, 1–11 (2008)
Framinan, J.M., Leisten, R.: An efficient constructive heuristic for flowtime minimisation in permutation flow shops. Omega 31(4), 311–317 (2003)
Framinan, J.M., Gupta, J.N.D., Leisten, R.: A review and classification of heuristics for permutation flow-shop scheduling with makespan objective. J. Oper. Res. Soc. 55(12), 1243–1255 (2005a)
Framinan, J.M., Leisten, R., Ruiz-Usano, R.: Comparison of heuristics for flowtime minimisation in permutation flowshops. Comput. Oper. Res. 32(5), 1237–1254 (2005b)
French, S.: Sequencing and Scheduling: An introduction to the mathematics of the job-shop. Ellis Horwood Ltd., Chichester (1982)
Gao, H., Liu, X.: Improved artificial immune algorithm and its applications on permutation flow shop sequencing problems. Inform. Technol. J. 6(6), 929–933 (2007)
Garey, M.R., Johnson, D.S., Sethi, R.: The complexity of flowshop and job-shop scheduling. Math. Oper. Res. 1(2), 117–129 (1976)
Gelders, L.F., Samdandam, N.: Four simple heuristics for scheduling a flowshop. Int. J. Prod. Res. 16, 221–231 (1978)
Gen, M., Cheng, R.: Genetic algorithms and engineering optimization. John Wiley&Sons, USA (2000)
Glover, F.: Tabu search: part I. ORSA. J. Comput. 1, 190–206 (1989)
Glover, F.: Tabu search: part II. ORSA. J. Comput. 2, 4–32 (1990)
Glover, F.W., Kochenberger, G.A.: Handbook of metaheuristics. Kluwer, Norwell (2003)
Gonzalez, T., Sahni, S.: Flowshop and jobshop schedules: Complexity and approximations. Oper. Res. 26(1), 36–52 (1978)
Grabowski, J., Pempera, J.: New block properties for the permutation flow shop problem with application in tabu search. J. Oper. Res. Soc. 52, 210–220 (2001)
Grabowski, J., Wodecki, M.: A very fast tabu search algorithm for the permutation flowshop problem with makespan criterion. Comput. Oper. Res. 31, 1891–1909 (2004)
Gupta, J.N.D.: An improved combinatorial algorithm for the flowshop problem. Oper. Res. 19, 1753–1758 (1971a)
Gupta, J.N.D.: A functional heuristic algorithm for the flowshop scheduling problem. Oper. Res. Quart. 22, 39–48 (1971b)
Gupta, J.N.D.: Heuristic algorithms for multistage flow shop problem. AIIE T. 4, 11–18 (1972)
Gupta, J.N.D., Palanimuthy, N., Chen, C.L.: Designing a tabu search algorithm for the two-stage flowshop problem with secondary criterion. Prod. Plan Control 10, 251–265 (1999)
Gupta, J.N.D., Stafford Jr., E.F.: Flowshop scheduling research after five decades. Eur. J. Oper. Res. 169, 699–711 (2006)
Haq, A.N., Saravanan, M., Vivekraj, A.R., Prasad, T.: A scatter search approach for general flowshop scheduling problem. Int. J. Adv. Manuf. Technol. 31, 731–736 (2007)
Haouari, M., Ladhari, T.: A branch-and-bound-based local search method for the flowshop problem. J. Oper. Res. Soc. 54(10), 1076–1084 (2003)
Hasija, S., Rajendran, C.: Scheduling in flowshops to minimize total tardiness of jobs. Int. J. Prod. Res. 42(11), 2289–2301 (2004)
Ho, J.C.: Flowshop sequencing with mean flowtime objective. Eur. J. Oper. Res. 81, 571–578 (1995)
Hundal, T.S., Rajgopal, J.: An extension of Palmer's heuristic for flowshop scheduling problem. Int. J. Prod. Res. 26, 1119–1124 (1988)

Ignall, E., Schrage, L.: Application of the branch and bound technique to some flow shop scheduling problems. Oper. Res. 13(3), 400–412 (1965)

Ishibuchi, H., Misaki, S., Tanaka, H.: Modified Simulated Annealing Algorithms for the Flow-Shop Sequencing Problem. Eur. J. Oper. Res. 81(2), 388–398 (1995)

Iyer, S.K., Saxena, B.: Improved genetic algorithm for the permutation flowshop scheduling problem. Comput. Oper. Res. 31(4), 593–606 (2004)

Jarboui, B., Ibrahim, S., Siarry, P., Rebai, A.: A combinatorial particle swarm optimization for solving permutation flowshop problems. Comput. Ind. Eng. 54, 526–538 (2008)

Jin, F., Song, S., Wu, C.: An improved version of the NEH algorithm and its application to large-scale flow-shop scheduling problems. IIE. Trans. 39, 229–234 (2007)

Johnson, S.M.: Optimal two- and three-stage production schedules with setup times included. Nav. Res. Logist. Q. 1(1), 61–68 (1954)

Kalczynski, P.J., Kamburowski, J.: An improved NEH heuristic to minimize makespan in permutation flowshops. Comput. Oper. Res. 35, 3001–3008 (2008)

Kennedy, J., Eberhart, R.C.: Particle swarm optimization. In: Proceedings of IEEE international conference on neural networks, pp. 1942–1948. Piscataway, New Jersey (1995)

Kim, Y.D.: Heuristics for flowshop scheduling problems minimizing mean tardiness. J. Oper. Res. Soc. 44(1), 19–28 (1993)

Kim, Y.D.: Minimizing total tardiness in permutation flowshops. Eur. J. Oper. Res. 85, 541–555 (1995)

Ladhari, T., Haouari, M.: A computational study of the permutation flowshop problem based on a tight lower bound. Comput. Oper. Res. 32, 1831–1847 (2005)

Laha, D., Chakraborty, U.K.: An efficient stochastic hybrid heuristic for flowshop scheduling. Eng. Appl. Artif. Intel. 20, 851–856 (2007)

Laha, D., Chakraborty, U.K.: An efficient heuristic approach to total flowtime minimization in permutation flowshop scheduling. Int. J. Adv. Manuf. Technol. 38, 1018–1025 (2008)

Lai, T.C.: The note on heuristics of flowshop scheduling. Oper. Res. 44, 648–652 (1996)

Lian, Z., Gu, X., Jiao, B.: A similar particle swarm optimization algorithm for permutation flowshop scheduling to minimize makespan. Appl. Math. Comput. 175, 773–785 (2006)

Lian, Z., Gu, X., Jiao, B.: A novel particle swarm optimization algorithm for permutation flowshop scheduling to minimize makespan. Chaos Soliton Fract. 35, 851–861 (2008)

Liao, C.J., Tseng, C.T., Luarn, P.: A discrete version of particle swarm optimization for flowshop scheduling problems. Comput. Oper. Res. 34, 3099–3111 (2007)

Liu, J., Reeves, C.R.: Constructive and composite heuristic solutions to the $P//\sum C_i$ scheduling problem. Eur. J. Oper. Res. 132, 439–452 (2001)

Lourenço, H.L.: Sevast' janos's algorithms for the flowshop scheduling problem. Eur. J. Oper. Res. 91, 176–189 (1996)

Low, C., Yeh, J.Y., Huang, K.I.: A robust simulated annealing heuristic for flowshop scheduling problems. Int. J. Adv. Manuf. Technol. 23, 762–767 (2004)

Manne, A.S.: On the jobshop scheduling problem. Oper. Res. 8(2), 219–223 (1960)

McMahon, G.B., Burton, B.: Flowshop scheduling with branch and bound method. Oper. Res. 15, 473–481 (1967)

Moccellin, J.V.: A new heuristic method for the permutation flowshop scheduling problem. J. Oper. Res. Soc. 46(7), 883–886 (1995)

Morton, T.E., Pentico, D.W.: Heuristic scheduling systems with applications to production systems and project management. John Wiley & Sons Inc., New York (1993)

Murata, T., Ishibuchi, H., Tanaka, H.: Genetic algorithms for flowshop scheduling problems. Comput. Ind. Eng. 30(4), 1061–1071 (1996)

Nagano, M.S., Moccellin, J.V.: A high quality solution constructive heuristic for flow shop sequencing. J. Oper. Res. Soc. 53(12), 1374–1379 (2002)

Nagano, M.S., Ruiz, R., Lorena, L.A.N.: A constructive genetic algorithm for permutation flowshop scheduling. Comput. Ind. Eng. 55, 195–207 (2008)

Nawaz, M., Enscore Jr., E., Ham, I.: A heuristic algorithm for the m-machine n-job flowshop sequencing problem. Omega 11, 91–95 (1983)

Nearchou, A.C.: A novel metaheuristic approach for the flowshop scheduling problem. Eng. Appl. Artif. Intel. 17, 289–300 (2004a)

Nearchou, A.C.: Flowshop sequencing using hybrid simulated annealing. J. Intell. Manuf. 15, 317–328 (2004b)

Neppalli, V.R., Chen, C.L., Aljaber, N.J.: An effective heuristic for the flowshop problem with weighted tardiness. In: Proceedings of the 3rd Industrial Engineering Research Conference, pp. 634–638 (1994)

Nowicki, E., Smutnicki, C.: A fast tabu search algorithm for the permutation flowshop problem. Eur. J. Oper. Res. 91, 160–175 (1996)

Nowicki, E., Smutnicki, C.: Some aspects of scatter search in the flowshop problem. Eur. J. Oper. Res. 169, 654–666 (2006)

Ogbu, F.A., Smith, D.K.: Simulated annealing for the permutation flowshop problem. Omega 19, 64–67 (1991)

Onwubolu, G.C., Mutingi, M.: Genetic algorithm for minimizing tardiness in flowshop scheduling. Prod. Plan Control 10(5), 462–471 (1999)

Onwubolu, G., Davendra, D.: Scheduling flowshops using differential evolution algorithm. Eur. J. Oper. Res. 171, 674–692 (2006)

Osman, I.H., Potts, C.N.: Simulated annealing for permutation flowshop scheduling. Omega 17(6), 551–557 (1989)

Ow, P.S.: Focused scheduling in proportionate flowshops. Manage. Sci. 31(7), 852–869 (1985)

Page, E.S.: An approach to the scheduling of jobs on machines. J. Roy. Stat. Soc. B. Met. 23(2), 484–492 (1961)

Palmer, D.S.: Sequencing jobs through a multi-stage process in the minimum total time – a quick method of obtaining near optimum. J. Oper. Res. Soc. 16, 101–107 (1965)

Pan, Q.K., Tasgetiren, M.F., Liang, Y.C.: A discrete differential evolution algorithm for the permutation flowshop scheduling problem. Comput. Ind. Eng. (2008) doi:10.1016/j.cie.2008.03.003

Pinedo, M.: Scheduling: Theory, algorithms, and systems, 2nd edn. Prentice-Hall Inc., New Jersey (2002)

Pinedo, M.L.: Planning and scheduling in manufacturing and services. Springer Science + Business Media, Inc., New York (2005)

Prabhaharan, G., Shahul Hamid Khan, B., Rakesh, L.: Implementation of grasp in flow shop scheduling. Int. J. Adv. Manuf. Technol. 30, 1126–1131 (2006)

Qian, B., Wang, L., Hu, R., Wang, W.L., Huang, D.X., Wang, X.: A hybrid differential evolution method for permutation flow-shop scheduling. Int. J. Adv. Manuf. Technol. 38, 757–777 (2008)

Qian, B., Wang, L., Huang, D.X., Wang, W.L., Wang, X.: An effective hybrid DE-based algorithm for multi-objective flowshop scheduling with limited buffers. Comput. Oper. Res. 36, 209–233 (2009)

Rad, S., Ruiz, R., Boroojerdian, N.: New high performing heuristics for minimizing makespan in permutation flowshops. Omega 37, 331–345 (2009)

Rajendran, C.: Heuristic algorithm for scheduling in a flowshop to minimize total flowtime. Int. J. Prod. Econ. 29, 65–73 (1993)

Rajendran, C., Chaudhuri, D.: An efficient heuristic approach to the scheduling of jobs in flowshop. Eur. J. Oper. Res. 61(3), 318–325 (1991)

Rajendran, C., Ziegler, H.: An efficient heuristic for scheduling in a flowshop to minimize total weighted flowtime of jobs. Eur. J. Oper. Res. 103, 129–138 (1997)

Rajendran, C., Ziegler, H.: Ant-colony algorithms for flowshop scheduling to minimize makespan/total flowtime of jobs. Eur. J. Oper. Res. 155(2), 426–438 (2004)

Reeves, C.R.: A genetic algorithm for flow shop sequencing. Comput. Oper. Res. 22(1), 5–13 (1995)

Reeves, C.R., Yamada, T.: Genetic algorithms, path relinking and the flow shop sequencing problem. Evol. Comput. 6(1), 230–234 (1998)

Reza Hejazi, S., Saghafian, S.: Flowshop scheduling problems with makespan criterion: A review. Int. J. Prod. Res. 43(14), 2895–2929 (2005)

Ruiz, R., Maroto, C.: A comprehensive review and evaluation of permutation flowshop heuristics. Eur. J. Oper. Res. 165, 479–494 (2005)

Ruiz, R., Maroto, C., Alcaraz, J.: Two new robust genetic algorithms for the flowshop scheduling problem. Omega 34, 461–476 (2006)

Ruiz, R., Stützle, T.: A simple and effective iterated greedy algorithm for the permutation flowshop scheduling problem. Eur. J. Oper. Res. 177, 2033–2049 (2007)

Saravanan, M., Haq, A.N., Vivekraj, A.R., Prasad, T.: Performance evaluation of the scatter search method for permutation flowshop sequencing problems. Int. J. Adv. Manuf. Technol. 37, 1200–1208 (2008)

Šeda, M.: Mathematical models of flow shop and job shop scheduling problems. Int. J. AM&CS 4(4), 241–246 (2007)

Smith, M.L., Dudek, R.A.: A general algorithm for solution of the n-job m-machine sequencing problem of the flowshop. Oper. Res. 15, 71–82 (1967)

Solimanpur, M., Vrat, P., Shankar, R.: A neurotabu search heuristic for the flow shop scheduling problem. Comput. Oper. Res. 31, 2151–2164 (2004)

Stinson, J.P., Smith, A.W.: A heuristic programming procedure for sequencing the static flowshop. Int. J. Prod. Res. 20, 753–764 (1982)

Storn, R., Price, K.: Differential evolution- A simple and efficient adaptive scheme for global optimization over continuous spaces, Technical Report TR-95-012, ICSI (1995)

Stützle, T.: An ant approach to the flow shop problem. In: Proceedings of the 6th European Congress on Intelligent Techniques and Soft Computing (EUFIT 1998), Verlag Mainz, Wissenschaftsverlag, Aachen, Germany, pp. 1560–1564 (1998a)

Stützle, T.: Applying iterated local search to the permutation flow shop problem. Technical Report, AIDA-98-04, Darmstadt University of Technology, Computer Science Department, Intellectics Group (1998b)

Szwarc, W.: Optimal elimination methods in mxn flowshop scheduling problem. Oper. Res. 21, 1250–1259 (1973)

Taillard, E.: Some efficient heuristic methods for the flowshop sequencing problem. Eur. J. Oper. Res. 47(1), 65–74 (1990)

Tang, L., Liu, J.: A modified genetic algorithm for the flow shop sequencing problem to minimize mean flow time. J. Intell. Manuf. 13, 61–67 (2002)

Tasgetiren, M.F., Liang, Y.C., Sevkli, M., Gencyilmaz, G.: Differential evolution algorithm for permutation flowshop sequencing problem with makespan criterion. In: Proceedings of the 4th International Symposium on Intelligent Manufacturing Systems (IMS 2004), Sakarya, Turkey, pp. 442–452 (2004)

Tasgetiren, M.F., Liang, Y.C., Sevkli, M., Gencyilmaz, G.: A particle swarm optimization algorithm for makespan and total flowtime minimization in the permutation flowshop sequencing problem. Eur. J. Oper. Res. 177, 1930–1947 (2007)

T'kindt, V., Billaut, J.C.: Multicriteria scheduling: Theory, models and algorithms. Springer, Berlin (2002)

Vallada, E., Ruiz, R.: Cooperative metaheuristics for the permutation flowshop scheduling problem. Eur. J. Oper. Res. (2008) doi:10.1016/j.ejor.2007.11.049

Wang, C., Chu, C., Proth, J.M.: Heuristic approaches for n/m/F/∑Ci scheduling problem. Eur. J. Oper. Res. 96, 636–644 (1997)

Wang, L., Zhang, L.: Determining optimal combination of genetic operators for flow shop scheduling. Int. J. Adv. Manuf. Technol. 30, 302–308 (2006)

Wang, L., Zhang, L., Zheng, D.Z.: A class of order-based genetic algorithm for flow shop scheduling. Int. J. Adv. Manuf. Technol. 22, 828–835 (2003)

Wang, L., Zhang, L., Zheng, D.Z.: The ordinal optimisation of genetic control parameters for flow shop scheduling. Int. J. Adv. Manuf. Technol. 23, 812–819 (2004)

Wang, L., Zheng, D.Z.: An effective hybrid heuristic for flow shop scheduling. Int. J. Adv. Manuf. Technol. 21(1), 38–44 (2003)

Werner, F.: On the heuristic solution of the permutation flow shop problem by path algorithms. Comput. Oper. Res. 20(7), 707–722 (1993)

Widmer, M., Hertz, A.: A new heuristic method for the flowshop sequencing problem. Eur. J. Oper. Res. 41, 186–193 (1989)

Woo, H.S., Yim, D.S.: A heuristic algorithm for mean total flowtime objective in flowshop scheduling. Comput. Oper. Res. 25(3), 175–182 (1998)

Yagmahan, B., Yenisey, M.M.: Ant colony optimization for multi-objective flow shop scheduling problem. Comput. Ind. Eng. 54(3), 411–420 (2008)

Yamada, T., Reeves, C.: Solving the Csum permutation flowshop scheduling problem by genetic local search. In: Proceedings of the 1998 IEE International Conference on Evolutionary Computing, pp. 230–234 (1998)

Ying, K.C., Liao, C.J.: An ant colony system for permutation flowshop sequencing. Comput. Oper. Res. 31(5), 791–801 (2004)

Yong, Z., Sannomiya, N.: An improvement genetic algorithm by search space reductions in solving large-scale flowshop problems. Trans. IEE. Japan 121-C(6), 1010–1015 (2001)

Zegordi, S.H., Itoh, K., Enkawa, T.: Minimizing makespan for flow shop scheduling by combining simulated with sequencing knowledge. Eur. J. Oper. Res. 85(3), 515–531 (1995)

Zhang, C., Sun, J., Zhu, X., Yang, Q.: An improved particle swarm optimization algorithm for flowshop scheduling problem. Inform Process Lett. (2008) doi:10.1016/j.ipl.2008.05.010

Zhang, L., Wang, L., Zheng, D.Z.: An adaptive genetic algorithm with multiple operators for flowshop scheduling. Int. J. Adv. Manuf. Technol. 27, 580–587 (2006)

Ziaee, M., Sadjadi, S.J.: Mixed binary integer programming formulations for the flow shop scheduling problems. A case study: ISD projects scheduling. Appl. Math. Comput. 185, 218–228 (2007)

Metaheuristics for Common due Date Total Earliness and Tardiness Single Machine Scheduling Problem

M. Fatih Tasgetiren[1], Quan-Ke Pan[2], P.N. Suganthan[3], Yun-Chia Liang[4], and Tay Jin Chua[5]

[1] Department of Operations Management and Business Statistics, Sultan Qaboos University, Muscat, Sultanate of Oman
mfatih@squ.edu.om
[2] College of Computer Science, Liaocheng University, Shandong Province, 252059, P.R.C
qkpan@lcu.edu.cn
[3] School of Electrical and Electronic Engineering, Nanyang Technological University, Singapore, 639798
epnsugan@ntu.edu.sg
[4] Department of Industrial Engineering and Management, Yuan Ze University, Taoyuan County, Taiwan, R.O.C
ycliang@saturn.yzu.edu.tw
[5] Singapore Institute of Manufacturing Technology, Singapore, 638075
tjchua@SIMTech.a-star.edu.sg

Summary. In this chapter, metaheuristic algorithms, namely, a binary particle swarm optimization, a discrete particle swarm optimization, and a discrete differential evolution algorithm, are presented to solve the common due date total earliness and tardiness single machine scheduling problem. Novel discrete versions of both particle swarm optimization and differential evolution algorithms are developed to be applied to all types of combinatorial optimization problems in the literature. The metaheuristic algorithms presented in this chapter employ a binary solution representation, which is very common in the literature in terms of determining the early and tardy job sets so as to implicitly tackle the problem. In addition, a constructive heuristic algorithm, here we call it MHRM, is developed to solve the problem. Together with the MHRM heuristic, a new binary swap mutation operator, here we call it BSWAP, is employed in the metaheuristic algorithms. Furthermore, metaheuristic algorithms are hybridized with a simple local search based on the BSWAP mutation operator to further improve the solution quality. The proposed metaheuristic algorithms are tested on 280 benchmark instances ranging from 10 to 1000 jobs from the OR Library. The computational results show that the metaheuristic algorithms with a simple local search generated either better or competitive results than those of all the existing approaches in the literature.

1 Introduction

Among all types of scheduling objectives, earliness and tardiness penalties are considered the most common and important ones in the Just-in-Time (JIT) environment. In a JIT production system, a job completing earlier than its due date incurs an earliness penalty (inventory cost) whereas a job completing later leads to a tardiness penalty (imposed by customers). If the optimal sequence cannot be constructed without

U.K. Chakraborty (Ed.): Comput. Intel. in Flow Shop and Job Shop Sched., SCI 230, pp. 301–340.
springerlink.com

considering the value of the due date, the common due date is called restrictive. In a single machine scheduling problem with common due date, all jobs are available to be processed at time zero. Each job j has a processing time p_j and a common due date d. Preemption is not allowed and the objective is to sequence jobs with a restrictive common due date such that the sum of weighted earliness and tardiness penalties is minimized. That is,

$$f(S)=\sum_{j=1}^{n}\left(\alpha_j E_j+\beta_j T_j\right) \tag{1}$$

When the job j completes its operation before the due date, its earliness is given by $E_j=\max(0,d-C_j)$, where C_j is the completion time of the job j. On the other hand, if the job finishes its operation after the due date, its tardiness is calculated by $T_j=\max(0,C_j-d)$. Earliness and tardiness penalties are also given by α_j and β_j, respectively. For convenience, S^E denotes the set of jobs completed before or at the due date whereas S^T represents the set of jobs completed after the due date.

It is well-known that for the case of restrictive common due date with general penalties, there exists an optimal schedule with the following properties:

1. No idle times are inserted between consecutive jobs [1]
2. The schedule is V-Shaped. In other words, jobs that are completed at or before the due date are sequenced in non-increasing order of the ratio p_j/α_j. On the other hand, jobs whose processing starts at or after the due date are sequenced in non-decreasing order of the ratio p_j/β_j [2]. Note that there might be a straddling job, which is started before the due date and completed after the due date [3].
3. There is an optimal schedule in which either the processing of the first job starts at time zero or one job is completed at the due date [4].

The complexity of the restrictive common due-date problem is proved to be NP-complete in the ordinary sense [5]. Therefore, only small-sized instances of the single machine scheduling problem with a common due date may be solved to optimality with reasonable computational time using exact algorithms. When the problem size increases, the computational time of exact methods grows explosively. On the other hand, heuristic algorithms require generally acceptable time and memory requirements to reach a near-optimal or optimal solution. In past decades, most research focused on developing metaheuristic algorithms such as tabu search (TS) [6, 7, 8], genetic algorithm (GA) [8, 9, 10], differential evolution (DE) [11], evolutionary strategy (ES), simulated annealing (SA) and threshold accepting (TA) [12]. Hybridization of heuristics is another trend of research track. For example, M'Hallah [13] proposed a hybrid algorithm that combines GA, hill climbing (HC), dispatching rules, and SA, and Hino et al. [8] proposed two hybrid methods HGT and HTG by combining TS, GA, and an efficient constructive heuristic HRM. Lastly, some effective heuristics are developed recently. Hendel & Sourd [14] employed neighborhood search based on the adjacent pairwise interchange (API) method, and Lin et al. [15] proposed a sequential exchange approach. In this study, following the HRM heuristic [8], we also present a modified version of the HRM heuristic, here we call it MHRM heuristic, by taking into account of the drawbacks in the HRM heuristic.

PSO is one of the latest evolutionary metaheuristic methods, which receives growing interest from the researchers in the literature. It is based on the metaphor of social interaction and communication such as bird flocking and fish schooling. PSO was first introduced to optimize various continuous nonlinear functions by Eberhart & Kennedy [16]. Distinctly different from other evolutionary-type methods such as GA and ES, PSO algorithms maintain the members of the entire population through the search procedure without considering the survival of fitness. In other words, selection is not employed in PSO algorithms. In a PSO algorithm, each individual is called a *particle*, and each particle moves around in the multi-dimensional search space with a velocity constantly updated by the particle's own experience, the experience of the particle's neighbors, or the experience of the whole swarm. That is, the search information is socially shared among particles to direct the population towards the best position in the search space. The comprehensive surveys of the PSO algorithms and applications can be found in [17, 18].

As well known, the original PSO is designed for solving the real-valued optimization problems. The PSO algorithm has already been extended to be applied to binary/discrete optimization problems. To cope with the binary variables, Kennedy and Eberhart [19] designed the velocity as a probability to determine whether or not the value of the positions x_{ij} will be 0 or 1. They squashed the velocity v_{ij} by using the sigmoid function $s(v_{ij}) = 1/(1+\exp(-v_{ij}))$ while the velocity is calculated with the traditional equation. If a random number within [0,1] is less than $s(v_{ij})$ then x_{ij} is set to 1, otherwise it is set to 0. The binary version of PSO outperformed several versions of GAs in all tested problems.

On the other hand, differential evolution (DE) is also one of the latest evolutionary optimization methods proposed by Storn & Price [20]. Like other evolutionary-type algorithms, DE is a population-based and stochastic global optimizer. In a DE algorithm, candidate solutions are represented by chromosomes based on floating-point numbers. In the mutation process of a DE algorithm, the weighted difference between two population members is added to a third member to generate a mutated solution. Then, a crossover operator follows to combine the mutated solution with the target solution so as to generate a trial solution. Thereafter, a selection operator is applied to compare the fitness function value of both competing solutions, namely, target and trial solutions to determine who can survive for the next generation. Since DE was first introduced to solve the Chebychev polynomial fitting problem by Storn & Price [20, 21], it has been successfully applied to a variety of applications that can be found in Corne et al. [22], Lampinen [23], Babu & Onwubolu [24], Price et al. [25], and Chakraborty [26].

The applications of PSO and DE on combinatorial optimization problems are still limited, but the past experiences of successfully applying PSO and DE algorithms to combinatorial problems in the literature [27, 28, 29, 30, 31, 32, 33, 34, 35] have shown the promising of PSO and DE on scheduling problems. Recently, the authors have also introduced a new and novel discrete version of the differential evolution algorithm in [36, 37], which is based on a discrete domain exploiting the basic features of its continuous counterpart. In this chapter, the discrete particle swarm algorithm and the discrete differential algorithm are given in very much detail, especially for their pure performance with and without a local search. We also show that a simple binary PSO of of Kennedy & Eberhart [19] can solve the problem on hand very efficiently when embedded with a local search. Furthermore, the MHRM heuristic is given in detail as to how it differs from its counterpart HRM heuristic with examples. The performance of the newly proposed binary mutation operator, BSWAP, is evaluated in detail too. Finally, a very detailed design of experiments is conducted to determine the parameters of the metaheuristics proposed. To sum up, this research presents discrete particle swarm optimization (DPSO) and discrete differential evolution (DDE) algorithms in detail as well as the binary PSO algorithm, here we denote it BPSO, of Kennedy and Eberhart [19] to solve the single machine total earliness and tardiness penalties with a common due date (E/T) problem.

The remainder of the chapter is organized as follows. Section 2 introduces the discrete particle swarm optimization together with the standard BPSO. The discrete differential evolution, local search employed, and the MHRM heuristic are discussed in Section 3. Section 4 presents the design of experiments for parameter setting, and the computational results over benchmark problems are discussed in Section 5. Finally, Section 6 summarizes the concluding remarks.

2 Discrete Particle Swarm Optimization Algorithm

In the standard PSO algorithm, all particles have their position, velocity, and fitness values. Particles fly through the n-dimensional space by learning from the historical information emerged from the swarm population. For this reason, particles are inclined to fly towards better search area over the course of evolution. Let NP denote the swarm size represented as $X^t = \left[X_1^t, X_2^t, \ldots, X_{NP}^t\right]$. Then each particle in the swarm population has the following attributes: A current position represented as $X_i^t = \left[x_{i1}^t, x_{i2}^t, \ldots, x_{in}^t\right]$; a current velocity represented as $V_i^t = \left[v_{i1}^t, v_{i2}^t, \ldots, v_{in}^t\right]$; a current personal best position represented as $P_i^t = \left[p_{i1}^t, p_{i2}^t, \ldots, p_{in}^t\right]$; and a current global best position represented as $G^t = \left[g_1^t, g_2^t, \ldots, g_n^t\right]$. Assuming that the function f is to be minimized, the current velocity of the jth dimension of the ith particle is updated as follows.

$$v_{ij}^t = w^{t-1} v_{ij}^{t-1} + c_1 r_1 \left(p_{ij}^{t-1} - x_{ij}^{t-1}\right) + c_2 r_2 \left(g_j^{t-1} - x_{ij}^{t-1}\right) \tag{2}$$

where w^t is the inertia weight which is a parameter to control the impact of the previous velocities on the current velocity; c_1 and c_2 are acceleration coefficients and r_1 and r_2 are uniform random numbers between [0,1]. The current position of the jth dimension of the ith particle is updated using the previous position and current velocity of the particle as follows:

$$x_{ij}^t = x_{ij}^{t-1} + v_{ij}^t \tag{3}$$

The personal best position of each particle is updated using

$$P_i^t = \begin{cases} P_i^{t-1} & if \quad f(X_i^t) \ge f(P_i^{t-1}) \\ X_i^t & if \quad f(X_i^t) < f(P_i^{t-1}) \end{cases} \tag{4}$$

Finally, the global best position found so far in the swarm population is obtained as

$$G^t = \begin{cases} \arg\min_{P_i^t} f(P_i^t) & if \quad \min f(P_i^t) < f(G^{t-1}) \\ G^{t-1} & else \quad 1 \le i \le NP \end{cases} \tag{5}$$

Regarding the BPSO algorithm, we follow Kennedy and Eberhart [19]. In the BPSO algorithm, the sigmoid function is used to force the real values between 0 and 1, and the velocities are restricted to the range of $\lfloor v_{\min}, v_{\max} \rfloor$. Once velocities are updated with the traditional equation (2), the sigmoid function is used to squash them to be within [0,1] as follows:

$$s(v_{ij}) = 1/(1+\exp(-v_{ij})) \tag{6}$$

Finally particles are updated such that:

$$x_{ij} = \begin{cases} 1 & if \quad r \le s(v_{ij}) \\ 0 & otherwise \end{cases} \tag{7}$$

where r is a uniform random number within 0 and 1. If r is less than $s(v_{ij})$, then position of the jth dimension of the ith particle is assigned to 1, otherwise it is assigned to 0.

Standard PSO equations cannot be used to generate discrete values since positions are real-valued. Pan et al. [30, 31, 33] have presented a DPSO optimization algorithm to tackle the discrete spaces, where particles are updated by using the temporary particles λ_i and δ_i as follows:

$$\lambda_i^t = \begin{cases} F_k(X_i^{t-1}) & if \quad r < w \\ X_i^{t-1} & otherwise \end{cases} \tag{8}$$

where w is the mutation probability, r is a random number between [0,1], and F_k is the mutation operator F_k with the mutation strength k. A uniform random number r is generated between 0 and 1. If r is less than the mutation probability w, then the mutation operator is applied to the particle X_i^{t-1} at the previous generation t-1 in order to produce the temporary particle by $\lambda_i^t = F_k(X_i^{t-1})$, otherwise the temporary particle is taken as $\lambda_i^t = X_i^{t-1}$.

$$\delta_i^t = \begin{cases} CR(\lambda_i^{t-1}, P_i^{t-1}) & if \quad r < c_1 \\ \lambda_i^t & Otherwise \end{cases} \tag{9}$$

where c_1 is the crossover probability, r is a random number between [0,1], and P_i^{t-1} is the personal best solution at the generation *t*-1. CR represents the crossover operator with the probability of c_1. Note that λ_i^t and P_i^{t-1} will be the first and second parents for the crossover operator, respectively. It results either in $\delta_i^t = CR\left(\lambda_i^t, P_i^{t-1}\right)$ or in $\delta_i^t = \lambda_i^t$ depending on the choice of a uniform random number.

$$X_i^t = \begin{cases} CR\left(\delta_i^{t-1}, G^{t-1}\right) & if \quad r < c_2 \\ \delta_i^{t-1} & Otherwise \end{cases} \tag{10}$$

where c_2 is the crossover probability, r is a random number between [0,1], G^{t-1} is the global best solution at the generation *t*-1. Again, CR represents the crossover operator with the probability of c_2. Note that δ_i^t and G^{t-1} will be the first and second parents for the crossover operator, respectively. It results either in $X_i^t = CR\left(\delta_i^t, G^{t-1}\right)$ or in $X_i^t = \delta_i^t$ depending on the choice of a uniform random number.

For the DPSO algorithm, the *gbest* (global neighborhood) model of Kennedy et al. [17] was followed. The basic idea behind the DPSO algorithm is to exploit the features of its continuous counterpart. Particles in the population are updated in such a way that they are guided to gather some information from their personal best solutions and the global best solution. Therefore, all population is ultimately directed towards the global best and personal best solutions during the search space without any selection procedure.

3 Discrete Differential Evolution

Currently, there exist several mutation variations of DE. The *DE/rand/1/bin* scheme of Storn & Price [20, 21] is presented below. The DE algorithm starts with initializing the target population in the size of *NP*. Each individual has an *n*-dimentional vector with parameter values determined randomly and uniformly between predefined search range. To generate a mutant individual, DE mutates vectors from the target population by adding the weighted difference between two randomly selected target population members to a third member as follows:

$$v_{ij}^t = x_{aj}^{t-1} + F\left(x_{bj}^{t-1} - x_{cj}^{t-1}\right) \tag{11}$$

where *a*, *b*, and *c* are three randomly chosen individuals from the target population such that $\left(a \neq b \neq c \in \left(1,.., NP\right)\right)$. $F > 0$ is a mutation scale factor which affects the differential variation between two individuals. Following the mutation phase, the crossover operator is applied to obtain the trial individual such that:

$$u_{ij}^t = \begin{cases} v_{ij}^t & if \quad r_{ij}^t \leq CR \quad or \quad j = D_j \\ x_{ij}^t & otherwise \end{cases} \tag{12}$$

where the D_j refers to a randomly chosen dimension $(j = 1,..,n)$, which is used to ensure that at least one parameter of each trial individual u_{ij}^t differs from its counterpart in the previous generation u_{ij}^{t-1}. *CR* is a user-defined crossover constant in the range [0,1], and r_{ij}^t is a uniform random number between 0 and 1. In other words, the trial individual is made up with some parameters of mutant individual, or at least one of the parameters randomly selected, and some other parameters of the target individual.

To decide whether or not the trial individual U_i^t should be a member of the target population for the next generation, it is compared to its counterpart target individual X_i^{t-1} at the previous generation. The selection is based on the survival of the fitness among the trial population and target population such that:

$$X_i^t = \begin{cases} U_i^t & \text{if} \quad f(U_i^t) \le f(X_i^{t-1}) \\ X_i^{t-1} & \text{otherwise} \end{cases} \tag{13}$$

Again the standard DE equations cannot be used to generate discrete values since positions are real-valued. Instead we propose a new and novel DDE algorithm whose solutions are based on discrete/binary values and therefore can be applied to discrete/binary combinatorial optimization problems. In the DDE algorithm for the E/T problem, the target population is constructed based on the binary 0-1 values as represented by $X_i = [X_1, X_2,,, X_{NP}]$. For the mutant population can be obtained as follows: the following equations can be used:

$$V_i^t = \begin{cases} F_k(G^{t-1}) & \text{if} \quad r < Pm \\ G^{t-1} & \text{else} \end{cases} \tag{14}$$

where G^{t-1} is the best solution found so far in the population; Pm is the mutation probability; and F_k is the mutation operator with the mutation strength of *k*. A uniform random number *r* is generated between [0,1]. If *r* is less than Pm then the mutation operator is applied to generate the mutant individual $V_i^t = F_k(G^{t-1})$, otherwise the global best solution is kept as the mutant individual $V_i^t = G^{t-1}$. Following the mutation phase, the trial individual is obtained such that:

$$U_i^t = \begin{cases} CR(X_i^{t-1}, V_i^t) & \text{if} \quad r < Pc \\ V_i^t & \text{else} \end{cases} \tag{15}$$

where CR is the crossover operator, and Pc is the crossover probability. In other words, if a uniform random number *r* is less than the crossover probability Pc, then the crossover operator is applied to generate the trial individual $U_i^t = CR(X_i^{t-1}, V_i^t)$. Otherwise the trial individual is chosen as $U_i^t = V_i^t$. By doing so, the trial individual is made up either from the outcome of mutation operator or from the crossover operator. Finally, the selection is based on the survival of the fitness among the trial population and target population such that:

$$X_i^t = \begin{cases} U_i^t & \text{if } f(U_i^t) \le f(X_i^{t-1}) \\ X_i^{t-1} & \text{otherwise} \end{cases} \tag{16}$$

In the proposed DDE algorithm, the basic idea is to direct the population towards the best solution so far in the population. In both algorithms, k represents the mutation strength. The lower the value of mutation strength k is, the lower the possibility that the algorithm would avoid getting stuck at the local minima. On the other hand, the higher the value of mutation strength k is, the higher the possibility that the algorithm would possess excessive randomness. So care must be taken in the choice of the value of the mutation strength.

3.1 Solution Representation

As mentioned before, a binary solution representation is employed for the problem in all algorithms. In the binary representation, x_{ij}^t, the position or individual value of the jth dimension of the ith particle or individual X_i^t, denotes a job. If $x_{ij}^t = 0$, the job j is said to complete before or at the due date, which belongs to the early job set S^E whereas if $x_{ij}^t = 1$, the job j is said to finish after the due date, which belongs to the tardy job set S^T. Binary solution representation is unique in terms of determining the early job set S^E and the tardy job set S^T. An example of solution representation is shown in Table 1. From Table 1, it is trivial to see that the jobs J_1, J_4 and J_6 belong to the early job set S^E; and the jobs J_2, J_3 and J_5 belong to the tardy job set S^T.

Table 1. Solution representation

j	1	2	3	4	5	6
x_{ij}	0	1	1	0	1	0
S^E	J_1			J_4		J_6
S^T		J_2	J_3		J_5	

3.2 Local Search

In this paper, we present a novel BSWAP mutation operator for all proposed meta-heuristics as well as in the local search algorithm presented. The BSWAP operator consists of two steps:

1. Generate two random integers, u and v, in the range $[1,n]$;
2. if $x_{iu}^t = x_{iv}^t$, then $x_{iu}^t = (x_{iu}^t + 1) \bmod 2$;
   else $x_{iu}^t = (x_{iu}^t + 1) \bmod 2$ and $x_{iv}^t = (x_{iv}^t + 1) \bmod 2$.

The main feature of the BSWAP mutation operator is to provide a balance between the early and tardy job sets in such a way that when a solution is determined by an early/tardy job set, the first part of the BSWAP mutation operator is possibly to find two jobs from the same set and assigning one of them to the early/tardy jobs or vice versa. On the other hand, if a solution is relatively balanced with the early and tardy jobs, the BSWAP mutation operator is more likely to find two jobs, one belonging to the early job set and the other belonging to the tardy job set, then swapping them from the early to tardy job set or vice versa.

After applying the BPSO, DPSO and DDE operators, the early job set S^E and the tardy job set S^T are determined from the binary representation. Then every fitness calculation follows the second property of optimality conditions. In other words, the V-Shaped schedule is constructed where jobs completed at or before the due dates are sequenced in non-increasing order of the ratio p_j / α_j whereas jobs whose processing starts at or after the due date are sequenced in non-decreasing order of the ratio p_j / β_j. Note that the set S^T might contain a straddling job. If there is a straddling job, the first job in the early job set S^E is started at time zero. After completing the last job of the early job set S^E, the straddling job and the jobs in the tardy job set S^T are sequenced. On the other hand, if there is no straddling job, the completion time of the last job in the early job set S^E is matched with the due date and the processing in the tardy job set S^T is followed immediately.

The local search in this study was based on the simple BSWAP neighborhood. It should be noted that the following local search was applied to the global best solution, G^t, at each iteration *t*. The pseudo code of the local search is given in Figure.1.

```
Procedure LocalSearch(G)
s:=perturbation(G)
for i:=1 to loopsize do
   flag:=true;
   while (flag=true) do
        s1 :=BSWAP(s);
        if f(s1) ≤ f(s) then
           s := s1;
        else
           flag:=false;
        endif
   endwhile
 endfor
 if f(s) ≤ f(G) then
   G:=s;
 else
   G:=G;
 endif
 return G
end
```

Fig. 1. Local Search Employed

In the local search algorithm, s refers to the perturbed global best or the best so far solution G^t at each generation t. That is, the global best or best so far solution is perturbed by swapping two jobs randomly; one from the tardy set S^T, and another from the early set S^E. Then the BSWAP operator was applied to the perturbed solution s. The size of the local search was carefully set to $loopsize = \min(30n, 6000)$ in order to obtain comparable results fair enough to the existing approaches in terms of CPU time requirements. For convenience, we denote all algorithms with the local search as $BPSO_{LS}$, $DPSO_{LS}$ and DDE_{LS}, respectively from now on throughout the chapter.

3.3 MHRM Heuristic

In a single-machine with n jobs, at most one job can be completed on the due date. For this reason, there will be two sets of jobs: an early job set denoted by S^E where the jobs are completed before or at the due date and a tardy job set denoted by S^T where the jobs are completed after the due date. Consistent with the HRM heuristic [8], the MHRM heuristic consists of: (i) determining these two sets, (ii) constructing a sequence for each set, and (iii) setting the final schedule S as the concatenation of both sequences. In order to ensure that S will satisfy properties (1) and (2), there will be no idle time between consecutive jobs, and the sequences of S^E and S^T will be "\-shaped" and "/-shaped", respectively.

At each generation, the non-scheduled jobs with the maximum ratios p_j / α_j and p_j / β_j are considered for inclusion in one of the two sets. According to the distance between each job's possible completion time and the due date, just one of the jobs is included. Adjustments in the inserted idle time at the beginning of the sequence are also considered. Finally, when all jobs are scheduled, an attempt to satisfy the property (3) is made. Following notation consistent with Hino et al. [8] is employed:

P : set of jobs to be allocated
g : idle time inserted at the beginning of the schedule
S^E : set of jobs completed before or at the due date
S^T : set of jobs completed after the due date
S : schedule representation $S = (g, S^E, S^T)$
e : candidate job for S^E
t : candidate job for S^T
E^e : distance between the possible completion time of the job e and the due date
T^t : distance between the possible completion time of the job t and the due date
d^T : time window available for inserting a job in the set S^T
d^E : time window available for inserting a job in the set S^E
p_j : processing time of job j
H : total processing time, $H = \sum_{j=1}^{n} p_j$

The computational flow of the MHRM heuristic is as follows:

Step 1: Let $P=\{1,2,..,n\}$; $S^E=S^T=\Phi$, $g=\left\lceil \max\left\{0, d-H\times\frac{1}{n}\sum_{j=1}^{n}\left(\frac{\beta_j}{\alpha_j+\beta_j}\right)\right\}\right\rceil$;

$d^E=d-g$ and $d^T=g+H-d$.

Step 2: Set $e=\arg\max_{j\in p}\{p_j/\alpha_j\}$ and $t=\arg\max_{j\in p}\{p_j/\beta_j\}$ (in case of a tie, select the job with the longest p_j).

Step 3: Set $E^e=d^E-p_e$ and $T^t=d^T$.

If $E^e\le 0$ then go to step 5.

If $T^t-p_t\le 0$ then go to step 6.

Step 4: Choose the job to be inserted:

- If $E^e>T^t$ then $S^E=S^E+\{e\}$, $d^E=d^E-p_e$ and $P=P-\{e\}$.
- If $E^e<T^t$ then $S^T=S^T+\{t\}$, $d^T=d^T-p_t$ and $P=P-\{t\}$.
- If $E^e=T^t$ then

  if $\alpha_e>\beta_t$ then $S^T=S^T+\{t\}$, $d^T=d^T-p_t$ and $P=P-\{t\}$;

  else $S^E=S^E+\{e\}$, $d^E=d^E-p_e$ and $P=P-\{e\}$.

Go to step 7.

Step 5: Adjustment of the idle time (end of the space before the due date):

- If $g+E_e<0$ then $S^T=S^T+\{t\}$, $d^T=d^T-p_t$ and $P=P-\{t\}$

  If $d^T<0$ then $g=0$
- Else

  $S^{E'}=S^E, S^{T'}=S^T\cup P, g'=d\sum_{j\in S^{E'}}p_j, S'=(g',S^{E'},S^{T'})$;

  $S^{E''}=S^E+\{e\}, S^{T''}=S^T\cup P-\{e\}, g''=d-\sum_{j\in S^{E''}}p_j$,

  $S''=(g'',S^{E''},S^{T''})$.

  If $f(S^{E'})\le f(S^{E''})$ then

  $S^T=S^T+\{t\}, d^E=0, d^T=d^T-p_t+g'-g$, $g=g'$ and $P=P-\{t\}$.

  Else $S^E=S^E+\{e\}, d^E=0$, $d^T=d^T+g''-g$, $g=g''$

  and $P=P-\{e\}$.

Go to step 7.

Step 6: Adjustment of the idle time (end of the space after the due date):

- If $g<T^t$ then $S^E=S^E+\{e\}, d^E=d^E-p_e$ and $P=P-\{e\}$
- Else

  $S^{T'}=S^T$, $S^{E'}=S^E\cup P$, $g'=d-\sum_{j\in S^{E'}}p_j, S'=(g',S^{E'},S^{T'})$;

$S^{T''} = S^T + \{t\}$, $S^{E''} = S^E \cup P - \{t\}$, $g'' = d - \sum_{j \in S^{E''}} p_j$,

$S'' = (g'', S^{E''}, S^{T''})$.

If $f(S') \le f(S'')$ then

$S^E = S^E + \{e\}, d^T = 0, d^E = d^E - p_e + g - g', g = g', P = P - \{e\}$;

Else $S^T = S^T + \{t\}, d^T = 0, d^E = d^E + g - g'', g = g''$, $P = P - \{t\}$.

***Step 7*:** If $P \neq \Phi$ then go to step 2.

***Step 8*:** If there is a straddling job (it must be the last job in S^T), then

- $S^{E'} = S^E, S^{T'} = S^T$, $g' = d - \sum_{j \in S^{E'}} p_j$, $S' = (g', S^{E'}, S^{T'})$.

  Solve $S = (g', S^{E'}, S^{T'})$
- If $f(S') < f(S)$ then $g = g'$. $S = S'$

Step 9: Stop.

As mentioned before, the MHRM heuristic is a modified version of HRM heuristic presented in Hino et al. [8]. The main difference between HRM and MHRM heuristics is due to the calculation of the inserted idle time in Step 1 such that

$$g = \left\lceil \max\left\{0, d - H \times \frac{1}{n} \sum_{j=1}^{n} \left(\frac{\beta_j}{\alpha_j + \beta_j} \right) \right\} \right\rceil \tag{17}$$

In Hino et al. [8], the inserted idle time is calculated by $g = \max\{0, d - 0.5 \times H\}$. Instead, in the MHRM heuristic, the inserted idle times are calculated based on the ratio of $\sum(\beta_j / (\alpha_j + \beta_i))$. By doing so, the inserted idle time completely depends on the particular instance considered to be solved. It implies that if the total tardiness penalty of a particular instance is greater than the total earliness penalty of that instance (i.e., $\sum \beta_j > \sum \alpha_j$), the inserted idle time would be larger for that particular instance. Hence more jobs would be completed before the due date. In other words, more jobs would be early. Since the total tardiness penalty is larger than the total earliness penalty, i.e., $\sum \beta_j > \sum \alpha_j$, the total penalty imposed on the fitness function would be less than the one used in the HRM heuristic.

In addition, the following modification is also made in Step 3. As shown in Figure 2, if the distance between the possible completion time of candidate job t and the due date is less than or equal to zero, both the start time and the completion time of the job t are before or at the due date, i.e., the job t is not a straddling job. In the MHRM algorithm, $T^t - p_t \le 0$ is employed instead of $T^t \le 0$ because $T^t - p_t \le 0$ implies that the job t is a straddling job. In this case, the adjustment of the idle time for the end of the space after the due date through Step 6 should be made. Accordingly, necessary modifications are also made in Steps 5, 6, and 8.

In order to justify the quality of the MHRM heuristic, an example is given in Appendix A by constructing an instance of 10 jobs with earliness and tardiness penalties as well as a common due date.

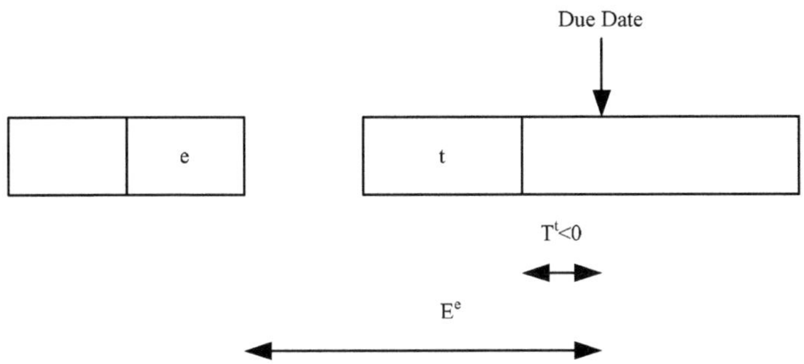

a. End of the space after the due date in the HRM heuristic

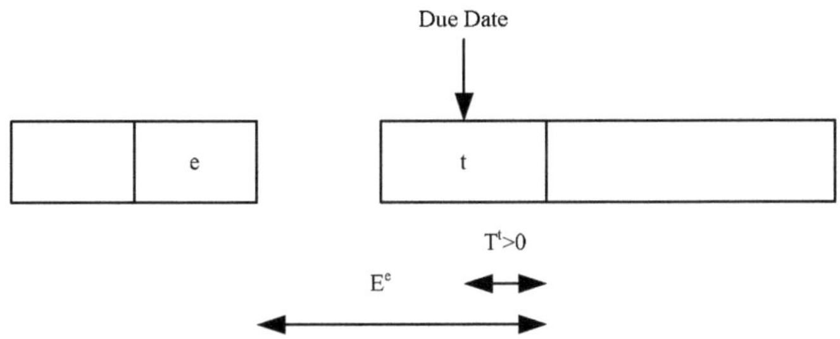

b. End of the space after the due date in the MHRM heuristic

Fig. 2. Difference between HRM and MHRM Heuristics

4 Design of Experiments

In this section, we present the Design of Experiments (DOE) approach [38] for parameter setting of the DPSO and DDE algorithms except for the BPSO algorithm for which the parameter setting for it is well-known in the literature. For this reason, we conduct the DOE for only the DPSO and DDE algorithms. we did not conduct the DOE for the $DPSO_{LS}$ and DDE_{LS} algorithms because the parameters given for the local search in Section 4 were quite effective based on our previous experience in Pan et al. [30, 31]. To conduct the initial runs for the DOE, traditional two-cut crossover and BSWAP mutation operators are used in both algorithms. In the DPSO and DDE algorithms, the mutation strength was only one swap of jobs from the early and tardy sets. Regarding the initial population, one of the solutions in the population is constructed with the MHRM heuristic, the rest is constructed randomly. In order to carry out the experiments, we randomly generated the E/T instances following the procedure in Biskup & Feldmann [3] where processing times are uniformly distributed in the range of [1,20], and the earliness and tardiness penalties were generated in the range of [1,10] and [1,15], respectively. The number of jobs, n, is considered as 10, 20, 50, 100, 200, 500 and 1000 whereas the restrictive factor h for determining the common due date, $d = \lfloor h \times \sum_{j=1}^{n} p_j \rfloor$, is considered to be 0.2, 0.4, 0.6

and 0.8. Four instances were generated for each combination of the number of jobs n and the restrictive factor h, thus resulting in $7\times4\times4=112$ problem instances as in Biskup & Feldmann [3]. Note that these instances are different than those in Biskup & Feldmann [3] since different seed numbers are used. However, they come from the same distribution. 112 instances were run for 10 replications for each treatment by the DPSO and DDE algorithms with a CPU time limit of $2\times n$ milliseconds. Setting the time limit with respect to the number of jobs provides the DPSO and DDE algorithms with more computation times as the number of jobs increases. All the experiments for the DOE are conducted on an Intel Pentium IV 3.0 GHz PC with 512 MB memory. The response variable was the average percentage relative deviation for R=1120 replications for each treatment and averaged as follows:

$$\Delta_{avg} = \sum_{i=1}^{R}\left(\frac{(F_i - F_{REF})\times 100}{F_{REF}}\right)/R \qquad (18)$$

where F_i, F_{REF}, and R were the fitness function value generated by each of three algorithms in each run, the reference fitness function value reported in Biskup & Feldmann [3], and R=1120 was the number of replications.

There are four parameters in the DPSO algorithm: population size (A), mutation probability of update equation (B = w), crossover probability (C = c_1), and crossover probability (D = c_2). Each factor has two levels and a full factorial design of $2^4=16$ treatments is employed. On the other hand, There are three parameters in the DDE algorithm: population size (A), mutation probability (B = Pm), crossover probability (C = Pc), and mutation equation (D). All factors have two levels and a general factorial design of $2^3=8$ treatments is employed. The details of the DOE analysis are given in Appendix B and Appendix C, respectively. Final parameter settings after the DOE analysis are given in Tables 2 and 3.

Table 2. Final Parameter setting for DPSO Algorithm

Factors	Levels	Description	Value
A	1	NP=high level	30
B	1	w =high level	0.8
C	-1	c_1 =low level	0.2
D	1	c_2 =high level	0.8

Table 3. Final Parameter Setting For DDE Algorithm

Factors	Levels	Description	Value
A	-1	NP=low level	10
B	1	P_m=high level	0.8
C	1	P_c=high level	0.8

The parameter setting for the BPSO algorithm was well studied in the literature [17]. The population size is taken as 30. Consistent with the literature, the initial inertia weight is taken as $w^0 = 0.9$ and decreased by $w = w^0 \times 0.975$. It was never decreased below 0.4. Acceleration coefficients c_1 and c_2 are taken as 2.0, respectively. Initial velocities are established uniformly within [-4,4]. The positions are randomly assigned to binary values either 0 or 1 with an equal probability in the initial population. Velocities after being updated by equation (2) are restricted to the range $[V_{\min}, V_{\max}] = [-4,4]$ to avoid having floating point error.

5 Computational Results

All the metaheuristic algorithms were coded in Visual C++ and run on an Intel Pentium IV 3.0 GHz PC with 512MB memory. Regarding the parameters of the DPSO and DDE algorithms, they were determined through DOE explained in Section 4. All the metaheuristics were applied to the benchmark problems that Biskup & Feldmann [3] developed a total of 280 instances ranging from 10 to 1000 jobs and restricting the common due date from 0.2 to 0.8 of sum of all processing times. These instances can be downloaded at the OR-Library web site http://www.ms.ic.ac.uk/jeb/orlib/schinfo.html. Ten runs (R=10) were carried out for each problem instance to report the statistics based on the percentage relative deviations (Δ) from the upper bounds in Biskup & Feldmann [3]. Again, Δ_{avg} was computed as follows:

$$\Delta_{avg} = \sum_{i=1}^{R} \left(\frac{(F_i - F_{REF}) \times 100}{F_{REF}} \right) / R \tag{19}$$

where F_i, F_{REF}, and R were the fitness function value generated by each of the three algorithms in each run, the reference upper bounds generated by Biskup & Feldmann [3], and the total number of runs, i.e., the number of runs for each instance times the number of instances for each problem category. In other words, $R = 10 \times 10 = 100$ runs are conducted for each combination of the number of jobs n and restrictive factor h. Note that Biskup & Feldmann [3] provided the optimal solutions for $n = 10$ problem instances. For this reason, we use the optimal solutions as upper bounds in our runs. For convenience, $\Delta_{\min}$, $\Delta_{\max}$, and Δ_{std} denote the minimum, maximum, and standard deviation of percentage relative deviation in fitness function value over R runs, respectively. For the computational effort consideration, t_{min}, t_{max}, t_{avg}, and t_{std} denote the minimum, maximum, average time and the standard deviation until termination of algorithms averaged over R runs in seconds. For the BPSO, DPSO, and DDE algorithms, the maximum number of generations is fixed to 1000. However, the maximum number of generations is fixed to 50 generations and the algorithms are terminated if the global best solution is not improved in 10 consecutive generations for the $BPSO_{LS}$, $DPSO_{LS}$, and DDE_{LS} algorithms.

The DOE presented in Section 4 was basically carried out for the parameter setting of the DPSO and DDE algorithms. Since the BSWAP mutation operator is newly presented and used in the update equations of the DPSO and DDE algorithms in this paper, its performance on the solution quality should be demonstrated. For this purpose, another simple design of experiments was carried out. Factors have been chosen as mutation and crossover operators with each having two levels. The design is shown in Table 4 consisting of $2^2 = 4$ experiments for the DPSO and DDE algorithms. In these experiments, the parameter values obtained during the DOE for the DPSO and DDE algorithms in Section 4 are used. No local search is employed and as mentioned before and our main goal was to see the impact of the BSWAP mutation operator on the solution quality.

Table 4. DOE for DPSO and DDE

	Factors	
Levels	A: Mutation Operator	B: Crossover Operator
-1	Single-Point Mutation	One-Cut Crossover
1	BSWAP Mutation	Two-Cut Crossover

Table 5. Comparison of Mutation and Crossover Operators

	DPSO (Δ)			
Mutation/Crossover	Min	Max	Avg	Std
BSWAP/One-Cut	-2.10	-1.96	-2.04	0.05
BSWAP/Two-Cut	-2.11	-1.96	-2.05	0.05
Single-Point/One-Cut	-2.09	-1.89	-2.01	0.08
Single-Point/Two-Cut	-2.10	-1.86	-2.03	0.08
	DDE (Δ)			
Mutation/Crossover	Min	Max	Avg	Std
BSWAP/One-Cut	-2.14	-2.00	-2.10	0.05
BSWAP/Two-Cut	-2.14	-2.01	-2.11	0.05
Single-Point/One-Cut	-2.05	-1.73	-1.92	0.12
Single-Point/Two-Cut	-2.07	-1.63	-1.91	0.16

Totally, $2^2 = 4$ experiments are run for 10 replications to get the response variable, which is the percentage relative deviation from the upper bounds. The experimental results are summarized in Table 5. From Table 5, it can be seen that the combination of the BSWAP mutation with two-cut crossover operator generated better results than those by other combinations. The impact of BSWAP mutation operator on the solution quality together with two-cut crossover operator was obvious that the minimum relative percentage deviation from the upper bounds of Biskup & Feldmann [3] was improved 2.11 percent and 2.14 percent by the DPSO and DDE algorithms, respectively. For this reason, the BSWAP mutation operator and two-cut crossover operator are employed in the $BPSO_{LS}$, $DPSO_{LS}$, and DDE_{LS} algorithms for the further runs.

Another contribution of this chapter is to present a novel MHRM construction heuristic inspired from the drawbacks of the HRM heuristic presented in Hino et al. [8]. In Section 3.3, the details of the MHRM heuristic were given and the examples of both constructive heuristics are also given in Appendix A. From two examples given, it was shown that the MHRM heuristic was superior to its counterpart HRM heuristic. However, a single instance would not be enough to judge on its solution quality. In order to see the performance of the MHRM heuristic on a wide range of problem instances, the benchmark suite of Biskup & Feldmann [3] is solved by the MHRM heuristic to be compared to its counterpart HRM heuristic. The computational results of both heuristics are given in Table 6. Note that the results for the HRM heuristic were adopted from Hino et al. [8]. From Table 6, it is clear that the MHRM heuristic was superior to its counterpart HRM heuristic in terms of relative percent deviations since the percentage relative deviation that the HRM heuristic presented in Hino et al. [8] was 2.42 percent worst than the upper bounds on average whereas the MHRM heuristic was able to improve the upper bounds by 0.65 percent on overall average. Especially, significant improvements over the HRM heuristic are observed on the problem instances having loose due date settings for h=0.6 and h=0.8.

Table 6. Computational Results for HRM and MHRM Heuristics (Δ)

	h	10	20	50	100	200	500	1000	Mean
HRM	0.2	1.53	-3.97	-5.33	-6.02	-5.63	-6.32	-6.68	-4.5
	0.4	8.68	0.46	-3.87	-4.42	-3.51	-3.46	-4.26	-1.48
	0.6	19.27	9.78	7.59	4.69	3.71	2.53	3.23	7.26
	0.8	22.97	13.52	8.1	4.7	3.71	2.53	3.23	8.39
	Mean	13.11	5.17	1.62	-0.26	-0.43	-1.18	-1.12	2.42
MHRM	h	10	20	50	100	200	500	1000	Mean
	0.2	1	-3.57	-5.45	-6.02	-5.62	-6.32	-6.69	-4.67
	0.4	5.91	-0.49	-4.03	-4.27	-3.52	-3.45	-4.27	-2.02
	0.6	2.77	2.02	1.51	1.5	1.71	1.41	1.55	1.78
	0.8	3.95	4.07	2.13	1.43	1.71	1.41	1.55	2.32
	Mean	3.41	0.51	-1.46	-1.84	-1.43	-1.74	-1.97	-0.65

Before getting into the detailed analysis of the metaheuristic algorithms against the recent metaheuristics in the literature, we again point out that our analysis is based on comparisons of our metaheuristics with and without a local search so as to make fair comparisons with all the existing algorithms in the literature. For comparison purposes, Avg I denotes the mean value for h=0.2, h=0.4, h=0.6, and h=0.8. Avg II denotes the mean value for h=0.2, and h=0.4 whereas Avg III represents the mean value for h=0.6 and h=0.8. The reason is because of the fact that an algorithm performs relatively good for a tight due date setting may not be so good for a loose due date setting.

In Table 7, an overall summary of the BPSO, DPSO and DDE algorithms is given in terms of Avg I. It is obvious from Table 7 that the DDE algorithm was superior to the BPSO and DPSO algorithms in terms of percentage relative deviations. Even its average performance was equal or better than the best performance of the DPSO and BPSO algorithms. In terms of CPU time requirements, the DPSO and DDE algorithms had similar speeds whereas the BPSO algorithm was much slower than both of them, which might be because of working on a continuous domain and using the sigmoid function to convert the velocity to binary values. Briefly, the DDE algorithm with this rough comparison was a clear winner.

Table 7. Comparison of Results with respect to Avg I: Without Local Search

					Time to Termination			
Alg.	$\Delta_{\min}$	$\Delta_{\max}$	Δ_{avg}	Δ_{std}	$t_{\min}$	$t_{\max}$	t_{avg}	t_{std}
DDE	-2.14	-2.01	-2.11	0.05	0.16	0.17	0.17	0.01
DPSO	-2.11	-1.96	-2.05	0.05	0.16	0.17	0.16	0.01
BPSO	-1.49	-1.42	-1.45	0.02	0.5	0.52	0.51	0.01

Table 8. Comparison of Results with respect to Avg I: With Local Search

					Time to Termination			
Alg.	$\Delta_{\min}$	$\Delta_{\max}$	Δ_{avg}	Δ_{std}	$t_{\min}$	$t_{\max}$	t_{avg}	t_{std}
DDE_{LS}	-2.15	-2.14	-2.15	0.01	0.45	1.15	0.77	0.24
$DPSO_{LS}$	-2.15	-2.13	-2.15	0.01	0.42	1.11	0.72	0.24
$BPSO_{LS}$	-2.15	-2.14	-2.15	0	0.42	1.1	0.72	0.23

However, the inclusion of a simple local search in all the metaheuristic algorithms led them to generate similar and improved results as seen in Table 8. All metaheuristics were able to improve the upper bounds by 2.15 percent with a CPU time of no more than 1.15 seconds at most on overall average since the maximum CPU time that the DDE_{LS} algorithm consumed was 1.15 seconds. Furthermore, the best, average, and the worst behavior of the $BPSO_{LS}$, $DPSO_{LS}$ and DDE_{LS} algorithms were very close to each other with very low standard deviations indicating the robustness of the metaheuristic algorithms presented.

Most recently, Hino et al. [8] developed a TS, GA and hybridization of both of them denoted as HTG and HGT. In addition, Nearchou [11] proposed a differential evolution approach (DEA) whereas a sequential exchange approach (SEA) is presented by Lin et al. [15]. It should be noted that Lin et al. [15] presented SEA1 and SEA2 algorithms and the best solution between SEA1 and SEA2 was reported as SEA in their paper. Since TS, GA, HTG, HGT, DEA, and SEA employed the same benchmark suite of Biskup & Feldmann [3] as we did in this chapter, we compare our results to those very recent approaches in the literature.

It should be noted that due to the stochastic nature of the metaheuristic algorithms, their minimum, maximum, average, and standard deviation of 10 runs for each instance should be given to evaluate their performance. However, except for Lin et al. [15], which is a deterministic algorithm, in Hino et al. [8], and Nearchou [11], 10 runs were conducted for each instance and the best out of 10 runs was picked up to be averaged over 10 instances even though they had some random components in their algorithms. It implies that no information was available at present about the average, and worst case behavior as well as the robustness of their algorithms. It led us to make comparisons with respect to minimum percentage relative deviation of the metaheuristic algorithms presented in this chapter. In addition, among the algorithms tested in Feldmann & Biskup [3], the TAR algorithm was superior to other algorithms namely, ES, SA, TA. However, recent approaches generated better results than the TAR algorithm. For this reason, the TAR algorithm was excluded in our comparisons even though they were the pioneering ones.

Table 9 presents the computational results of best performing algorithms for the E/T problem in the literature together with the metaheuristic algorithms presented in

Table 9. Comparison of Results with respect to $\Delta_{\min}$: Without Local Search

n	*h*	BPSO	DPSO	DDE	TS	GA	HTG	HGT	SEA	DEA
10	0.2	0.00	0.00	0.00	0.25	0.12	0.12	0.12	0.01	0.00
	0.4	0.00	0.00	0.00	0.24	0.19	0.19	0.19	0.00	0.00
	0.6	0.00	0.00	0.00	0.10	0.03	0.03	0.01	0.01	0.00
	0.8	0.00	0.00	0.00	0.00	0.00	0.00	0.00	0.00	0.00
20	0.2	-3.84	-3.84	-3.82	-3.84	-3.84	-3.84	-3.84	-3.79	-3.84
	0.4	-1.63	-1.63	-1.63	-1.62	-1.62	-1.62	-1.62	-1.58	-1.63
	0.6	-0.72	-0.72	-0.70	-0.71	-0.68	-0.71	-0.71	-0.64	-0.72
	0.8	-0.41	-0.41	-0.41	-0.41	-0.28	-0.41	-0.41	-0.39	-0.41
50	0.2	-5.45	-5.66	-5.69	-5.70	-5.68	-5.70	-5.70	-5.58	-5.69
	0.4	-4.04	-4.62	-4.65	-4.66	-4.60	-4.66	-4.66	-4.42	-4.66
	0.6	0.93	-0.27	-0.27	-0.32	-0.31	-0.27	-0.31	-0.31	-0.32
	0.8	1.23	-0.24	-0.24	-0.24	-0.19	-0.23	-0.23	-0.24	-0.24
100	0.2	-6.02	-6.17	-6.18	-6.19	-6.17	-6.19	-6.19	-6.21	-6.17
	0.4	-4.27	-4.85	-4.91	-4.93	-4.91	-4.93	-4.93	-4.85	-4.89
	0.6	1.50	-0.14	-0.15	-0.01	-0.12	0.08	0.04	-0.15	-0.13
	0.8	1.40	-0.17	-0.18	-0.15	-0.12	-0.08	-0.11	-0.18	-0.17
200	0.2	-5.62	-5.75	-5.77	-5.76	-5.74	-5.76	-5.76	-5.76	-5.77
	0.4	-3.52	-3.66	-3.72	-3.74	-3.75	-3.75	-3.75	-3.73	-3.72
	0.6	1.71	-0.14	-0.15	-0.01	-0.13	0.37	0.07	-0.15	0.23
	0.8	1.71	-0.14	-0.15	-0.04	-0.14	0.26	0.07	-0.15	0.20
500	0.2	-6.32	-6.38	-6.41	-6.41	-6.41	-6.41	-6.41	-6.43	-6.43
	0.4	-3.45	-3.48	-3.52	-3.57	-3.58	-3.58	-3.58	-3.57	-3.57
	0.6	1.41	-0.06	-0.11	0.25	-0.11	0.73	0.15	-0.11	1.72
	0.8	1.41	-0.06	-0.11	0.21	-0.11	0.73	0.13	-0.11	1.01
1000	0.2	-6.69	-6.71	-6.73	-6.73	-6.75	-6.74	-6.74	-6.77	-6.72
	0.4	-4.27	-4.28	-4.31	-4.39	-4.40	-4.39	-4.39	-4.40	-4.38
	0.6	1.55	0.22	-0.06	1.01	-0.05	1.28	0.42	-0.06	1.29
	0.8	1.55	0.22	-0.06	1.13	-0.05	1.28	0.40	-0.06	2.79
Avg I		-1.49	-2.11	-2.14	-2.01	-2.12	-1.94	-2.06	-2.13	-1.87
Avg II		-3.94	-4.07	-4.10	-4.08	-4.08	-4.09	-4.09	-4.08	-4.11
Avg III		0.95	-0.14	-0.19	0.06	-0.16	0.22	-0.03	-0.18	0.38

this paper. Note that no local search is used in these results. As seen in Table 9, the BPSO algorithm was the worst algorithm whereas the DDE, SEA, GA, DPSO, HGT, TS, HTG algorithms were the best performing algorithms in terms of Avg. I. When considering Avg. II, i.e., tight due date settings, DEA and DDE were the best with the fact that other algorithms compared has also generated almost similar results except for the BPSO algorithm. However, when considering Avg. III, i.e., loose due date settings, the best performing ones were the DDE, SEA, and GA algorithms whereas the worst ones were the BPSO, DEA, HTG, TS, respectively. Briefly, the best results were obtained by the DDE, SEA, GA, and DPSO algorithms, respectively. So the performance of the DDE algorithm without a local search was better than all the algorithms compared.

As seen in Table 10, the inclusion of the local search in all the metaheuristic algorithms led them to be the best performing algorithms in the literature. As seen in Table 10, the $BPSO_{LS}$, $DPSO_{LS}$ and DDE_{LS} algorithms generated better results than those of all the existing approaches in the literature in terms of Avg. I, Avg. II and Avg. III. Even their worst case performances in Tables 7 and 8 were better or equivalent to all the existing approaches compared. It is interesting to compare the algorithms in terms of Avg. I, Avg. II and Avg. III because when the due date becomes loose, i.e., h=0.6 and h=0.8, the performance of some algorithms was deteriorated except for the $BPSO_{LS}$, $DPSO_{LS}$ and DDE_{LS} algorithms. For instance, the performance of TS, HTG, HGT and DEA for h=0.6 and h=0.8 was deteriorated while they performed relatively well for h=0.2 and h=0.4 instances. Especially, the DEA algorithm performed one of the best for h=0.2 and h=0.4 instances whereas it failed for h=0.6 and h=0.8 instances. The best algorithms can be ranked with respect to Avg. I, Avg. II and Avg. III as the $BPSO_{LS}$, $DPSO_{LS}$, DDE_{LS}, SEA, GA, HGT, TS, HTG and DEA algorithms, respectively. However, the best results so far in the literature were presented by the $BPSO_{LS}$, $DPSO_{LS}$ and DDE_{LS} algorithms, respectively, in this chapter.

Table 11 summarizes the CPU time requirements for all the algorithms compared. It is difficult to compare the algorithms in terms of the CPU time requirements since different machines were used. However, Table 11 provides some clues about the speed of the algorithms compared. It is very obvious that the DEA algorithm was the most time consuming one amongst them since its average CPU time performance was 1815.53 seconds. Even with some fair correction factors, it was clearly the most expensive one in terms of consuming CPU time. As seen in Table 11, the fastest algorithms were $BPSO_{LS}$, $DPSO_{LS}$, DDE_{LS} and SEA since their average CPU times was 0.42, 0.45, 0.42 and 4.64 seconds, respectively. Owing to the fact that we used a machine approximately three times faster than the one used in SEA, a fair comparison should be made. However, even with a correction factor of 3, $0.42 \times 3 = 1.26$ seconds, $0.45 \times 3 = 1.35$ seconds and $0.42 \times 3 = 1.26$ seconds were still much less than 4.64 seconds that the SEA algorithm spent on average. For this reason, it can be concluded that the fastest algorithm so far in the literature were also $BPSO_{LS}$, $DPSO_{LS}$ and DDE_{LS} together with the best percentage relative deviations reported so far in the literature.

In order to statistically test the performance of the BPSO, DPSO and DDE algorithms with and without the local search, a series of the paired t-test at the 95% significance level was carried out after checking the normality assumption of the differences in the algorithms [39]. In the paired t-test, we are interested in the differences in two

Table 10. Comparison of Results with respect to Δ_{min} : With Local Search

n	h	BPSO	DPSO	DDE	TS	GA	HTG	HGT	SEA	DE
10	0.2	0.00	0.00	0.00	0.25	0.12	0.12	0.12	0.01	0.00
	0.4	0.00	0.00	0.00	0.24	0.19	0.19	0.19	0.00	0.00
	0.6	0.00	0.00	0.00	0.10	0.03	0.03	0.01	0.01	0.00
	0.8	0.00	0.00	0.00	0.00	0.00	0.00	0.00	0.00	0.00
20	0.2	-3.84	-3.84	-3.84	-3.84	-3.84	-3.84	-3.84	-3.79	-3.84
	0.4	-1.63	-1.63	-1.63	-1.62	-1.62	-1.62	-1.62	-1.58	-1.63
	0.6	-0.72	-0.72	-0.72	-0.71	-0.68	-0.71	-0.71	-0.64	-0.72
	0.8	-0.41	-0.41	-0.41	-0.41	-0.28	-0.41	-0.41	-0.39	-0.41
50	0.2	-5.69	-5.69	-5.70	-5.70	-5.68	-5.70	-5.70	-5.58	-5.69
	0.4	-4.66	-4.66	-4.66	-4.66	-4.60	-4.66	-4.66	-4.42	-4.66
	0.6	-0.34	-0.34	-0.34	-0.32	-0.31	-0.27	-0.31	-0.31	-0.32
	0.8	-0.24	-0.24	-0.24	-0.24	-0.19	-0.23	-0.23	-0.24	-0.24
100	0.2	-6.19	-6.19	-6.19	-6.19	-6.17	-6.19	-6.19	-6.21	-6.17
	0.4	-4.94	-4.94	-4.94	-4.93	-4.91	-4.93	-4.93	-4.85	-4.89
	0.6	-0.15	-0.15	-0.15	-0.01	-0.12	0.08	0.04	-0.15	-0.13
	0.8	-0.18	-0.18	-0.18	-0.15	-0.12	-0.08	-0.11	-0.18	-0.17
200	0.2	-5.78	-5.77	-5.78	-5.76	-5.74	-5.76	-5.76	-5.76	-5.77
	0.4	-3.75	-3.75	-3.75	-3.74	-3.75	-3.75	-3.75	-3.73	-3.72
	0.6	-0.15	-0.15	-0.15	-0.01	-0.13	0.37	0.07	-0.15	0.23
	0.8	-0.15	-0.15	-0.15	-0.04	-0.14	0.26	0.07	-0.15	0.20
500	0.2	-6.43	-6.42	-6.43	-6.41	-6.41	-6.41	-6.41	-6.43	-6.43
	0.4	-3.57	-3.57	-3.57	-3.57	-3.58	-3.58	-3.58	-3.57	-3.57
	0.6	-0.11	-0.11	-0.11	0.25	-0.11	0.73	0.15	-0.11	1.72
	0.8	-0.11	-0.11	-0.11	0.21	-0.11	0.73	0.13	-0.11	1.01
1000	0.2	-6.77	-6.76	-6.77	-6.73	-6.75	-6.74	-6.74	-6.77	-6.72
	0.4	-4.39	-4.38	-4.39	-4.39	-4.40	-4.39	-4.39	-4.40	-4.38
	0.6	-0.06	-0.06	-0.06	1.01	-0.05	1.28	0.42	-0.06	1.29
	0.8	-0.06	-0.06	-0.06	1.13	-0.05	1.28	0.40	-0.06	2.79
Avg I		**-2.15**	**-2.15**	**-2.15**	-2.01	-2.12	-1.94	-2.06	-2.13	-1.87
Avg II		**-4.12**	**-4.11**	**-4.12**	-4.08	-4.08	-4.09	-4.09	-4.08	-4.11
Avg III		**-0.19**	**-0.19**	**-0.19**	0.06	-0.16	0.22	-0.03	-0.18	0.38

observations within the pairs. Let $\mu_D = \mu_1 - \mu_2$ denote the true average difference between the percentage relative deviations generated by two different algorithms, the null hypothesis is given by $H_0 : \mu_D = \mu_1 - \mu_2 = 0$ saying that there is no difference between the average percentage relative deviations generated by two algorithms compared. On the other hand, the alternative hypothesis is given by $H_1 : \mu_D = \mu_1 - \mu_2 \neq 0$ saying that there is a difference between the average percentage relative deviations

Table 11. Comparison of CPU Times in Seconds

h/n	10	20	50	100	200	500	1000	**Avg**
SEA	0.00	0.00	0.01	0.05	0.25	3.65	28.77	
	0.00	0.00	0.01	0.05	0.28	3.99	31.67	PIII
	0.00	0.00	0.01	0.05	0.23	3.35	27.09	1 GHz
	0.00	0.00	0.01	0.05	0.23	3.36	26.72	
Avg	**0.00**	**0.00**	**0.01**	**0.05**	**0.25**	**3.59**	**28.56**	**4.64**
DEA	0.23	1.02	2.44	23.21	242.09	3941.17	8561.02	
	0.21	1.13	3.01	24.61	230.09	3925.08	8609.22	PIV
	0.19	1.18	2.38	17.23	216.39	3950.76	8441.70	1.2GHz
	0.20	1.00	2.11	18.02	240.91	3912.82	8465.37	
Avg	**0.21**	**1.08**	**2.49**	**20.77**	**232.37**	**3932.46**	**8519.33**	**1815.53**
$DPSO_{LS}$	0.00	0.00	0.02	0.10	0.21	0.61	1.37	
	0.00	0.00	0.02	0.10	0.26	0.66	1.68	PIV
	0.00	0.00	0.02	0.11	0.24	0.72	1.75	3 GHz
	0.00	0.00	0.02	0.11	0.24	0.73	1.75	
Avg	**0.00**	**0.00**	**0.02**	**0.11**	**0.24**	**0.68**	**1.64**	**0.38**
DDE_{LS}	0.00	0.00	0.02	0.10	0.22	0.63	1.52	
	0.00	0.00	0.03	0.11	0.25	0.84	2.03	PIV
	0.00	0.00	0.02	0.11	0.24	0.75	1.83	3 GHz
	0.00	0.00	0.02	0.11	0.24	0.76	1.83	
Avg	**0.00**	**0.00**	**0.02**	**0.11**	**0.24**	**0.75**	**1.80**	**0.42**
$BPSO_{LS}$	0.00	0.00	0.02	0.10	0.23	0.68	1.62	
	0.00	0.00	0.03	0.11	0.25	0.73	2.12	PIV
	0.00	0.00	0.03	0.11	0.24	0.77	1.83	3 GHz
	0.00	0.00	0.03	0.11	0.24	0.77	1.83	
Avg	**0.00**	**0.00**	**0.03**	**0.11**	**0.24**	**0.74**	**1.85**	**0.42**
GA	PIV 1.7 GHz							0.21
HTG	PIV 1.7 GHz							7.80
HGT	PIV 1.7 GHz							7.80

generated by two algorithms compared. The paired t-test results based on the percentage relative deviations in Tables 9 and 10 are given in Table 12.

Table 12 indicates the poor performance of the BPSO algorithm against all the algorithms compared since the null hypothesis was rejected on the behalf of the algorithms compared. It means that the differences between them were meaningful at the significance level of 0.95. An important indication of Table 12 is that the performance of the DPSO, was equivalent to GA, SEA and DEA since the null hypothesis was failed to be rejected implying that the differences between these algorithms were not meaningful at the significance level of 0.95. However, the null hypothesis was rejected on the behalf of the DPSO algorithm against the TS, HTG, and HGT algorithms indicating that the differences were meaningful at the significance level of 0.95. When considering the DDE algorithm versus the GA and SEA algorithms, the null hypothesis was failed to be rejected indicating that differences were not meaningful at the significant level of 0.95. In other words, they were equivalent. However, the null hypothesis was rejected on the behalf of the DDE algorithm when compared to the TS, HTG, HGT, and DEA algorithms. It indicates that the differences were meaningful at the significance level of 0.95. Briefly, BPSO algorithm was not competitive to all the algorithms compared and the best performing algorithms were the DDE, SEA, and DPSO, GA algorithms, respectively when considering no local search.

Table 12. Paired t-Test at Significance level of 0.95

H_0	t	p	$p < 0.05$	H_0
BPSO=DPSO	4.52	0.00	Yes	R
BPSO=DDE	4.61	0.00	Yes	R
DPSO=DDE	2.59	0.02	Yes	R
BPSO=TS	4.18	0.00	Yes	R
BPSO=GA	4.45	0.00	Yes	R
BPSO=HTG	4.19	0.00	Yes	R
BPSO=HGT	4.63	0.00	Yes	R
BPSO=SEA	4.50	0.00	Yes	R
DPSO=TS	-2.11	0.04	Yes	R
DPSO=GA	**0.89**	**0.38**	**No**	**FR**
DPSO=HTG	-2.58	0.02	Yes	R
DPSO=HGT	-2.08	0.05	Yes	R
DPSO=SEA	**1.36**	**0.38**	**No**	**FR**
DPSO=DEA	**-2.01**	**0.05**	**No**	**FR**
DDE=TS	-2.29	0.03	Yes	R
DDE=GA	**-1.78**	**0.09**	**No**	**FR**
DDE=HTG	-2.68	0.01	Yes	R
DDE=HGT	-2.78	0.01	Yes	R
DDE=SEA	**-1.03**	**0.31**	**No**	**FR**
DDE=DEA	-2.12	0.04	Yes	R

Next we compare the local search version of our metaheuristics to the best performing algorithms in the literature. Table 13 gives the paired t-test results for the $BPSO_{LS}$, DDE_{LS}, and $DPSO_{LS}$ algorithms against the best performing ones in the literature. As seen in Table 13, the null hypothesis was rejected on the behalf of the DDE_{LS} algorithm against all the algorithms compared. In other words, the differences between the DDE_{LS} and those of all the algorithms compared were meaningful at the significance level of 0.95. We do not report the $BPSO_{LS}$ and $DPSO_{LS}$ since the null hypothesis was rejected on the behalf of them too. From this statistical analysis, it can be concluded that the DDE_{LS}, $DPSO_{LS}$ and $BPSO_{LS}$ algorithms were statistically proved to be the best algorithms so far in the literature.

Finally, we wanted to evaluate the peak performance of the DDE_{LS} algorithm by running 500 generations in order to see if there is still some room for future researchers to improve the results. The computational results for 500 generations are given in Table 14. We were able to improve the results a little bit, which may be conjectured that those solutions might be optimal ones. However, it would never be said so unless

Table 13. Paired t-Test at Significance level of 0.95

H_0	t	p	$p < 0.05$	H_0
$BPSO_{LS}$=$DPSO_{LS}$	-2.12	0.043	Yes	R
$BPSO_{LS}$=DDE_{LS}	1.00	0.326	No	FR
$DPSO_{LS}$=DDE_{LS}	2.42	0.022	Yes	R
DDE_{LS} =TS	-2.6	0.015	Yes	R
DDE_{LS} =GA	-3.87	0.001	Yes	R
DDE_{LS} =HTG	-2.93	0.007	Yes	R
DDE_{LS} =HGT	-3.53	0.002	Yes	R
DDE_{LS} =SEA	-2.52	0.018	Yes	R
DDE_{LS} =DEA	-2.25	0.032	Yes	R

Table 14. Peak Performance of DDE_{LS} Algorithm with 500 generations

		Δ				Time to Termination			
n	h	$\Delta_{\min}$	$\Delta_{\max}$	Δ_{avg}	Δ_{std}	$t_{\min}$	$t_{\max}$	t_{avg}	t_{std}
10	0.2	0.00	0.00	0.00	0.00	0.08	0.10	0.09	0.01
	0.4	0.00	0.00	0.00	0.00	0.09	0.11	0.09	0.01
	0.6	0.00	0.00	0.00	0.00	0.09	0.10	0.09	0.00
	0.8	0.00	0.00	0.00	0.00	0.09	0.11	0.10	0.01
20	0.2	-3.84	-3.84	-3.84	0.00	0.26	0.34	0.28	0.02
	0.4	-1.63	-1.63	-1.63	0.00	0.26	0.29	0.27	0.01
	0.6	-0.72	-0.72	-0.72	0.00	0.26	0.29	0.27	0.01
	0.8	-0.41	-0.41	-0.41	0.00	0.26	0.28	0.27	0.01
50	0.2	-5.70	-5.68	-5.69	0.01	1.26	1.32	1.28	0.02
	0.4	-4.66	-4.66	-4.66	0.00	1.30	1.34	1.32	0.02
	0.6	-0.34	-0.34	-0.34	0.00	1.37	1.41	1.39	0.02
	0.8	-0.24	-0.24	-0.24	0.00	1.36	1.40	1.38	0.02
100	0.2	-6.19	-6.19	-6.19	0.00	4.34	4.36	4.35	0.01
	0.4	-4.94	-4.94	-4.94	0.00	4.82	4.85	4.83	0.01
	0.6	-0.15	-0.15	-0.15	0.00	5.15	5.18	5.17	0.01
	0.8	-0.18	-0.18	-0.18	0.00	5.15	5.19	5.17	0.01
200	0.2	-5.78	-5.78	-5.78	0.00	8.73	8.79	8.76	0.02
	0.4	-3.75	-3.75	-3.75	0.00	10.06	10.16	10.11	0.03
	0.6	-0.15	-0.15	-0.15	0.00	10.85	10.91	10.87	0.02
	0.8	-0.15	-0.15	-0.15	0.00	10.85	10.90	10.88	0.02
500	0.2	-6.43	-6.43	-6.43	0.00	23.15	23.35	23.25	0.06
	0.4	-3.58	-3.57	-3.58	0.00	27.49	27.79	27.65	0.10
	0.6	-0.11	-0.11	-0.11	0.00	29.38	29.47	29.42	0.03
	0.8	-0.11	-0.11	-0.11	0.00	29.40	29.48	29.44	0.02
1000	0.2	-6.77	-6.77	-6.77	0.00	48.58	49.23	48.88	0.20
	0.4	-4.40	-4.39	-4.39	0.00	59.14	59.67	59.42	0.17
	0.6	-0.06	-0.06	-0.06	0.00	61.87	62.06	61.96	0.06
	0.8	-0.06	-0.06	-0.06	0.00	61.85	62.04	61.94	0.07
Mean		**-2.16**	**-2.15**	**-2.15**	**0.00**	**14.55**	**14.66**	**14.61**	**0.04**

proved mathematically. It should be noted that we also run them for 1000 generations too. However, we were unable to further improve the results. This is last to say that all solution details would be available upon request.

6 Conclusions

PSO and DE are recent evolutionary optimization methods. It has been widely used in a wide range of applications. Besides the standard versions, we presented a new and novel discrete version of both promising algorithms, so called here DPSO and DDE, in this paper together with the standard binary PSO. To the best of our knowledge, these are the first reported applications of both DPSO and DDE algorithm to the single-machine total earliness and tardiness penalties with a common due date problem in the literature.

Unlike the standard PSO and DE, the DPSO and DDE algorithms are novel algorithms, which are based on a discrete domain exploiting the basic features of its continuous counterpart. They employ a binary solution representation for the problem on hand. It indicates that both algorithms can be applied to other binary/discrete combinatorial optimization problems with some modifications in the literature. Another contribution of this chapter is to a presentation of a novel MHRM constructive heuristic algorithm in such a way that the MHRM heuristic is given in detail as to how it differs from its counterpart HRM heuristic with examples. We have also presented a BSWAP mutation operator to be used for binary solution spaces. Furthermore, all the metaheuristic algorithms are hybridized with a simple local search to further improve the solution quality. Finally, a very detailed design of experiments is conducted to determine the parameters of the metaheuristics proposed.

The proposed metaheuristic algorithms were applied to the benchmark problems in Biskup and Feldmann [3]. The computational results statistically show that the proposed algorithms with the local search have generated better results than all of the existing approaches in the literature.

As a final note, it is obvious that the proposed algorithms can be easily extended to solve the flowshop scheduling problems as well as other other discrete/combinatorial optimization problems.

Appendix A: MHRM Heuristic

Table A1. An Example Instance with A Common Due Date: d=103

Job	1	2	3	4	5	6	7	8	9	10
p_j	6	19	20	16	11	11	5	11	10	20
α_j	5	8	5	8	3	6	9	7	10	5
β_j	9	12	1	15	12	1	13	1	2	1

HRM Solution

Step1. Let $H = 129$, $d = 103$, $g = \max\{0, d - 0.5H\} = 39$, $d^E = d - g = 64$, $d^T = g + H - d = 65$, $S^E = S^F = \Phi$, $P = \{1,2,3,4,5,6,7,8,9,10\}$.

Step 2. Set $e = \arg\max_{j\in P}\{p_j / \alpha_j\} = 3$ and $t = \arg\max_{j\in P}\{p_j / \beta_j\} = 3$.

Step 3. Set $E^e = d^E - p_e = 44$ and $T^t = d^T = 65$; Since $E^e > 0$ and $T^t > 0$, go to step 4.

Step 4. Since $E^e < T^t$, $S^T = S^T + \{t\} = \{3\}$, $d^T = d^T - p_t = 45$, $P = P - \{t\} = \{1,2,4,5,6,7,8,9,10\}$. Go to step 7.

Step 7. Since $P \neq \{\phi\}$, go to step 2.

Step 2. Set $e = 10$ and $t = 10$.

Step 3. Set $E^e = 44$ and $T^t = 45$; Since $E^e > 0$ and $T^t > 0$, go to step 4.

Step 4. Since $E^e < T^t$, $S^T = \{10,3\}$, $d^T = 25$, $P = \{1,2,4,5,6,7,8,9\}$. Go to Step 7.

Step 7. Since $P \neq \{\phi\}$, go to step 2.

Step 2. Set $e = 5$, $t = 8$.

Step 3. Set $E^e = 53$, and $T^t = 25$; Since $E^e > 0$ and $T^t > 0$, go to step 4.

Step 4. Since $E^e > T^t$, $S^E = S^E + \{e\} = \{5\}$, $d^E = d^E - p_e = 53$, and $P = P - \{e\} = \{1,2,4,6,7,8,9\}$. Go to step 7.

Step 7. Since $P \neq \{\phi\}$, go to step 2.

Step 2. Set $e = 2$, $t = 8$.

Step 3. Set $E^e = 34$, and $T^t = 25$; Since $E^e > 0$ and $T^t > 0$, go to step 4.

Step 4. Since $E^e > T^t$, $S^E = \{5,2\}$, $d^E = 34$, and $P = \{1,4,6,7,8,9\}$. Go to step 7.

Step 7. Since $P \neq \{\phi\}$, go to step 2.

Step 2. Set $e = 4$, $t = 8$.

Step 3. Set $E^e = 18$, and $T^t = 25$; Since $E^e > 0$ and $T^t > 0$, go to step 4.

Step 4. Since $E^e < T^t$, $S^T = \{8,10,3\}$, $d^T = 14$, $P = \{1,4,6,7,9\}$. Go to Step 7.

Step 7. Since $P \neq \{\phi\}$, go to step 2.

Step 2. Set $e = 4$, $t = 6$.

Step 3. Set $E^e = 18$, and $T^t = 14$; Since $E^e > 0$ and $T^t > 0$, go to step 4.
Step 4. Since $E^e > T^t$, $S^E = \{5,2,4\}$, $d^E = 18$, and $P = \{1,6,7,9\}$. Go to step 7.
Step 7. Since $P \neq \{\phi\}$, go to step 2.
Step 2. Set $e = 6$, $t = 6$.
Step 3. Set $E^e = 7$, and $T^t = 14$; Since $E^e > 0$ and $T^t > 0$, go to step 4.
Step 4. Since $E^e < T^t$, $S^T = \{6,8,10,3\}$, $d^T = 3$, $P = \{1,7,9\}$. Go to Step 7.
Step 7. Since $P \neq \{\phi\}$, go to step 2.
Step 2. Set $e = 1$, $t = 9$.
Step 3. Set $E^e = 12$, and $T^t = 3$; Since $E^e > 0$ and $T^t > 0$, go to step 4.
Step 4. Since $E^e > T^t$, $S^E = \{5,2,4,1\}$, $d^E = 12$, and $P = \{7,9\}$. Go to step 7.
Step 7. Since $P \neq \{\phi\}$, go to step 2.
Step 2. Set $e = 9$, $t = 9$.
Step 3. Set $E^e = 2$, and $T^t = 3$; Since $E^e > 0$ and $T^t > 0$, go to step 4.
Step 4. Since $E^e < T^t$, $S^T = \{9,6,8,10,3\}$, $d^T = -7$, $P = \{7\}$. Go to Step 7.
Step 7. Since $P \neq \{\phi\}$, go to step 2.
Step 2. Set $e = 7$, $t = 7$.
Step 3. Set $E^e = 7$, and $T^t = -7$; Since $T^t \leq 0$, go to step 6.
Step 6. End of the space after the due date

- Set $S^{E'} = S^E \cup P = \{5,2,4,1,7\}$, $S^{T'} = S^T = \{9,6,8,10,3\}$,
  $g' = d - \sum_{i \in S^{E'}} p_i$, $g' = 103 - 57 = 46$.
  $S' = (g', S^{E'}, S^{T'})$; then $f(S') = 664$.
- Set $S^{E''} = (S^E \cup P) - \{t\} = \{5,2,4,1\}$, $S^{T''} = S^T + \{t\} = \{7,9,6,8,10,3\}$,
  $g'' = d - \sum_{i \in S^{E''}} p_i$, $g'' = 103 - 52 = 51$, $S'' = (g'', S^{E''}, S^{T''})$, then $f(S') = 639$.
- Since $f(S'') < f(S')$, $S^T = S^T + \{t\} = (7,9,6,8,10,3)$, $S^E = \{5,2,4,1\}$, $g'' = 51$, and $P = (\phi)$.
- Since there is no straddling job, no g-test is required.
- The solution is $S = (g, S^E, S^T)$ where $g = 51$, $S^E = \{5,2,4,1\}$, $S^T = (7,9,6,8,10,3)$ and $f(S') = 639$.

MHRM Solution

Step 1. Let $H = 129$, $d = 103$, $g = \left\lceil \max\left\{0, d - H \times \frac{1}{n}\sum_{j=1}^{n}\left(\frac{\beta_j}{\alpha_j + \beta_j}\right)\right\} \right\rceil = 51$,

$d^E = d - g = 52$, $d^T = g + H - d = 77$, $S^E = S^F = \Phi$, $P = \{1,2,3,4,5,6,7,8,9,10\}$.
Step 2. Set $e = \arg\max_{j \in P}\{p_j / \alpha_j\} = 3$, $t = \arg\max_{j \in P}\{p_j / \beta_j\} = 3$.

Step 3. Set $E^e = d^E - p_e = 32$, $T^t = d^T = 77$, $T^t - p_t = 57$. Since $E^e > 0$ and $T^t - p_t > 0$, go to step 4.

Step 4. Since $E^e < T^t$, $S^T = S^T + \{t\} = \{3\}$, $d^T = d^T - p_t = 57$, $P = P - \{t\} = \{1,2,4,5,6,7,8,9,10\}$. Go to step 7.

Step 7. Since $P \neq \{\phi\}$, go to step 2.

Step 2. Set $e = 10$, $t = 10$.

Step 3. Set $E^e = d^E - p_e = 32$, $T^t = d^T = 57$, $T^t - p_t = 37$. Since $E^e > 0$ and $T^t - p_t > 0$, go to step 4.

Step 4. Since $E^e < T^t$, $S^T = \{10,3\}$, $d^T = d^T - p_t = 37$, $P = \{1,2,4,5,6,7,8,9\}$. Go to step 7.

Step 7. Since $P \neq \{\phi\}$, go to step 2.

Step 2. Set $e = 5$, $t = 8$.

Step 3. Set $E^e = d^E - p_e = 41$, $T^t = d^T = 37$, $T^t - p_t = 26$. Since $E^e > 0$ and $T^t - p_t > 0$, go to step 4.

Step 4. Since $E^e > T^t$, $S^E = \{5\}$, $d^E = d^E - p_e = 41$, $P = \{1,2,4,6,7,8,9\}$. Go to step 7.

Step 7. Since $P \neq \{\phi\}$, go to step 2.

Step 2. Set $e = 2$, $t = 8$.

Step 3. Set $E^e = d^E - p_e = 22$, $T^t = d^T = 37$, $T^t - p_t = 26$. Since $E^e > 0$ and $T^t - p_t > 0$, go to step 4.

Step 4. Since $E^e < T^t$, $S^T = \{8,10,3\}$, $d^T = d^T - p_t = 26$, $P = \{1,2,4,6,7,9\}$. Go to step 7.

Step 7. Since $P \neq \{\phi\}$, go to step 2.

Step 2. Set $e = 2$, $t = 6$.

Step 3. Set $E^e = d^E - p_e = 22$, $T^t = d^T = 26$, $T^t - p_t = 15$. Since $E^e > 0$ and $T^t - p_t > 0$, go to step 4.

Step 4. Since $E^e < T^t$, $S^T = \{6,8,10,3\}$, $d^T = d^T - p_t = 15$, $P = \{1,2,4,7,9\}$. Go to step 7.

Step 7. Since $P \neq \{\phi\}$, go to step 2.

Step 2. Set $e = 2$, $t = 9$.

Step 3. Set $E^e = d^E - p_e = 22$, $T^t = d^T = 15$, $T^t - p_t = 5$. Since $E^e > 0$ and $T^t - p_t > 0$, go to step 4.

Step 4. Since $E^e > T^t$, $S^E = \{5,3\}$, $d^E = d^E - p_e = 22$, $P = \{1,4,7,9\}$. Go to step 7.

Step 7. Since $P \neq \{\phi\}$, go to step 2.

Step 2. Set $e = 4$, $t = 9$.

Step 3. Set $E^e = d^E - p_e = 6$, $T^t = d^T = 15$, $T^t - p_t = 5$. Since $E^e > 0$ and $T^t - p_t > 0$, go to step 4.

Step 4. Since $E^e < T^t$, $S^T = \{9,6,8,10,3\}$, $d^T = d^T - p_t = 5$, $P = \{1,4,7\}$. Go to step 7.

Step 7. Since $P \neq \{\phi\}$, go to step 2.

Step 2. Set $e = 4$, $t = 4$.

Step 3. Set $E^e = d^E - p_e = 6$, $T^t = d^T = 5$, $T^t - p_t = -11$. Since $T^t - p_t \leq 0$, go to step 6.

Step 6. End of the space after the due date.

Since $g > T^t$,

- $S^{E'} = S^E \cup P = \{5,2,4,1,7\}$, $S^{T'} = S^T = \{9,6,8,10,3\}$, $g' = d - \sum_{j \in S^{E'}} p_j = 46$, $S' = (g', S^{E'}, S^{T'})$. Then $f(S') = 664$.
- Set $S^{E''} = S^E \cup P - \{t\} = \{5,2,1,7\}$, $S^{T'} = S^T + \{t\} = \{4,9,6,8,10,3\}$, $g'' = d - \sum_{j \in S^{E''}} p_j = 62$, $S'' = (g'', S^{E''}, S^{T''})$. Then $f(S'') = 736$.

  Since $f(S') \leq f(S'')$,

  - $S^E = S^E + \{e\} = \{5,2,4\}$, $d^T = 0$, $d^E = d^E - p_e + g - g'$, $d^E = 22 - 16 + 51 - 46 = 11$, $g = g' = 46$. $P = P - \{e\} = \{1,7\}$.

Step 7. Since $P \neq \{\phi\}$, go to step 2.

Step 2. Set $e = 1$, $t = 1$.

Step 3. Set $E^e = d^E - p_e = 5$, $T^t = d^T = 0$, $T^t - p_t = -6$. Since $T^t - p_t \leq 0$, go to step 6.

Step 6. End of the space after the due date.

Since $g > T^t$,

- $S^{E'} = S^E \cup P = \{5,2,4,1,7\}$, $S^{T'} = S^T = \{9,6,8,10,3\}$, $g' = d - \sum_{j \in S^{E'}} p_j = 46$, $S' = (g', S^{E'}, S^{T'})$. Then $f(S') = 664$.
- Set $S^{E''} = S^E \cup P - \{t\} = \{5,2,4,7\}$, $S^{T''} = S^T + \{t\} = \{1,4,9,6,8,10,3\}$, $g'' = d - \sum_{j \in S^{E''}} p_j = 52$, $S'' = (g'', S^{E''}, S^{T''})$. Then $f(S'') = 615$.
- Since $f(S') > f(S'')$,

  $S^T = S^T + \{t\} = \{1,9,6,8,10,3\}$, $d^T = 0$, $d^E = d^E + g - g''$, $d^E = 11 + 46 - 52 = 5$, $g = g'' = 52$. $P = P - \{t\} = \{7\}$

Step 7. Since $P \neq \{\phi\}$, go to step 2.

Step 2. Set $e = 7$, $t = 7$.

Step 3. Set $E^e = d^E - p_e = 0$, $T^t = d^T = 0$, $T^t - p_t = -5$. Since $E^e \leq 0$, go to step 5.

Step 5. End of the space after the due date.

Since $g + E^e \geq 0$,

- $S^{E'} = S^E = \{5,2,4\}$, $S^{T'} = S^T \cup P = \{7,1,9,6,8,10,3\}$, $g' = d - \sum_{j \in S^{E'}} p_j = 46$, $S' = (g', S^{E'}, S^{T'})$. Then $f(S') = 660$.
- $S^{E''} = S^E + \{e\} = \{5,2,4,7\}$, $S^{T''} = S^T \cup P - \{e\} = \{1,9,6,8,10,3\}$, $g' = d - \sum_{j \in S^{E''}} p_j = 52$, $S'' = (g'', S^{E''}, S^{T''})$. Then $f(S'') = 615$.

Since $f(S') > f(S'')$,

- $S^E = S^E + \{e\} = \{5,2,4,7\}$, $d^E = 0$, $d^T = d^T + g'' - g = 0$, $g = g'' = 52$. $P = P - \{e\} = \{\phi\}$,
- Go to Step 7.

Step 7. Since $P = \{\phi\}$, go to step 8.

Step 8. Since there is no straddling job. Solve $S' = (g', S^E, S^T)$ **with** $S^E = \{5,2,4,7\}$ **and** $S^T = \{7,1,9,6,8,10,3\}$. $f(S) = 615$ **and**

Appendix B: Design of Experiments for the DPSO Algorithm

There are four parameters in the DPSO algorithm: population size (A), mutation probability of update equation (B = w), crossover probability (C = c_1), and crossover probability (D = c_2). Each factor has two levels and a full factorial design of $2^4 = 16$ treatments is employed. Table B1 shows the factors and their levels whereas Table B2 illustrates DOE for the DPSO algorithm.

Table B1. Factors and Their levels for DPSO

	Factors			
Level	A	B	C	D
Low (-1)	10	0.2	0.2	0.2
High (1)	30	0.8	0.8	0.8

After the response variable was determined for each treatment as given in Table B2, the following statistical analysis were made to determine the level of parameters. In order to screen and identfy the key factors influencing the response variable, the Normal Probability Plot of Effects is used to compare the relative magnitude of the effects. As well known, points in the normal probability plot of effects falling near the fitted line usually indicate important effects. In other words, important effects are larger and further from the fitted line whereas unimportant effects tend to be smaller and centered around zero. To sum up, the Normal Probability Plot of Effects provides a very good screening of important factors in the design.

As seen in the Normal Probability Plot of the Effects in Figure B1, the parameters and their interactions having significant effects on the response variable can easily be determined based on how far they are from the fitted line. For this reason, from

Table B2. Full Factorial Design for DPSO Algorithm

A	B	C	D	R_1	R_2	..	R_{1120}	Response
-1	1	1	1					0.04
1	-1	1	1					0.07
1	-1	1	-1					0.04
1	-1	-1	1					0.03
-1	1	-1	1					0.03
-1	-1	1	-1					0.03
-1	-1	-1	1					0.03
1	1	-1	-1					0.09
-1	-1	1	1					0.07
-1	-1	-1	-1					0.04
-1	1	-1	-1					0.07
1	1	1	-1					0.04
1	-1	-1	-1					0.04
1	1	-1	1					0.03
1	1	1	1					0.03
-1	1	1	-1					0.03

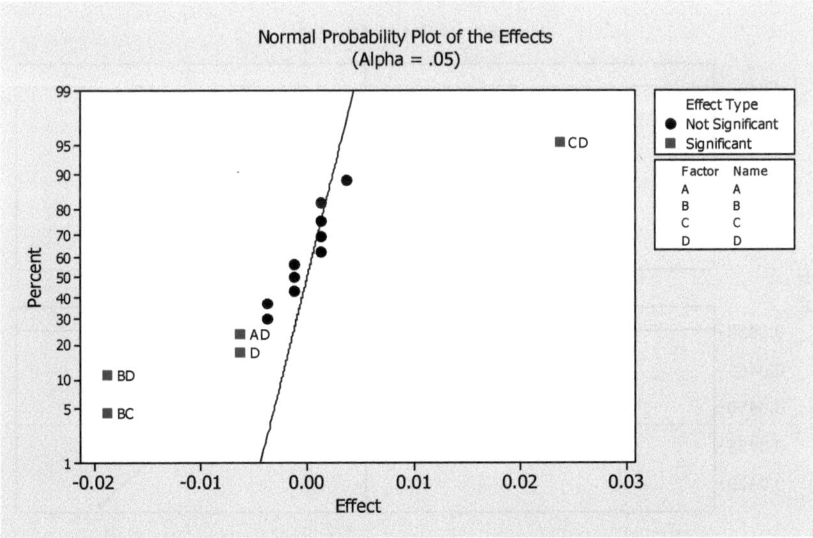

Fig. B1. Normal Probability Plot of the Effects for DPSO

Figure B1, it can be seen that the most significant factor was D, and the most significant interactions were BC, BD, AD, and CD, respectively.

In order to justify the interpretation resulted from the Normal Probability Plot of Effects, a statistical analysis is needed. The Generalized Linear Model (GLM) was used to conduct the Analysis of Variance (ANOVA). To apply ANOVA, three main hypothesis should be checked: normality, homogenity, and independence of residuals. The residuals from the experimental results were analyzed and all three hypothesis could be accepted. The ANOVA results are given in Table B3. Table B3 justifies the significancy of the factor D and the interactions BC, BD, AD and CD since the *F* values were high enough and the *p*-values are less than 0.05.

Table B3. Analysis of Variance for DPSO

Source	DF	Seq SS	Adj SS	Adj MS	F	P
A	1	0.0000562	0.0000562	0.0000562	2.74	0.14
B	1	0.0000062	0.0000062	0.0000062	0.3	0.6
C	1	0.0000062	0.0000062	0.0000062	0.3	0.6
D	1	0.0001562	0.0001563	0.0001563	7.61	0.03
BC	1	0.0014063	0.0014063	0.0014063	68.48	0
BD	1	0.0014063	0.0014062	0.0014062	68.48	0
AD	1	0.0001562	0.0001562	0.0001562	7.61	0.03
CD	1	0.0022563	0.0022563	0.0022563	109.87	0
Error	7	0.0001437	0.0001437	0.0000205		
Total	15	0.0055938				

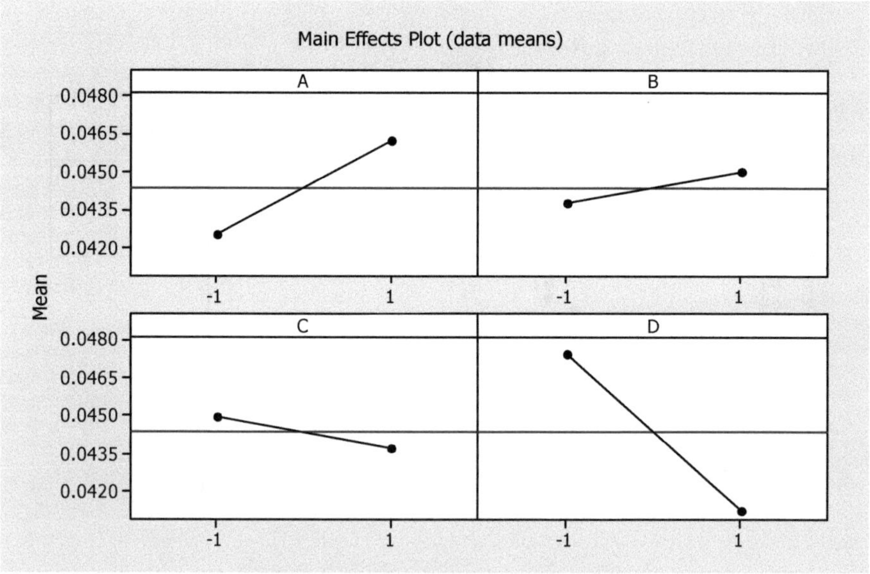

Fig. B2. Main Effects Plot for DPSO

In order to determine the level of each factor, the Main Effects Plot can be used. A main Effect Plot shows the mean values of each level of a factor considered in the design. A main effects happens if the mean response varies accross different levels of a factor considered. It is generally used to asses the relative strength of the effects across different levels of a factor in the design.The Main Effect Plot is shown in Figure B2. If only the main effects were to be considered, it would be suitable to run all the factors at the following levels in Table B4.

Table B4. Parameter settings of DPSO from Main Effect Plot

Factors	Levels	Description	Value
A	-1	NP=low level	10
B	-1	w =low level	0.2
C	1	c_1 =high level	0.8
D	1	c_2 =high level	0.8

However, it is always necessary to look into any interaction that is significant due to the fact that main effects do not have much meaning when they are involved in significant interactions. For this purpose, Interaction Plots can be used, which show the mean values for each level of a factor with the level of a second factor held constant. An interaction between factors happens if the change in the response from the low level to the high level of one factor is not the same as the change in the response at the same levels of a second factor considered. It indicates that the effect of one factor is dependent on a second factor. For this reason, the effect of interactions should be analyzed on deciding the levels of parameters. The BC interaction is illustrated in Figure B3 where BC interaction does not give a clear picture about the level of the parameters since both levels seem to have similar effect on the response variable.

BD interaction is given in Figure B4 where it can be seen that the best results are obtained with both high levels of B and D. It suggests that the mutation probability w and the crossover probability c_2 should be 0.8 and 0.8, respectively.

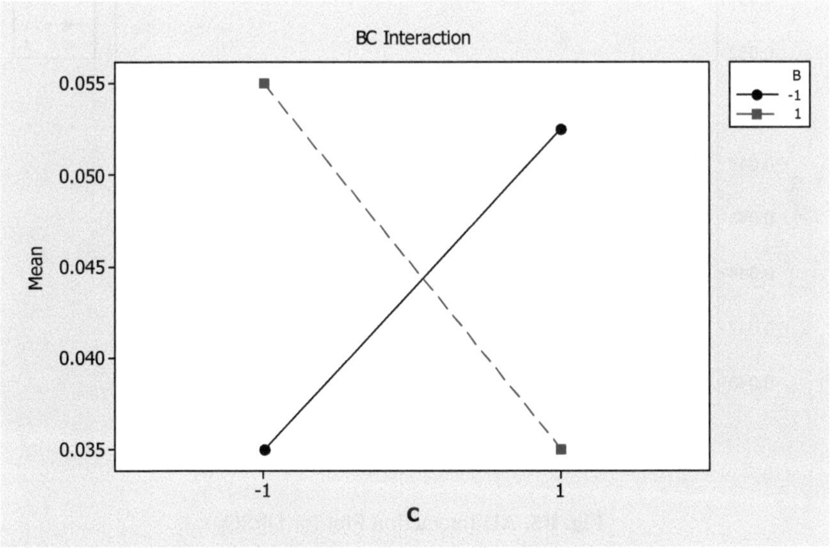

Fig. B3. BC Interaction Plot for DPSO

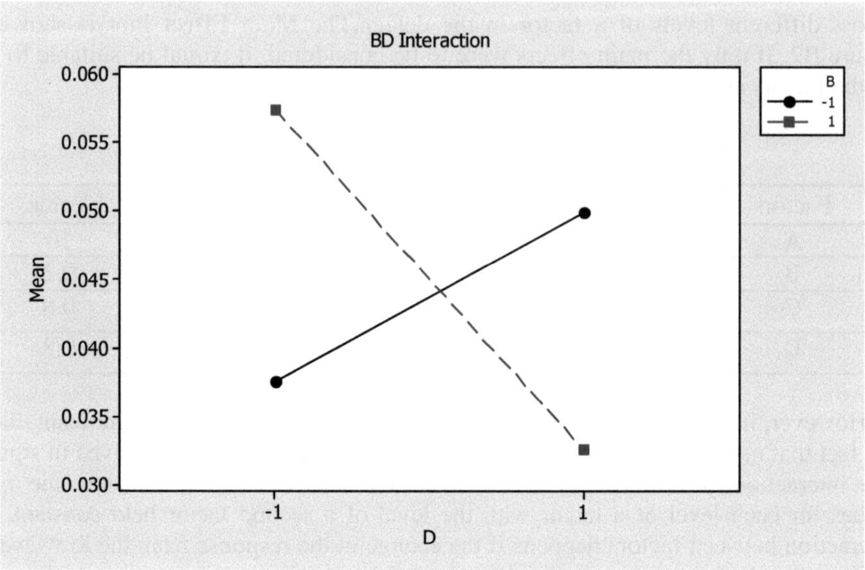

Fig. B4. BD Interaction Plot for DPSO

AD interaction is given in Figure B5 where the best results are obtained with both high level of A and D justifying again the higher effect of D in Figure B4. For this reason, the population size is taken as *NP*=30.

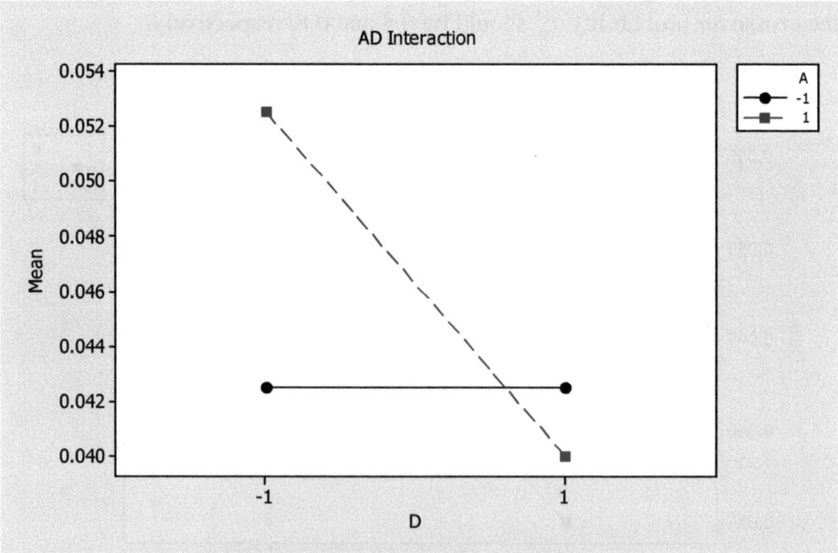

Fig. B5. AD Interaction Plot for DPSO

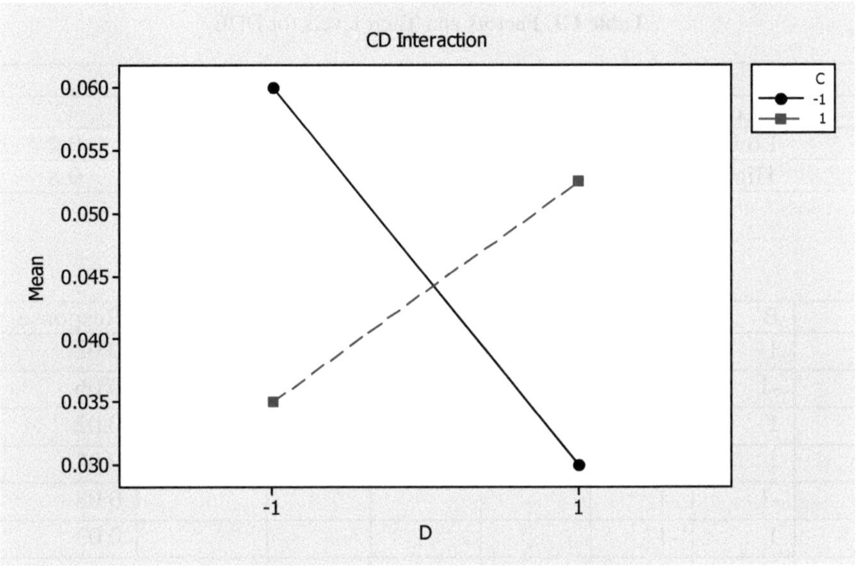

Fig. B6. CD Interaction Plot for DPSO

Finally, the CD interaction is illustrated in Figure B6 where we conclude that the best results are obtained with again a high level of D and a low level of C. For this reason, crossover probabilities c_1 and c_2 were taken as 0.8 and 0.8, respectively. The final parameter setting for the DPSO algorithm is given in Table B5.

Table B5. Final Parameter settings for DPSO Algorithm

Factors	Levels	Description	Value
A	1	NP=high level	30
B	1	w =high level	0.8
C	-1	c_1 =low level	0.2
D	1	c_2 =high level	0.8

Appendix C: Design of Experiments for the DDE Algorithm

There are three parameters in the DDE algorithm: population size (A), mutation probability (B = P_m), crossover probability (C = P_c), and mutation equation (D). All factors have two levels and a general factorial design of $2^3 = 8$ treatments is employed. Table C1 shows the factors and their levels whereas Table C2 illustrates DOE for the DDE algorithm.

Table C1. Factors and Their levels for DDE

	Factors		
Level	A	B	C
Low (-1)	10	0.2	0.2
High (1)	30	0.8	0.8

Table C2. General Design for DDE Algorithm

A	B	C	R_1	R_2	..	R_{1120}	Response
-1	1	1					0.02
-1	-1	1					0.06
1	1	1					0.02
1	1	-1					0.05
1	-1	-1					0.08
-1	1	-1					0.03
-1	-1	-1					0.05
1	-1	1					0.06

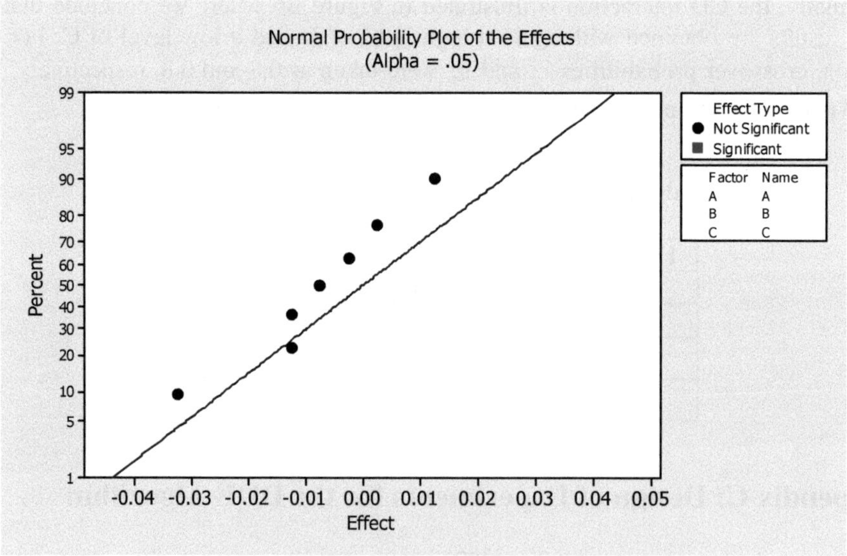

Fig. C1. Normal Probability Plot of the Effects for DDE

Again, once the response variable was determined for each treatment, a similar statistical analysis was made to determine the level of parameters. From the Normal Probability Plot of the Effects in Figure C1, it can be seen that no parameters and interaction were significant. Then the Generalized Linear Model (GLM) was used to conduct the Analysis of Variance (ANOVA) once again. The residuals from the experimental results were also analyzed and all three hypothesis could be accepted for this design too. The ANOVA results are given in Table C3 where no factor had significantly less than p=0.05. For this reason, it can be concluded that the Main Effects Plot would be enough to judge on the level of parameters.

Table C3. Analysis of Variance for DDE

Source	DF	Seq SS	Adj SS	Adj MS	F	P
A	1	0.0003125	0.0003125	0.000312	25.00	0.13
B	1	0.0021125	0.0021125	0.0021125	169.00	0.05
C	1	0.0003125	0.0003125	0.0003125	25.00	0.13
AB	1	0.0000125	0.0000125	0.0000125	1.00	0.50
AC	1	0.0003125	0.0003125	0.0003125	25.00	0.13
BC	1	0.0001125	0.0001125	0.0001125	9.00	0.21
Error	1	0.0000125	0.0000125	0.0000125		
Total	7	0.0031875				

The Main Effects Plot is given in Figure C2. Following the Main Effects Plot, it can easily be seen that best results are obtained with a low level of A, and both high levels of B and C. For this reason, the level of parameters is determined as shown in Table C4.

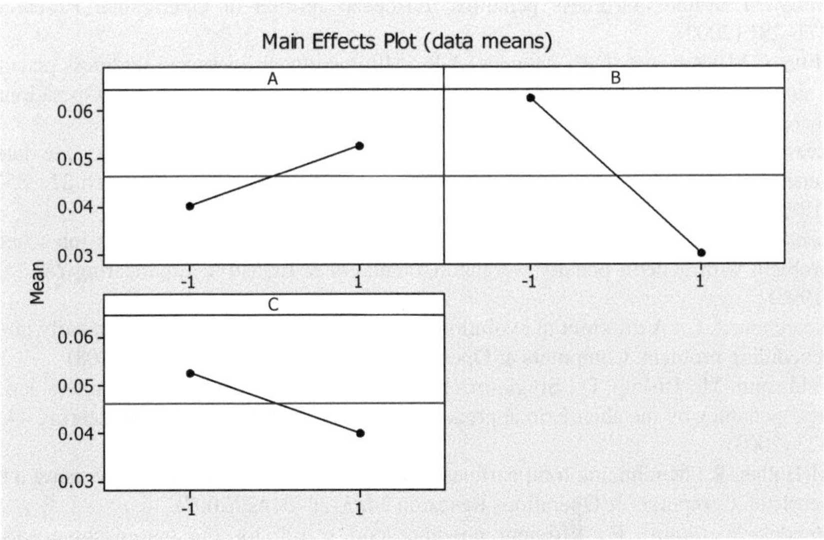

Fig. C2. Main Effects Plot for DDE

Table C4. Final Parameter Settings for DDE Algorithm

Factors	Levels	Description	Value
A	-1	*NP*=low level	10
B	1	P_m=high level	0.8
C	1	P_c=high level	0.8

References

1. Cheng, T.C.E., Kahlbacher, H.G.: A proof for the longest/job/first policy in one/machine scheduling. Naval Research Logistics 38, 715–720 (1991)
2. Baker, K.R., Scudder, G.D.: On the assignment of optimal due dates. Journal of the Operational Research Society 40, 93–95 (1989)
3. Biskup, D., Feldmann, M.: Benchmarks for scheduling on a single machine against restrictive and unrestrictive common due dates. Computers & Operations Research 28, 787–801 (2001)
4. Hoogeveen, J.A., van de Velde, S.L.: Scheduling around a small common due date. European Journal of Operational Research 55, 237–242 (1991)
5. Hall, N.G., Kubiak, W., Sethi, S.P.: Earliness-tardiness scheduling problems II: weighted deviation of completion times about a restrictive common due date. Operations Research 39(5), 847–856 (1991)
6. James, R.J.W., Buchanan, J.T.: Using tabu search to solve the common due date early/tardy machine scheduling problem. Computers & Operations Research 24, 199–208 (1997)
7. Wan, G., Yen, B.P.C.: Tabu search for single machine with distinct due windows and weighted earliness/tardiness penalties. European Journal of Operational Research 142, 271–281 (2002)
8. Hino, C.M., Ronconi, D.P., Mendes, A.B.: Minimizing earliness and tardiness penalties in a single-machine problem with a common due date. European Journal of Operational Research 160, 190–201 (2005)
9. Lee, C.Y., Choi, J.Y.: A genetic algorithm for jobs sequencing with distinct due dates and general early-tardy penalty weights. Computers & Operations Research 22, 857–869 (1995)
10. Lee, C.Y., Kim, S.J.: Parallel genetic algorithms for the earliness/tardiness job scheduling problem with general penalty weights. Computers & Industrial Engineering 28, 231–243 (1995)
11. Nearchou, A.C.: A differential evolution approach for the common due date early/tardy job scheduling problem. Computers & Operations Research 35, 1329–1343 (2008)
12. Feldmann, M., Biskup, D.: Single-machine scheduling for minimizing earliness and tardiness penalties by metaheuristic approaches. Computers & Industrial Engineering 44, 307–323 (2003)
13. M'Hallah, R.: Minimizing total earliness and tardiness on a single machine using a hybrid heuristic. Computers & Operations Research 34, 3126–3142 (2007)
14. Hendel, Y., Sourd, F.: Efficient neighborhood search for the one-machine earliness-tardiness scheduling problem. European Journal of Operational Research 173, 108–119 (2006)

15. Lin, S.-W., Chou, S.-Y., Ying, K.-C.: A sequential exchange approach for minimizing earliness-tardiness penalties of single-machine scheduling with a common due date. European Journal of Operational Research 177, 1294–1301 (2007)
16. Eberhart, R.C., Kennedy, J.: A new optimizer using particle swarm theory. In: Proceedings of the Sixth International Symposium on Micro Machine and Human Science, pp. 39–43 (1995)
17. Kennedy, J., Eberhart, R.C., Shi, Y.: Swarm Intelligence. San Mateo, Morgan Kaufmann (2001)
18. Clerc, M.: Particle Swarm Optimization. ISTE Ltd., France (2006)
19. Kennedy, J., Eberhart, R.C.: A discrete binary version of the particle swarm algorithm. In: Proceedings of the World Multiconference on Systemics, Cybernetics and Informatics 1997, Piscataway, NJ, pp. 4104–4109 (1997)
20. Storn, R., Price, K.: Differential evolution – a simple and efficient adaptive scheme for global optimization over continuous spaces. ICSI, Technical Report TR-95-012 (1995)
21. Storn, R., Price, K.: Differential evolution - a simple and efficient heuristic for global optimization over continuous space. Journal of Global Optimization 11, 341–359 (1997)
22. Corne, D., Dorigo, M., Glover, F.: Part Two: Differential Evolution. In: New Ideas in Optimization, pp. 77–158. McGraw-Hill, New York (1999)
23. Lampinen, J.: A bibliography of differential evolution algorithm. Lappeenranta University of Technology, Department of Information Technology, Laboratory of Information Processing, Technical Report (2001)
24. Babu, B.V., Onwubolu, G.C. (eds.): New Optimization Techniques in Engineering. Springer, Heidelberg (2004)
25. Price, K., Storn, R., Lampinen, J.: Differential Evolution – A Practical Approach to Global Optimization. Springer, Heidelberg (2006)
26. Chakraborty, U.K.: Advances in Differential Evolution. Springer, Berlin (2008)
27. Tasgetiren, M.F., Liang, Y.-C.: A binary particle swarm optimization algorithm for lot sizing problem. Journal of Economic and Social Research 5(2), 1–20 (2003)
28. Tasgetiren, M.F., Liang, Y.-C., Sevkli, M., Gencyilmaz, G.: Particle swarm optimization and differential evolution for single machine total weighted tardiness problem. International Journal of Production Research 44(22), 4737–4754 (2006)
29. Tasgetiren, M.F., Sevkli, M., Liang, Y.-C., Yenisey, M.M.: Particle swarm optimization and differential evolution algorithms for job shop scheduling problem. International Journal of Operations Research 3(2), 120–135 (2006)
30. Pan, Q.-K., Tasgetiren, M.F., Liang, Y.-C.: A discrete particle swarm optimization algorithm for single machine total earliness and tardiness problem with a common due date. In: Proceedings of the World Congress on Evolutionary Computation, CEC 2006, Vancouver, Canada, pp. 3281–3288 (2006)
31. Pan, Q.-K., Tasgetiren, M.F., Liang, Y.-C.: Minimizing total earliness and tardiness penalties with a common due date on a single machine using a discrete particle swarm optimization algorithm. In: Dorigo, M., Gambardella, L.M., Birattari, M., Martinoli, A., Poli, R., Stützle, T. (eds.) ANTS 2006. LNCS, vol. 4150, pp. 460–467. Springer, Heidelberg (2006)
32. Tasgetiren, M.F., Liang, Y.-C., Sevkli, M., Gencyilmaz, G.: Particle swarm optimization algorithm for makespan and total flowtime minimization in permutation flowshop sequencing problem. European Journal of Operational Research 177(3), 1930–1947 (2007)
33. Pan, Q.-K., Tasgetiren, M.F., Liang, Y.-C.: A discrete particle swarm optimization algorithm for the no-wait flowshop scheduling problem with makespan and total flowtime criteria. Computers & Operations Research 35, 2807–2839 (2008)

34. Al-Anzi, F.S., Allahverdi, A.: A self adaptive differential evolution heuristic for two-stage assembly scheduling problem to minimize maximum lateness with setup times. European Journal of Operational Research 182, 80–94 (2007)
35. Liao, C.-L., Tseng, C.-T., Luarn, P.: A discrete version of particle swarm optimization for flowshop scheduling problems. Computers & Operations Research 34, 3099–3111 (2007)
36. Tasgetiren, M.F., Pan, Q.-K., Liang, Y.-C., Suganthan, P.N.: A discrete differential evolution algorithm for the total earliness and tardiness penalties with a common due date on a single machine. In: Proceedings of the 2007 IEEE Symposium on Computational Intelligence in Scheduling (CISched 2007), Hawaii, USA, pp. 271–278 (2007)
37. Tasgetiren, M.F., Pan, Q.-K., Liang, Y.-C., Suganthan, P.N.: A discrete differential evolution algorithm for the no-wait flowshop scheduling problem with total flowtime criterion. In: Proceedings of the 2007 IEEE Symposium on Computational Intelligence in Scheduling (CISched 2007), Hawaii, USA, pp. 251–258 (2007)
38. Montgomery, D.C.: Design and Analysis of Experiments., 5th edn. John Wiley and Sons, Chichester (2000)
39. Devore, J.L.: Probability and Statistics for Engineering and the Sciences, 5th edn., Duxbury, Thomson Learning (2000)

Author Index

Author Index

Index

MIX
Papier aus verantwortungsvollen Quellen
Paper from responsible sources
FSC® C105338

If you have any concerns about our products,
you can contact us on
ProductSafety@springernature.com

In case Publisher is established outside the EU,
the EU authorized representative is:
Springer Nature Customer Service Center GmbH
Europaplatz 3, 69115 Heidelberg, Germany

Printed by Libri Plureos GmbH
in Hamburg, Germany